Innovation and Growth

CHASING A MOVING FRONTIER

Edited by

Vandana Chandra, Deniz Eröcal, Pier Carlo Padoan
and Carlos A. Primo Braga

OECD
WORLD BANK

ORGANISATION FOR ECONOMIC CO-OPERATION AND DEVELOPMENT

The OECD is a unique forum where the governments of 30 democracies work together to address the economic, social and environmental challenges of globalisation. The OECD is also at the forefront of efforts to understand and to help governments respond to new developments and concerns, such as corporate governance, the information economy and the challenges of an ageing population. The Organisation provides a setting where governments can compare policy experiences, seek answers to common problems, identify good practice and work to co-ordinate domestic and international policies.

The OECD member countries are: Australia, Austria, Belgium, Canada, the Czech Republic, Denmark, Finland, France, Germany, Greece, Hungary, Iceland, Ireland, Italy, Japan, Korea, Luxembourg, Mexico, the Netherlands, New Zealand, Norway, Poland, Portugal, the Slovak Republic, Spain, Sweden, Switzerland, Turkey, the United Kingdom and the United States. The Commission of the European Communities takes part in the work of the OECD.

OECD Publishing disseminates widely the results of the Organisation's statistics gathering and research on economic, social and environmental issues, as well as the conventions, guidelines and standards agreed by its members.

ISBN 978-92-64-07396-8 (print)
ISBN 978-92-64-07397-5 (PDF)

Corrigenda to OECD publications may be found on line at: *www.oecd.org/publishing/corrigenda*.

Foreword

Innovation is the cornerstone of sustained economic growth and prosperity. We often think of innovation in terms of breakthrough inventions – but it can also be linked to organisational changes and technology diffusion. In a globalised world, in which countries and firms compete fiercely to buy and sell their products and services, innovation is a key driver of competitiveness.

We see this today in the critical role innovation plays in the rapid growth of emerging economies, as well as lagging growth when innovation is absent. There are strong signs, for instance, of increasing innovation activity in China and other fast-growing emerging economies, and it is certainly playing a role in their convergence with more advanced economies.

At the same time, the global financial crisis has given focus to the issue of the sustainability of existing growth and innovation models, while increasing the relevance of better understanding the role that innovation can play in restoring sustainable growth.

The OECD and the World Bank Group are adding to this better understanding.

Currently, the OECD is working on an "Innovation Strategy" to help countries develop policies to boost innovation (www.oecd.org/innovation/strategy). This project is built around evidence-based analysis and benchmarking. It will include policy principles to guide the design and implementation of policies for innovation, new indicators on the innovation-economic performance link, initiatives for innovation-friendly business environments, and the development of best practices and policy recommendations. This work is also expected to contribute to the G20 initiative to enhance sustainable growth.

The Growth Commission's recent report (*www.growthcommission.org*), supported by the World Bank Group, highlights the need for country-specific policies to jump-start technological learning in developing countries to fuel innovation. The World Bank Group has also developed numerous other studies addressing the economics of innovation and technological change (such as the 2008 Global Economic Prospects report, *Technology Diffusion in the Developing World*) and financed many science and technology projects over the years. The Bank Group's Poverty Reduction and Economic Management (PREM) Network, in particular, is committed to further exploring the implications of innovation policies for inclusive growth.

Based on this work, as well as the practical experience of advanced and emerging economies, we know that markets alone cannot always offer the private sector sufficient incentives to innovate. Governments can play an essential role in fostering innovation. The importance of a multilateral framework to co-ordinate these efforts only adds to the value of a shared understanding of global best practices in the area of innovation policy.

The OECD and the World Bank Group are partnering in an effort to advance our understanding of the economics of innovation. This volume introduces some of the key issues and information directly relevant to this agenda. The ideas explored in the following chapters were initially presented at a joint conference on **Innovation and Growth in a Globalised World** (*www.oecd.org/innovation/wbconference*), held in Paris in November 2008. The OECD and the World Bank Group will continue to address these themes as we further explore the role of innovation in promoting sustainable growth in developed and developing economies.

Angel Gurría	Robert B. Zoellick
Secretary-General	President
OECD	The World Bank Group

Acknowledgements

The conference on Innovation and Sustainable Growth in a Globalised World (November 2008) and this volume would not have been possible without the support of Angel Gurría, Secretary-General, OECD and Danny M. Leipziger, Vice President & Head of Network for Poverty Reduction and Economic Management (PREM), The World Bank, at the time of the conference. The Editors, who were also the principal organisers of the conference, are grateful for the leading role they played in advancing OECD-World Bank co-operation in the area of innovation and growth.

Thanks are also due to a number of high-level participants who contributed to the success of the conference by sharing their knowledge and helping guide the two organisations as they continue to develop their joint work in this area. They include (with affiliations as of November 2008): Dr. V.S. Arunachalam, Chairman, Center for Study of Science Technology and Policy (CSTEP), Bangalore; H.E. Debapriya Bhattacharya, Ambassador, Permanent Representative of Bangladesh to the WTO and UN Offices in Geneva; Professor G.K. Chadha, Chief Executive Officer, South Asian University, and Member, PM's Economic Advisory Council, India; Jean-Philippe Cotis, Director-General, National Institute for Statistics and Economic Studies (INSEE), France; Jorgen Elmeskov, Director of Policy Studies Branch, OECD Economics Department; Professor Dr. Fang Xin, Vice President, Chinese Academy of Sciences; Professor Yuko Harayama, Graduate School of Engineering, Tohoku University, Japan; H.E. Chris Hoornaert, Ambassador, Permanent Representative of Belgium to the OECD; Professor Michael Kahn, University of Stellenbosch, South Africa; Ilkka Lakaniemi, Corporate Affairs, Nokia Siemens Networks; Jeffrey D. Lewis, Senior Advisor, Office of Vice President & PREM (Poverty Reduction Economic Management Network), The World Bank; Douglas Lippoldt, Acting Head, Development Division, Trade and Agriculture Directorate, OECD; Dawn Nafus, Senior Global Policy Specialist, Intel Corporation; Jérôme Saulnier, Economic Analyst, Co-ordination of Structural Reforms and of the Economic Service, Directorate General for Economic and Financial Affairs, European Commission; Jennifer Schenker, Correspondent, *Business Week*; Klaus Schmidt-Hebbel, OECD Chief Economist and Head of Economics Department; Professor Luc Soete, Director of UNU-MERIT (United Nations University, Maastricht Economic Research Institute on Innovation and Technology); Alfred Watkins, The World Bank Science and Technology Program Coordinator; Andrew Wyckoff, Deputy Director, Directorate for Science, Technology and Industry, OECD. All individuals participated in the Conference in their personal capacity.

Many others have our gratitude for contributing in various ways to the conference sessions and making them enjoyable for all. We want to thank especially the participants from the OECD and developing countries who took the time to attend the conference over two days and broadened the scope of its discussions. Special thanks go to Sandrine Abrazian (OECD) who prepared the conference website, and the support staff of both organisations, in particular Maureen Nash, Karine Normant and Isabelle Renaud of the OECD and Sarah Lipscomb and Ivana Ticha of the World Bank, who made the event and this book possible.

Table of Contents

Chapter 1

Introduction: why innovation matters

Vandana Chandra, Deniz Eröcal, Pier Carlo Padoan and Carlos A. Primo Braga[1]

> *In this volume, the OECD and the World Bank jointly take stock of how globalisation is posing new challenges for innovation and growth in both developed and developing countries, and of how countries are coping with them. The authors discuss options for national and global policy initiatives that can foster technological innovation in the pursuit of faster and sustainable growth.*

[1] Vandana Chandra is Senior Economist, The World Bank; Deniz Eröcal is Project Co-ordinator, Enhanced Engagement with Major Non-Member Economies, Office of the Secretary-General, OECD; Pier Carlo Padoan is Deputy Secretary-General, OECD; and Carlos A. Primo Braga is Director of the Economic Policy and Debt Department, PREM Network, The World Bank.

In the past few decades, as the international flows of trade, capital and labour have expanded across the global marketplace, the competitiveness and prosperity of high-income economies has come to rely increasingly on their innovative capability. Unlike OECD countries, developing countries' competitiveness and prosperity remains largely tied to their endowments of natural resources. Their governments have been less successful in fostering technological innovation. Moreover, low productivity levels continue to constrain their competitiveness in the global market.

The unique nature of innovative activity and the growing interconnectedness of the world economy call, however, for greater attention to the interplay of openness and technological innovation not only in OECD countries, but also in developing economies. Innovation systems increasingly rely on "open" platforms and collaboration side by side with competition. At the same time, the geography of innovation is being redrawn as economic interdependence grows, emerging economies accumulate immaterial assets, and modern communication networks redefine opportunities for "leapfrogging". The experience of the so-called "BRICs" (Brazil, Russia, India and China) is illustrative in this context.

This publication comes at a time of global financial and economic crisis. The extent of the crisis is such that many elements of national and global innovation systems are being affected and may be severely compromised. The crisis may affect innovative activities through several channels: lower R&D spending (as private-sector R&D is usually pro-cyclical); loss of human capital (as protracted unemployment may erode existing skills); lower risk taking (as a result of disruptions in financial markets); and weaker international diffusion of technology (reflecting declines in trade and foreign direct investment). While the following chapters do not specifically address these issues, their analyses of the policy challenges faced by countries trying to pursue sustainable innovation-led growth provide useful information for the debate about how innovation can play a role in the process of economic recovery.

The various chapters highlight how the emergence of an integrated global market affects the impact of national innovation policy. What seemed like effective innovation strategies (*e.g.* policies designed to strengthen the R&D capacity of domestic firms) are no longer sufficient for effective catch-up. This is partly due to the changing nature of innovation. The more open and global nature of innovation makes innovation policies more difficult to design, implement and monitor at the national scale alone. These challenges are further complicated by new and still unfolding phenomena, such as the emergence of global value chains and the fragmentation of production, the growing role of global corporations, and the information and communications technology (ICT) revolution. Where a global corporation chooses to anchor its production and why are, in different ways, affecting the playing field for OECD and developing economies alike.

This volume starts by discussing existing growth theories and their implications for innovation policies. The "neo-Schumpeterian model", for example, highlights the role of competition in fostering innovation depending on a country's distance from the technological frontier. The setting is a national economy in which firms innovate to remain competitive. The scene changes as the setting widens to the global marketplace in which countries, especially emerging economies, compete with each other and a country's competitiveness is measured with respect to the industry leader in the global economy. The innovation policy implications for emerging markets such as the BRICs vary quite significantly from those for OECD countries in view of the greater dispersion of their firms' productivity. When developing countries, especially low-income ones, are added to the analysis, what they need to do to grow faster and catch up with the OECD countries gains additional facets as the differences between technological diffusion and innovation become sharper.

Bringing entry barriers into the picture shows that competition can have mixed effects on innovation. New entrants will offer new possibilities for innovation, but at the same time they may discourage established firms from investing in innovation for fear that they may be driven out of business anyway. Theory suggests that the threat of the entry of foreign firms on the world technology frontier will discourage R&D and innovation by domestic firms that are far from the frontier but will encourage R&D by domestic firms that are closer to the frontier, as their best defence against a frontier entrant is to be on the frontier as well. Attention is also drawn to the fact that firms in the ICT industry compete not only for market share, but also that their successes and failures help shape the national innovation system. ICT is a general purpose technology, *i.e.* a technology that can serve as a fundamental input to other technologies and applications, and thus act as a multiplier of innovation. This can have significant spillover effects and influence the overall catching-up process.

Each chapter addresses a distinct aspect of innovation policy for growth in different settings. Chapter 2 starts with an analytical framework from the new Schumpeterian theory of growth which sets the stage for thinking about innovation and growth. The conventional wisdom from neoclassical theory is that competition exerts downward pressure on costs, reduces slack and provides incentives for efficient organisation of production. Rethinking Schumpeterian theory in an OECD country setting, Howitt shows both theoretically and empirically that product market competition is essential to growth, but that how much it affects innovation and hence, growth itself, depends upon how far the incumbent firm is from the technological frontier, as measured by comparing the productivity levels of the incumbent and entrant. The efficiency-improving policy recommendation consists of pro-competition policies combined with policies that facilitate the re-allocation or mobility of workers from the lagging to the technologically leading firms in a country. However, as noted by Klaus Schmidt-Hebbel, a former OECD Chief Economist, at the conference where these papers were initially presented, while the distance to frontier model suggests that the relation between competition and innovation is "U-shaped" and may vary by industry and country: "OECD evidence seems to suggest that the positive segment of the relation is the dominant one"-- *i.e.* on the whole, framework conditions characterised by competitive product and labour markets are associated with greater innovation.

When Howitt's model is extended to the global marketplace in which emerging market countries such as the BICs (Brazil, India, China) compete to catch up with Korea, the importance of industry-level global competition comes to the fore. In Chapter 3, Chandra, Osorio-Rodarte and Primo Braga validate the model presented in Chapter 2, but show that to catch up, emerging market countries need, in addition to framework conditions (competition policy), to consider the design of distance-shortening innovation polices. They underscore the role of industry- and country-specific innovation policies for catch-up.

Chapters 4 and 5 examine empirical evidence from OECD countries and rapidly evolving transnational economic phenomena. The challenge is to link macroeconomic performance to structural factors. In Chapter 4, Blöndal and Dougherty expand on the OECD's "Going for Growth" framework, developed to assess the impact of structural reforms and policies on long-term growth, to take detailed stock of trends in OECD's innovation indicators and in catch-up through productivity growth in countries like Turkey and New Zealand; they also note cases in which the impact of recommendations based on the indicators has been disappointing. They conclude that for the "leaders" among OECD countries, product market competition and investments in education are more important now than ever before. In studying the trends in the global innovation networks of transnational corporations which foster the fragmentation of production and global value chains, Pilat, De Backer, Basri, Box and Cervantes argue in Chapter 5 that globalisation dampens the returns from national investments in innovation. This implies that the policy implications for large and small countries are different, even though both are subject to the same level and degree of global competition. Clearly, this observation has implications for innovation policies in the OECD and developing countries alike. The level and pace of innovation need not be limited by the level and quality of education in a country. In addition to national investments in education, the authors note that policies that attract global talent can play an important role in supplementing and boosting R&D, innovation and growth.

Chapters 6, 7, and 8 turn to more country-specific innovation experiences and their links with growth. Their main contribution is to introduce the reader to the "how to" of innovation. In Chapter 6, Guinet, Hutschenreiter and Keenan discuss lessons from OECD countries in a globalising world. In Chapter 7, Dahlman does the same for the emerging market economies, analysing how they have successfully leveraged catch-up. In Chapter 8, Burns introduces a metric for innovation in developing countries in general, highlighting the enormous challenges facing them. The onus of innovation policy in these three chapters falls largely on national governments.

Chapters 9, 10 and 11 turn to how the ICT revolution has transformed the pace of innovation and the process of economic catch-up. These chapters discuss regional experiences as well as ICT as a unique instrument for innovation. In Chapter 9, de Laiglesia describes how ICT has affected development patterns in Latin America. In Chapter 10 Kelly does the same for Asia. Paltridge examines the role of ICT and innovation in Chapter 11 and discusses how network externalities can influence innovation in countries at any stage of development.

In sum, this volume provides insights into the relations between innovation and growth in developed and developing countries. The theoretical framework, developed by Aghion and Howitt and summarised in Chapter 2 is rich and flexible enough to accommodate both general and country-specific aspects of innovation and growth. Perhaps more importantly, useful policy implications that go beyond traditional policy recommendations can be derived from such a framework. While country-specific factors remain relevant, the analysis presented here underscores that innovation and innovation policies are useful at all stages of development. After all, innovation needs not necessarily to be identified with the introduction of new technologies but can result as well from non-technological innovation (such as new business practices) and mechanisms to foster technological diffusion. This is even more the case if the role of general purpose technologies such as ICT is taken into account; their positive impact on GDP growth in poor countries still needs to be better understood.

Of course much more needs to be done, both in refining the theoretical framework and putting it to test in different country environments. The OECD and the World Bank are well positioned to continue to explore this agenda with a view to better understanding the role of innovation in fostering growth and economic development.

Chapter 2

Competition, innovation and growth: theory, evidence and policy challenges[1]

Peter Howitt[2]

Product market competition is essential to the growth process but how much it affects innovation depends upon how technologically competitive the incumbent firm is. What are the guiding principles for pro-competition policies to spur growth?

- Conventional wisdom: competition exerts downward pressure on costs, reduces slack, provides incentives for efficient organisation of production.

- The Schumpeterian alternative: the only effective form of competition is innovation; antitrust measures reduce the reward to innovation.

- Rethinking Schumpeterian theory – an inverted U.

- Policy lessons from the ABHV model.

- Six principles for promoting competition in developing countries.

Conclusion: as you approach the frontier, appropriate policy changes.

1. Keynote Address to the OECD-World Bank Joint Conference on Innovation and Sustainable Growth in a Globalised World, held in Paris, 18-19 November, 2008.
2. Peter Howitt is Lyn Crost Professor of Social Sciences, Department of Economics, Brown University.

Introduction

How is a country's growth rate affected by competition policies? The conventional wisdom on this subject, as expressed for example in the Spence report (World Bank, 2008), is that competition exerts downward pressure on costs, reduces slack, and provides incentives for efficient organisation of production, thereby favouring economic growth. An alternative view, sometimes ascribed to Schumpeter, claims the opposite – that rigorous pursuit of antitrust policies is detrimental to growth because anything that reduces the scope for monopoly profit also reduces the reward to innovation, which in turn is the mainspring of long-run growth.

For more than two decades now I have been working, together with my colleague Philippe Aghion, to develop and test a version of growth theory based on Schumpeter's idea of creative destruction (Aghion and Howitt, 1992, 1998, 2009). My purpose in this lecture is to describe what we have learned from this investigation about the effects of competition on growth and innovation. As it turns out, theory and evidence do not provide unequivocal support for either the conventional wisdom or the alternative "Schumpeterian" view.

Schumpeterian growth theory

Schumpeterian theory starts from the same premise as almost every other growth theory, namely that long-run growth is driven by productivity growth, which in turn is driven by technological progress. It differs from neoclassical theory by treating technological progress as an economic phenomenon. And it differs from other endogenous growth theories in emphasising that the main force driving technological progress is industrial innovation, the same force that is central to the competitive process of any market economy.

The theory also emphasises that successful technology strategies vary from country to country, depending on such factors as institutions, geography, educational levels, environmental conditions, and especially distance to the world technology frontier. Countries that are on or near the frontier tend to produce leading-edge innovations, whereas countries that are further from the frontier tend to implement technologies that have been developed elsewhere. It thus produces a context-dependent theory of what we call "appropriate growth policy" (Aghion and Howitt, 2006).

The earliest version of this theory seemed to vindicate the Schumpeterian view of competition and growth; increased competition that reduced monopoly profits did indeed reduce growth by reducing the reward to successful innovation. It did not take us long, however, to recognise that this particular implication of the theory is contradicted by several empirical studies. Porter (1990) observed that Japanese firms in more competitive prefectures tended to grow more rapidly, and econometric work on UK manufacturing firms by Blundell *et al.* (1995) and by Nickell (1996) showed that a linear regression of either productivity growth or the frequency of patenting on various determinants, including an industry-specific measure of the degree of competition, indicated that firms are more innovative, and productivity grows more rapidly, in industries that are more competitive, other things equal.

More generally, many economic historians have come to the conclusion (for example Crafts, 1996) that rapid growth is associated with openness and competition in product markets rather than with protected monopolies. And of course the "modern miracle" growth countries of East Asia have all had export-oriented policies under which firms were forced to submit to competitive market pressures. Even Korea, with its *chaebol* and its import restrictions, has made successful competition in foreign markets a necessary condition for support and for permission to import key inputs.

In the face of this evidence it was clear that we had to modify the first generation models of Schumpeterian growth theory to include some of the mechanisms by which competition spurs growth, mechanisms that would, at least in some circumstances, counteract the Schumpeterian appropriability effect.

Ex ante barriers to entry

Introducing *ex ante* barriers to entry is one such modification. A policy of strengthening competition by removing barriers will allow more potential innovators to be active. This obviously should raise an economy's overall rate of innovation. These barriers to entry, as measured by the cost of starting a business, can be quite substantial; they vary greatly across countries and are highly negatively correlated with a country's level of development. Table 2.1 shows, for example, that to start a business in India requires the entrant to undertake ten procedures and is estimated to take a total of 77 days and to require an expenditure equal to 58% of Indian per capita income. The table indicates that these costs are significantly higher in BRIC countries (Brazil, the Russian Federation, India and China) than in the average country in the top quartile of the world distribution of per capita income, by enough to constitute significant barriers to entry and to help explain why the BRICs remain poorer than the top quartile.

Table 2.1. Entry barriers in select countries

	Procedures	Time	Cost
Brazil	15	63	20%
Russia	20	57	20%
India	10	77	58%
China	12	92	14%
Richest quartile	7	24	10%

Source: Djankov, S., R. La Porta, F.Lopez-de-Silanes, and A. Shleifer (2002). "The regulation of entry", *Quarterly Journal of Economics*, Vol. 117, pp. 1-37.

Incorporating entry barriers into the theory shows however that effects on innovation can be mixed. On the one hand, the new entrants will bring new possibilities for innovation; on the other hand, established firms might be discouraged from undertaking R&D projects for fear that an entrant will arrive with such an advanced technology that the firm is driven out of business anyway. The theory suggests that the threat of foreign entry by firms on the world technology frontier will indeed discourage R&D by domestic firms that are far from the frontier to begin with but that it will encourage R&D by domestic firms that are close to the frontier; they will realise that the best defense against a frontier entrant is to get onto the frontier themselves.

Figure 2.1 shows that this theoretical prediction seems to be borne out by UK manufacturing data. Both curves show a partial non-linear regression of domestic UK establishment productivity growth on lagged foreign entry in the establishment's industry over the period 1987-93. The upper curve shows the regression curve when the sample is restricted to observations in which the industry's distance to the frontier (as measured by productivity relative to US productivity in the same industry) is above the median for the whole sample, while the lower line shows the regression for observations below the median distance to the frontier. It is clear that domestic establishments in industries that are closer to the frontier tend to have faster productivity growth when there has been a lot of entry recently while the reverse is true for establishments in industries that are further from the frontier.

Figure 2.1. Entry, incumbent TFP growth and distance to frontier, 1987-93

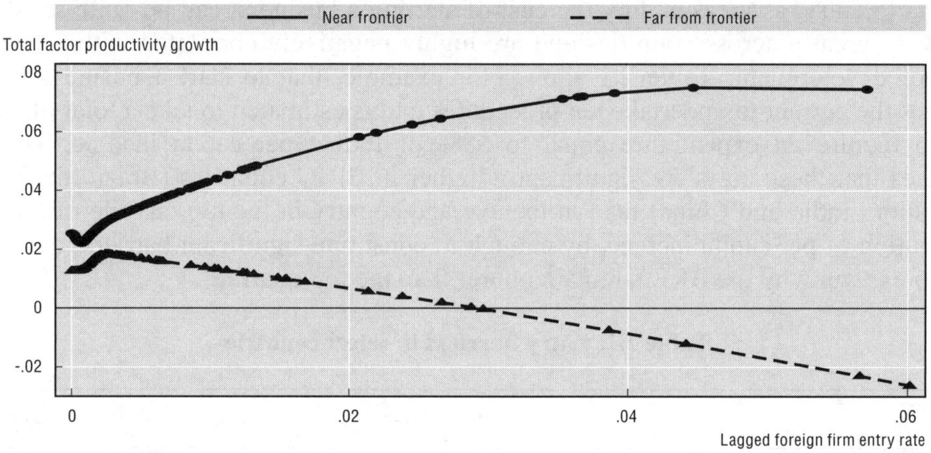

Source: Aghion, P., R. Blundell, R. Griffith, P. Howitt and S. Prantl (2009), "The Effects of Entry on Incumbent Innovation and Productivity", *Review of Economics and Statistics*, 91, forthcoming.

One of the questions that this conference will be addressing is the extent to which competition might help to spur innovation in fast-growing middle-income countries (BRICs). The results of Figure 2.1 suggest that, to the extent that many of the industries in the BRICs are relatively far from the frontier, increased trade liberalization might discourage innovation on the whole. This is where the context-dependence of Schumpeterian theory helps to make sharper predictions. Specifically, the theory predicts that if combined with appropriate labour market regulations which allow firms to respond adequately, the threat of foreign entry might even act as a spur to domestic innovation in the relatively more advanced sectors of the middle-income countries. Indeed, this is what Aghion *et al.* (2006) found when exploring the effects of delicensing in India. On the whole, delicensing had a negligible effect on domestic innovation. But in regions with pro-firm labour market regulations the effect was positive, implying that even in India it is possible to use competition as a spur to innovation in many sectors, provided that it is accompanied by appropriate regulatory reforms.

Corporate governance problems

The original versions of Schumpeterian growth theory assumed that firms were profit maximisers. When agency problems arise, however, and firms are run in the interests of managers rather than of shareholders, increased competition may lead to faster growth by reducing the slack available to managers. Aghion *et al.* (1999) provided a variant of Schumpeterian theory in which managers seeking a quiet life want to avoid the stress of innovation and therefore innovate only as much as needed to keep the firm solvent in the face of competition. In such a setting, increased competition will reduce the profit flow that would otherwise allow managers to remain solvent even when their technology has fallen far below the frontier, and thus forces them to innovate or die, to use Porter's (1990) expression. A good example of this phenomenon is provided by the US Saving and Loans industry, which until the mid-1970s was well protected by regulations and was so non-innovative that it failed to come up with the adjustable rate mortgages that protected other countries' mortgage lenders when inflation rose. It therefore found itself locked into mortgages at low rates that had to be financed by high-rate term deposits of shorter duration than the mortgages.

To the extent that corporate governance is an even bigger problem in low- and middle-income countries than in rich countries, this mechanism by which competition fosters innovation and growth is probably even more relevant for BRICs than for OECD countries. Maloney's (2002) analysis indicates that one of the most significant problems of industries that were protected from competition in Latin America under import substitution policies was that they were not innovative and fell far behind the world technology frontier.

In theory it could be argued that the positive effect of competition on growth will be reversed if managers choose to use their cash flow not to enjoy a quiet life but to build scientific empires. Bell Labs in the United States is a case in point. When AT&T was broken up, Bell Labs was reduced to a more prosaic supporting role and soon ceased to be the source of fundamental scientific and technological progress that it had been when the company had enough slack to support the basic research that led to such breakthroughs as the transistor, the laser, and even the discovery of the background cosmic radiation that confirmed the big-bang theory of the origin of the universe. However I doubt that such cases are common even in rich countries, let alone in the poor and middle-income countries that account for a small fraction of the world's fundamental research.

Escape competition and the inverted U

In principle, the incentive to innovate should depend not on the absolute level of monopoly profits that a successful innovator could earn, but on the incremental profits that would result from an innovation. This distinction between absolute and incremental profit is irrelevant for a small start-up firm whose profit would be zero in the absence of a successful innovation. But for many firms that are already producing and earning profits, a successful innovation will to some extent cannibalise existing profits, making incremental profits significantly smaller than absolute profits.

Moreover, product market competition affects incremental profits quite differently from the way it affects absolute profits. As Boone (2000) has explained in detail, more intense competition, measured in any of a variety of ways, gives a big advantage to a more productive firm, by allowing it to take better advantage of its superior technology. This can easily result in an increase in incremental profit from an innovation. Put differently, although an increase in competition may reduce the absolute profit of a successful innovator, it will reduce even more the absolute profit of an unsuccessful innovator.

Thus firms can have an incentive to innovate in order to escape competition from rivals with superior or even identical technologies. This incentive will be larger the more competitive the environment. In the absence of strict enforcement of antitrust regulations, even firms with no technological advantage over their actual and potential rivals can make substantial profits from collusive agreements, perhaps backed up by regulatory authority that helps protect their market from intruders. But expose them to competition and they will innovate or disappear.

Aghion *et al.* (2001) developed a growth model in which all innovations are assumed to be conducted by incumbent firms whose pre-existing technology would be devalued by an innovation. In this model, the Schumpeterian appropriability effect is present in industries with a large gap between the technology leader and the laggards, but the escape competition effect is present in industries in which the firms are all at a similar technology level. At the level of the whole economy this model implies that an increase in competition will have either a monotonically increasing effect or an inverted-U effect on the pace of innovation and productivity growth.

To see how this inverted U works, consider first what happens in an economy with very weak competition. Firms in neck-and-neck industries have relatively little incentive to innovate, whereas firms that are laggards in uneven sectors have a strong incentive to catch up with the leaders and share in their collusive profits. In such an economy most industries will end up being neck-and-neck. But this is the situation in which most firms would be stimulated to innovate more rapidly if competition were more intense. Thus, starting from a low level of competition, an increase in competition is likely to result in faster economic growth.

But now consider what happens if the economy becomes extremely competitive. In this case the incentive to innovate becomes so strong in neck-and-neck industries, where no one is able to earn any rents because no one has a technological advantage, that the industry will not long remain neck-and-neck. Thus most industries would soon be in the uneven state, and the dominant effect of a further increase in competition would be to reduce the rate of economic growth.

Figure 2.2 shows that, across different UK manufacturing firms, more competition (*i.e.* a reduction of the Lerner index) does indeed have an inverted-U effect on the rate of innovation, as measured by a citation-weighted patent count. Moreover, most UK firms are in the upward sloping part of the curve, where competition has a positive effect on growth. This suggests that the economy-wide effect on UK productivity growth is probably positive.

Even more than the results shown in Figure 2.1, it suggests that the economy-wide effect in most middle-income countries, where generally speaking competition is not as intense as in OECD countries, should be positive. However, one must qualify this with the realisation that to the extent that the market served by the typical BRIC is an international market and that the local firm is far behind the frontier, the dominant effect of trade liberalisation on domestic innovation may be the negative Schumpeterian effect that applies to distant laggards.

Two other important lessons come from this consideration of the escape competition effect. First, what really matters is *ex ante* competition, not *ex post*. Indeed, more *ex ante* competition is likely to result in less *ex post* competition. That is, a small Lerner effect will intensify innovation so strongly in neck-and-neck industries that most industries will end up being dominated by the small number of firms that have survived that intense competition. Thus, somewhat paradoxically, more competitive sectors are likely to become being more highly concentrated. This is why we use a Lerner index rather than a concentration ratio as our measure of competition.

Figure 2.2. Innovation and competition: the neck and neck split

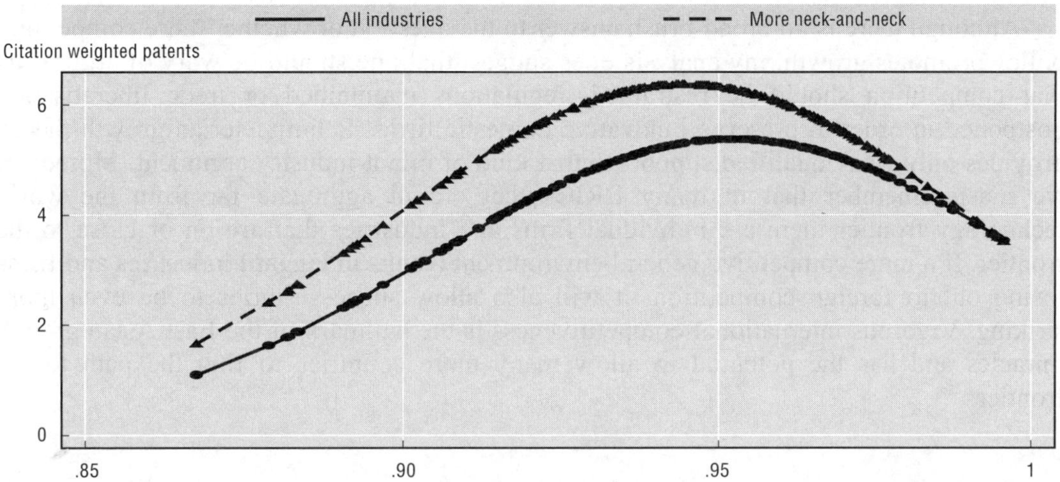

Source: Aghion, P., N. Bloom, R. Blundell, R. Griffith, R. and P. Howitt (2005), "Competition and Innovation: An Inverted U-Relationship", *Quarterly Journal of Economics*, Vol. 120, pp. 701-728.

The other lesson has to do with the interaction between competition policy and patent policy. Competition policy and patent policy are complementary, in the sense that antitrust is more effective when patent policies are relatively lax. More specifically, as Aghion *et al.* (2001) show, growth is more likely to be stimulated by more intense competition when patent laws allow laggards occasionally to catch up with the leader in their industry than when this is never possible. This is because the *ex post* concentration that results from a fiercely competitive environment can lead to a lot of satisfied leaders that see little reason to innovate and a lot of discouraged laggards that have given up. But allowing patents to lapse sooner or to be broken more easily can bring some of the latter back into the R&D-intense neck-and-neck state, at relatively little cost of discouraging innovation on the part of the neck-and-neck firms. Once again we see that what matters for economic growth is not just competition policy but the whole package of policies.

Lessons for competition policy

What are the lessons of this brief and selective tour of Schumpeterian growth theory for policy makers trying to use competition policy to promote growth in rapidly growing middle-income countries? The main lesson is the very general one that appropriate growth policy is context-dependent. What works in one country does not necessarily work in another, and even within a country what works in one industry does not necessarily work in another. In particular, distance to the frontier is a key criterion. Generally speaking, the further from the frontier a country (industry) is, the less likely it is, other things being equal, that more intense competition will promote economic growth.

Firms that are far behind their rivals technologically are more likely to be discouraged from innovating by the prospect of more intense competition than firms that are already close to the frontier.

Another closely related lesson is that we miss a lot if we think of competition policy alone rather than thinking of combinations of policies that involve a change in the competitive environment. For example we have seen that it may take more relaxed labour-market regulation or more relaxed patent laws in order for more competition to stimulate growth.

Although there is no broad-brush answer to the question of whether more competition policy promotes growth, my analysis does suggest that one should be wary of arguments that competition should be restrained, regulations maintained or trade liberalisation postponed in order to preserve innovative domestic firms. Schumpeterian growth theory provides only weak qualified support to that kind of infant-industry argument. Moreover, we must remember that in many BRICs that are in aggregate far from the world technology frontier there are individual firms and industries that are on or close to the frontier. If a more competitive general environment results in laggard industries and firms losing out to foreign competition, it will also allow success stories to be even more striking. Vigorous international competitiveness is the hallmark of the East Asian growth miracles and has the potential to allow many more countries to find the path to the frontier.

References

Aghion, P., N. Bloom, R., Blundell, R. Griffith and P. Howitt (2005), "Competition and Innovation: An inverted U-Relationship", *Quarterly Journal of Economics*, Vol. 120, pp. 701-728.

Aghion, P., R. Blundell, R. Griffith, P. Howitt and S. Prantl (2009), "The Effects of Entry on Incumbent Innovation and Productivity", *Review of Economics and Statistics*, Vol. 91, forthcoming.

Aghion, P., R. Burgess, S. Redding and F. Zilibotti (2006), "The Unequal Effects of Liberalization: Evidence from Dismantling the License Raj in India", CEPR Discussion Paper No. 5492.

Aghion, P., M. Dewatripont and P. Rey (1999), "Competition, Financial Discipline, and Growth", *Review of Economic Studies*, Vol. 66, pp. 825-852.

Aghion, P., C. Harris, P. Howitt and J. Vickers (2001), "Competition, Imitation and Growth with Step-by-Step Innovation", *Review of Economic Studies*, Vol. 68, pp. 467-492.

Aghion, P. and P. Howitt (1992), "A Model of Growth through Creative Destruction", *Review of Economic Studies*, Vol. 60, pp. 323-351.

Aghion, P. and P. Howitt (1998), *Endogenous Growth Theory*, MIT Press, Cambridge, MA.

Aghion, P. and P. Howitt (2006), "Appropriate Growth Policy: An Integrating Framework", *Journal of the European Economic Association*, Vol. 4, pp. 69-314.

Aghion, P. and P. Howitt (2009), *The Economics of Growth*. MIT Press, Cambridge, MA.

Blundell, R., R. Griffith and J. Van Reenen (1995), "Dynamic Count Data Models of Technological Innovation", *Economic Journal*, Vol. 105, pp. 333-344.

Boone, J. (2000), "Measuring Product Market Competition", CEPR Discussion Paper No. 2636.

Crafts, N.B. (1996), "Economic History and Endogenous Growth", in D. Kreps and K. Wallis (eds.), *Advances in Economics and Econometrics: Theory and Applications*, Cambridge University Press, Cambridge.

Djankov, S., R. La Porta, F. Lopez-de-Silanes and A. Shleifer (2002), "The Regulation of Entry", *Quarterly Journal of Economics*, Vol. 117, pp. 1-37.

Maloney, William F. (2002) "Technology, Growth and Resource Abundance: Observations from Latin America", unpublished, The World Bank.

Nickell, S.J. (1996), "Competition and Corporate Performance", *Journal of Political Economy*, Vol. 104, pp. 724-746.

Porter, M. (1990), *The Competitive Advantage of Nations*, Free Press, New York.

World Bank (2008), *The Growth Report: Strategies For Sustained Growth and Inclusive Development*, The World Bank, Washington, DC.

Chapter 3

Korea and the BICs (Brazil, India and China): catching-up experiences

V. Chandra, I. Osorio-Rodarte and C.A. Primo Braga[1]

This chapter tests the neo-Schumpeterian model of Aghion et al. *with industry-level data to analyse how Brazil, India and China are catching up to Korea's technological frontier in a globalised world. It validates Aghion* et al.*'s inverted-U hypothesis (industries closer to the technological frontier innovate to escape competition while longer distances discourage innovating). It suggests that for effective catch-up, distance-shortening (or innovation-enhancing) policies may be a necessary complement to liberalisation. Korea and China combined a variety of distance-shortening policies with financial subsidies to promote high-technology industries and an export-led growth strategy. Following liberalisation, they leveraged competition to spur catch-up. Brazil, which in 1980was as rich as Korea, and India, which was as rich as China, are catching up more slowly. Import-substitution industrialisation strategies saddled Brazil and India with a large anti-export bias, and unfocused attention to innovation-enhancing policies dampened global competitiveness. Post-liberalisation, many of their industries were too far from the technological frontier to benefit effectively from competition. The catch-up experiences of the BICs show that distance from the technological frontier matters and that the design of country-specific distance-shortening policies can be an important complement to trade liberalisation in promoting catch-up with richer countries.*

1. V. Chandra is Senior Economist, I. Osorio-Rodarte is a consultant, and C.A. Primo Braga is Director of the Economic Policy and Debt Department, PREM Network, The World Bank. The authors are indebted to N. Cooke, C. Dahlman, H. Hesse, P. Howitt, V. Thomas, V. Nehru, B. Hoekman, J. Daly, L. Garrido, J. Crespo-Cuaresma, G. Prennushi and participants at the OECD November 2008 Conference and World Bank March 2009 seminar for comments. The findings, interpretations and conclusions are those of the authors, and do not necessarily represent the views of The World Bank, its Board of Directors, or any of its member countries.

Introduction

The story of how Korea leveraged its growth strategy to begin closing the income gap with OECD countries and in less than three decades transformed itself into an innovation leader is well known (*e.g.* Commission on Growth and Development, 2008; Suh *et al.*, 2007; World Bank, 1998). The catch-up game is also being pursued by other emerging market economies such as Brazil, India, China (BICs), South Africa, Chile and the Russian Federation. Interested readers are invariably drawn to two questions with important implications for public policy. Are trade liberalisation and global integration a hindrance or a help in catching up with high-income economies? Why is it that after implementing trade reforms, only a handful of countries are in fact catching up?

Developing countries have good reasons to study the Korean growth experience. It is true that even though income levels doubled (in 2000 constant USD) in Korea, Brazil, and other middle-income countries between 1965 and 1980, a distinct divide persisted between this group and the OECD countries (Figure 3.1). Around 1981, however, in a break from its steady growth trajectory, Korean growth accelerated and began catching up with the high-income non-OECD countries. By 2007, its income per capita was USD 19 690 in nominal terms. In comparison, Brazil's income per capita was USD 5 910.

Figure 3.1. Korea and Brazil – catching up with the OECD

GDP per capita in constant 2000 USD

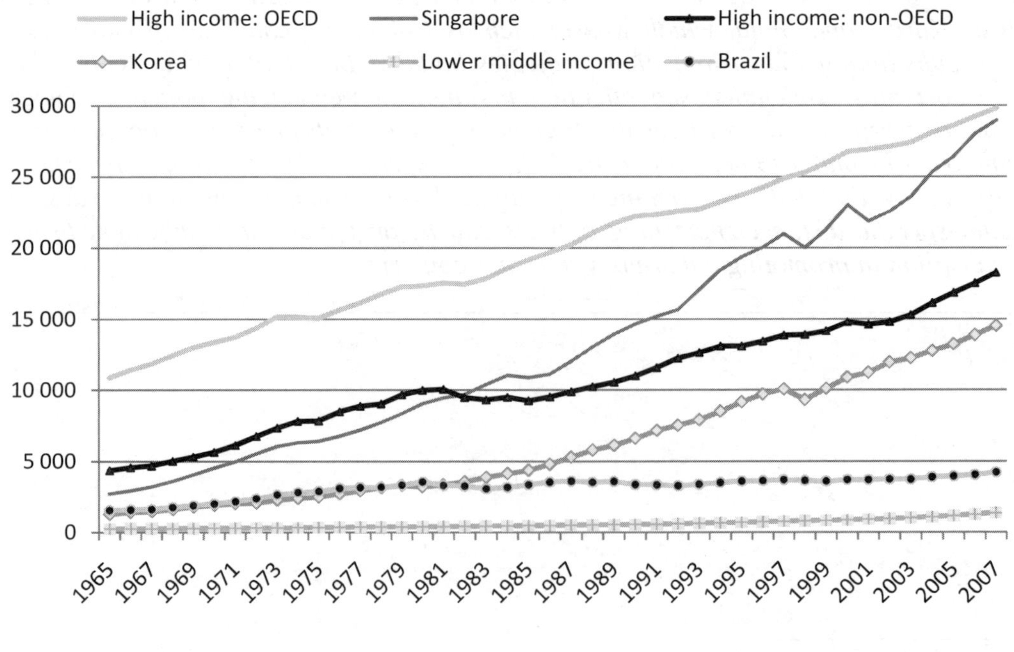

Source: World Bank, *World Development Indicators*.

China's tale of catch-up is equally impressive. Around the mid-1980s, when its per capita income was still below USD 200 (in 2000 constant USD), China began closing the gap with India (Figure 3.2). By 2007, its per capita income had risen to USD 2 360 while

India's was only USD 950 in nominal terms. In sum, starting at levels of income per capita typical of a low-income country in the 1950s, Korea was already a middle-income country by the early 1980s, joined the OECD in 1996, and has been catching up with the OECD average income per capita in spite of the temporary setback associated with the Asian financial crisis (1997-98). At the same time, the income gap between Korea and the BICs has widened.

Figure 3.2. China and India in the catch up game

GDP per capita in constant 2000 USD

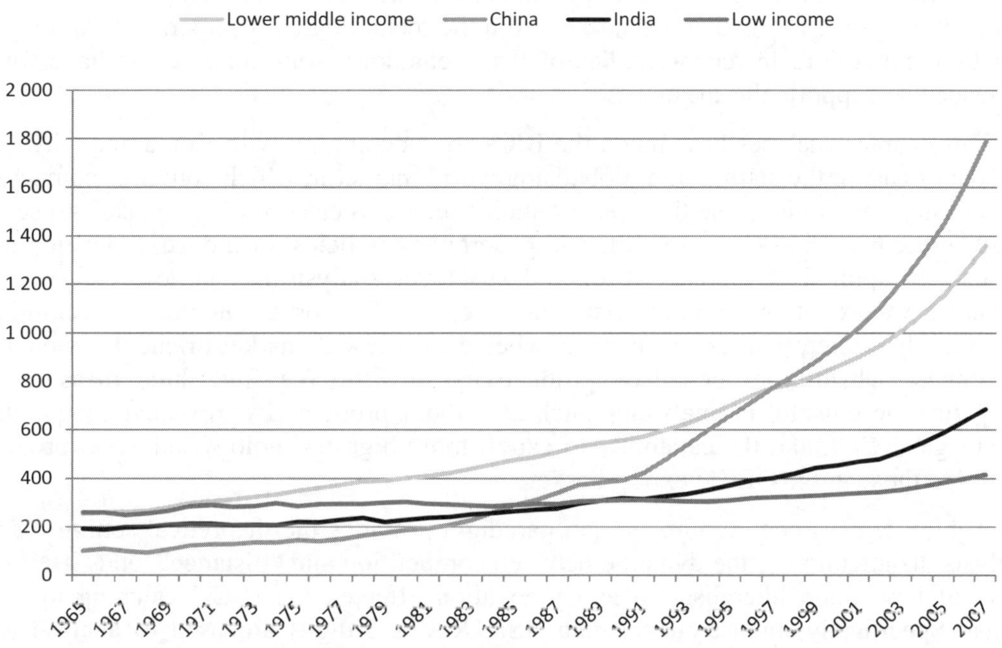

Source: World Bank, *World Development Indicators*.

Since the 1980s, many low- and middle-income developing countries have liberalised their economies and pursued greater integration with global markets with the expectation of becoming more prosperous, but only a handful are catching up with OECD economies. What can they learn from the experience of the BICs and Korea? And, among the BICs, what is China doing differently to overtake Brazil and India?

Standard neoclassical economic theory does not offer conclusive answers to these questions. It asserts that increasing the level of competition through trade liberalisation, for example, improves efficiency, spurs innovation and increases productivity growth. Empirical evidence confirms that productivity growth can contribute as much as 50% of overall economic growth (Loayza *et al.*, 2005; De Ferranti *et al.*, 2003); it also suggests that productivity growth is powered by innovation. However, it does not indicate that liberalisation automatically breeds innovation. In fact, in many developing countries, trade liberalisation has not been followed by sustained innovation-powered productivity gains that have led to higher per capita income in the long term.

Economists have proposed two alternative paradigms to explain why trade liberalisation or increased competition may not lead to a sustained increase in productivity. Schumpeter contended that economic rents were essential for innovation to occur. From this perspective, by dissipating rents, competition can adversely affect innovation. Recently, in advancing the new growth theory which is grounded in a Schumpeterian framework, Aghion *et al.* (2004, 2005, 2006 and 2008; Acemoglu *et al.*, 2006) hypothesised that competition can have variable effects on productivity.[2] It need not benefit all firms in an industry uniformly. Firms that are relatively technologically sophisticated and hence close to the most technologically sophisticated firms (*i.e.* the proxy for the innovation frontier) are able to "escape" competition from new entrants when faced with threat of entry. They innovate to become more competitive and reap productivity gains. Firms that are farther from the technological frontier, in turn, may be hurt by competition. In recent studies of firm behaviour, some researchers have found evidence that supports this argument.

This chapter analyses how much the BICs are catching up with Korea and why. The analysis is cast in the setting of a globalising world market in which countries compete at the industry level to increase their per capita incomes and catch up. This makes sense. As tariffs, other trade barriers to imports and government policies for industrial development are usually applied at industry level, industry-level analysis has policy relevance. In particular, we are interested in understanding the role of export orientation in catching up. We are also interested in evaluating whether domestic market-oriented production (*i.e.* import substitution) penalised productivity growth. We find three measures of competitiveness useful in analysing catch-up: labour productivity, revealed comparative advantage (RCA), and the capability to export more high-technology industrial products similar to those in the OECD export basket.

Aghion *et al.*'s new Schumpeterian paradigm provides the theoretical setting for the analysis. Its attention to the dynamic between competition and "distance" lends itself to a study of how trade liberalisation and innovation affected the BICs' catching up with Korea. Specifically, industry-level data (ISIC Rev. 2 3-digit) are used to analyse why some industries in some BICs are catching up with Korea while others are not. Catch-up is measured by growth in labour productivity and by the level of labour productivity which serves as a proxy for total factor productivity (TFP).[3] Conventionally, the latter is considered a proxy for innovation. As Korea is the benchmark, its labour productivity is treated as the technological frontier. The distance from the frontier for any given industry is defined as the ratio of one of the BICs' labour productivity levels to Korea's in that specific industry. The lower its labour productivity, the greater its distance from the Korean technological frontier. Catch-up is faster if the country is able to shorten a particular industry's distance to the frontier. Trade liberalisation, relying on a reduction in weighted tariffs, is used as a proxy for industry-level competition.

---

2. Unless otherwise indicated, in the following discussion the term "Aghion *et al.*" will be used to refer to this body of literature.

3. In the absence of data on the stock of capital at the industry level, labour productivity is used as a proxy for total factor productivity. This is a reasonable assumption and can be verified from the Investment Climate Assessment surveys of enterprises for Brazil (2002; 1 485 firms), China (2002: 1 153 firms) and India (2004: 1 420 firms). An analysis of the survey data by industry as well as for the whole manufacturing sector using various definitions of TFP indicates that there is a strong correlation between value added per worker and TFP in the BICs.

The analysis indicates several interesting characteristics of the global catch-up phenomenon as well as a new result with policy implications that complement those of the Aghion *et al.*'s model:

- First, on the eve of trade liberalisation, the timing and pace of which varied significantly among the BICs, most of the BICs' leading industries (which account for the bulk of their GDP) were a "very long" distance from the Korean technological frontier. In a set of about 30 industries, there were eight exceptions[4] for which Brazil was within 80-100% of the technological frontier.
- Second, distance to the frontier is not fixed. In most industries, the frontier – the Korean industry's labour productivity – is growing very fast. This makes it more challenging for the BICs to catch up unless their industries are growing even more rapidly. With distinctly higher rates of productivity growth, certain Chinese industries have acquired and maintained superior performance and had more success in shortening the distance to the technological frontier. The fact that this result holds even after controlling for the efficiency-improving effects of competition in all industries supports the hypothesis that some other factor such as innovation spurred these productivity gains. It is not possible to determine at the industry level which innovation strategies are enabling China's success because of data limitations, but the remarkable variation in their performance suggests that they are not industry-neutral.
- Third, a test of Aghion *et al.*'s (2006) hypothesis in our model (conducted at the industry level) finds empirical support for their findings.
- Fourth, we also find that regardless of competition, if an industry is too far from the frontier, it has no incentive to innovate and become competitive. As a result, it is unlikely to improve its productivity and catch up. This result has interesting policy implications. If policies that can shorten the distance to the technological frontier are implemented, it is possible to boost productivity. This in turn should boost competitiveness. This result directly puts the onus of distance-shortening or innovation-enhancing (these terms are used interchangeably here) policies on the BIC country which desires to catch up in the global market. In contrast to the Aghion model, which emphasises competition policy as the source of catch-up, our result suggests that in industries which are relatively far from the technological frontier, there may be a need to implement innovation-enhancing policies in addition to trade liberalisation.

This chapter first briefly discusses Aghion *et al.*'s theoretical framework and then looks at data limitations. The following section sets the stage for benchmarking Korea's technological frontier and briefly discusses Korea's catch-up. The following sections examine China's catch-up with Korea, Brazil and India. On the basis of that discussion, the balance between competition and innovation is considered in a comparative context and the empirical results are presented. A final section concludes.

4. These were transport equipment, industrial chemicals, non-electrical machinery, paper and paper products, printing and publishing products, professional and scientific equipment, non-ferrous metals, and petroleum refineries.

Theory and some evidence

The framework developed by Aghion *et al.* introduces potential entry and the effects of the threat of entry on the incumbent's investment in innovation to boost productivity growth. Foreign firms are assumed to represent the technological frontier. If the foreign firm enters and competes with a local firm which has lower productivity, it takes over the market and becomes the new dominant firm in that industry. If it competes with the local firm that has equivalent productivity, competition drives the profits of both firms to zero. If the potential entrant knows in advance the post-innovation productivity level of the incumbent, it will enter only if the latter's post-innovation productivity level is below the technological frontier. Of course, assuming perfect foresight, the foreign firm will never enter if the local firm has achieved frontier innovation levels.

An increased threat of a new product has effects that are similar to those of liberalisation. Increasing the threat of entry encourages innovation in advanced firms, and discourages it in backward firms. The intuition for this comparative statics is as follows: the higher the threat of entry, the more instrumental innovations will be in helping incumbent firms that are already close to the technological frontier to retain the local market. However, if they are too far away from the frontier, they have no possibility to win against a potential entrant. In that case, a higher threat of entry will only lower the expected net gain from innovation. This translates into an inverted-U relationship between competition and innovation.

Aghion and Griffith (2008) note that according to the Schumpeterian model, growth results from entrepreneurial innovation. Each new innovation destroys monopoly rents generated by previous innovators. Those in favour of competition – which is fostered by trade liberalisation, free entry by foreign investors, and monetary integration – underscore that competition is necessary for innovation because it encourages new entry and provides an incentive for incumbent firms to innovate to survive competition. In contrast, the proponents of the endogenous growth paradigm argue that more intense competition may dissipate the monopoly rents that are necessary for a firm to innovate by investing in R&D and other factors that propel innovation. This can discourage entry and the emergence of new industries and have a negative effect on productivity growth (Dixit and Stiglitz, 1977; Romer, 1990; Aghion and Howitt, 1992; Grossman and Helpman, 1991). As emphasised in Aghion and Griffith (2008, p. 16), "These models predict that property rights protection is growth enhancing; however, for exactly the same reason they also predict that competition policy is unambiguously detrimental to growth. Patent protection protects monopoly rents from innovation, whereas increased product market competition destroys these rents. Thus if we were to take these models at face value, when making policy prescriptions, we would never advocate that patent policy and antitrust be pursued at the same time, at least not from the point of view of promoting dynamic efficiency."[5]

In the theoretical framework of Aghion *et al.*, when firms innovate to compete and retain or even expand their market share because of threat of entry, they engage in "escape competition" tactics; when firms fail to innovate and compete, the authors argue that they are falling prey to Schumpeterian effects. In their early models, price costs margins are used as the competition indicator and citation-weighted patents as a measure of innovation. In their more recent work, the authors find that technologically advanced

---

5. For a discussion of the role of intellectual property rights in the process of economic development see Primo Braga *et al.* (2000).

entry by foreign firms has a positive effect on innovation if the domestic firms are close to the frontier. The effect of entry on total factor productivity growth interacts negatively with respect to the distance to the frontier. Aghion *et al.* (2008) find support for the inverted-U hypothesis by testing it with firm level data from Indian states for 1980-97, as do Aghion *et al.* (2004) for the United Kingdom between 1970 and 1994. They also examine de-licensing and its effect on within-industry inequality in productivity.

Carlin *et al.* (2004) test the inverted-U hypothesis using BEEPS data from transition economies. They conclude that innovation is greater in monopolistic industries when there is little competition. Gorodnichenko *et al.* (2008) also conclude the same with additional data from BEEPS 2002 and 2005. Sabirianova *et al.* (2004, 2005) find support for the heterogeneous effect of firm entry on firm performance using data from Russian and Czech firms. In their most recent paper, Gorodnichenko *et al.* (2008) use data from 11 500 firms from over 27 countries and find that product market competition has a negative effect on innovation. However, their findings do not validate the inverted-U relation between competition and innovation as proposed in Aghion and Griffith (2008). They support, however, the basic Schumpeterian view that monopolist market structures boost innovation.

The remainder of this chapter uses Aghion *et al.*'s framework to discuss how much the BICs have succeeded in catching up with Korea at the industry level.

Data limitations

UNIDO's ISIC Rev. 2 3-digit industry level data are used for about 30 industries for Brazil, India, China and Korea to test the predictions of the Schumpeterian model at the country level. A finer level of disaggregation, *i.e.* ISIC Rev. 2 4-digit, is available for Brazil and India but not for China. Moreover, Brazilian data in the UNIDO dataset are available only for 1996-2005. Trade data are available for all countries and all industries for the 1976-2005 period. Data for weighted tariffs are available for industries and most countries from the 1980s. Unfortunately, a full time series for non-tariff barriers is not available. Data from the Penn World Tables is used to calculate TFP growth at the national level for all four countries for 1964-2000.

Catch-up is defined as the increase in an industry's competitiveness in the global market with respect to the Korean technological frontier. Labour productivity, revealed comparative advantage from world trade data, and the technological content of exports are used as proxies. Competition is marked by industry-level tariff reduction. Ideally, innovation is measured by changes in innovation indicators, but as the latter are unavailable at industry level, innovation is not explicitly measured. There is ample national-level data on innovation indicators in the *OECD Science, Technology and Industry Outlook* (2008a) and the *OECD Science, Technology and Industry Scoreboard* (2007). This information is not repeated here.

Korea's economic performance was powered by uninterrupted and rapid growth in labour productivity from the 1960s. Korea's ability to outstrip middle-income countries like Brazil after 1980 and catch up with the OECD was fuelled by a sustained increase in technological innovation as illustrated by trends in labour productivity (Figure 3.3) (see footnote 3). This record was unbroken until the mid-1980s when similar trends emerged in Chinese labour productivity. They explain most of the miracle associated with China's catch-up with Korea. While the trend in labour productivity in the Indian economy since the early 1990s pales in comparison with China's, it is impressive by the standards of many other developing economies.

A collapse in productivity marked Brazil's "lost decade" and affected its ability to start catching up in the 1990s. An upturn in the trend in Brazil's labour productivity since 2004 suggests that Brazil may be beginning to benefit (in terms of innovation performance) from market-oriented reforms in the 1990s.

Figure 3.3. Benchmark: trends in catch-up with the Korean technological frontier

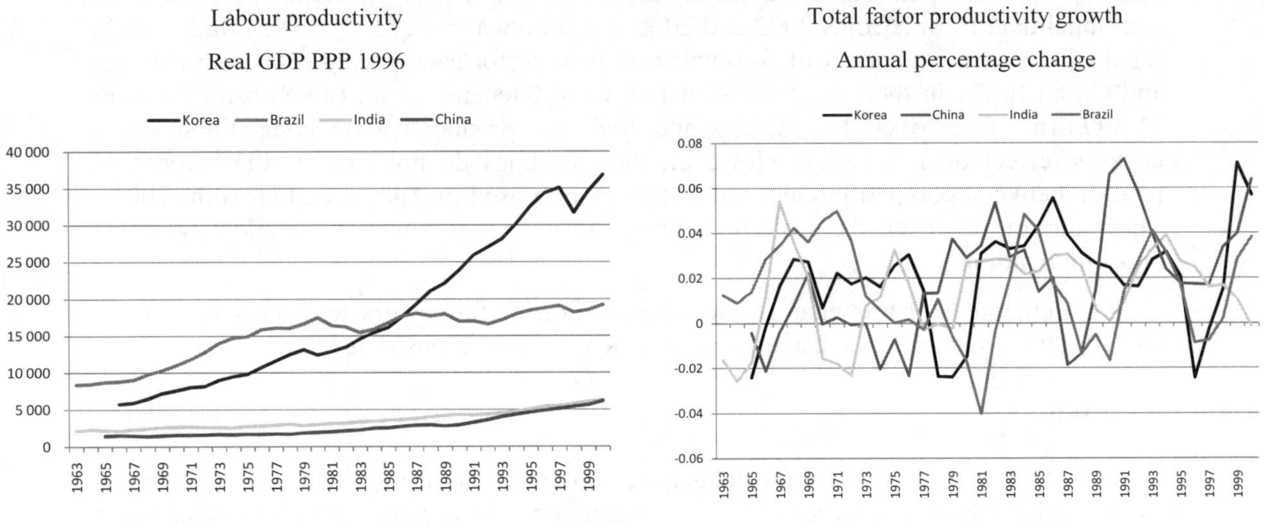

Source: Penn World Tables.

How did Korea catch up with the OECD?

Figure 3.1 illustrates the rapid pace of Korea's catch-up with the OECD and contrasts it with the experience of Brazil, which had a similar level of per capita income around 1980. Figure 3.4 displays a set of two figures each for the eight industries that became Korea's industries with the highest share of manufacturing value added (MVA). The left-hand figures present the levels of labour productivity in the BICs relative to the Korean frontier, *i.e.* Korean labour productivity is set equal to 100. The right-hand figures display the trend in labour productivity in Korea and the BICs. The first point to note is that the Korean frontier is far above the relative productivity levels in the BICs in most industries, with the exception of a few in Brazil. The second point to note is that the technological frontier is a moving target as Korean labour productivity was growing much faster than that of the BICs.

Korean competitiveness is also captured by its evolving revealed comparative advantage (RCA) in an industry. This measure is relevant in the global market place in which countries must compete if they rely on exports as a driver of growth. The eight figures in Figure 3.6 display the trends in the RCA of Korea's leading industries. Most notable are the trends in relatively high-technology exports such as electrical machinery, non-electrical machinery, transport equipment, and iron and steel. The declining trend in exports of the textiles and wearing apparel industry is consistent with the diminishing importance of these industries in its growth strategy (Table 3.1).

Figure 3.4. The BICs and Korea: distance of largest industries to the Korean technological frontier, 1980-2005

Product 321: Textiles

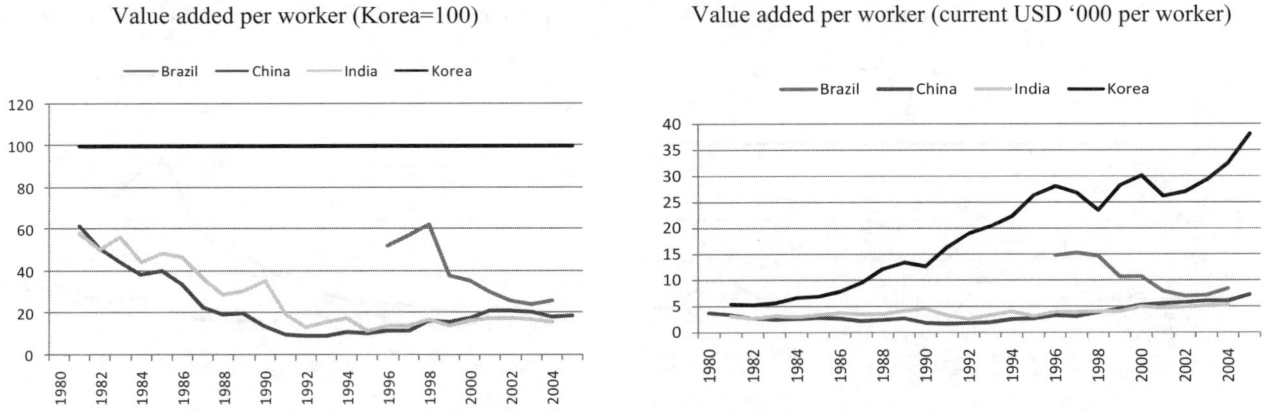

Product 311: Food products

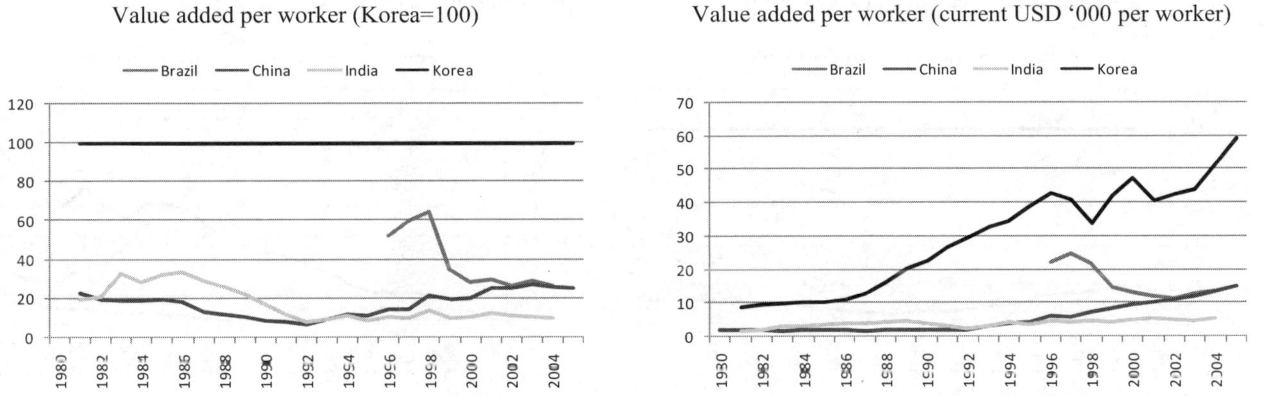

Product 382: Machinery, except electrical

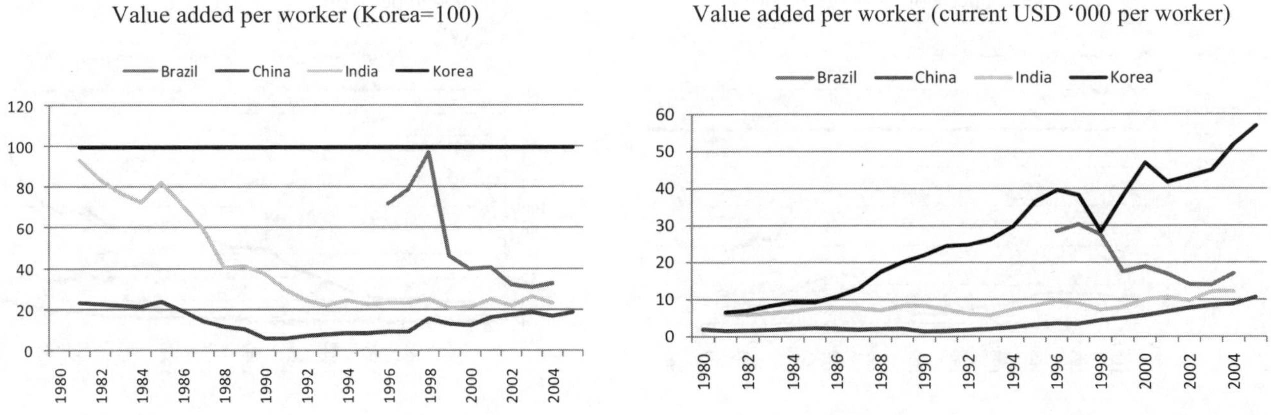

Figure 3.4. The BICs and Korea: distance of largest industries to the Korean technological frontier, 1980-2005 (cont'd.)

Product 383: Machinery, electric

Value added per worker (Korea=100)

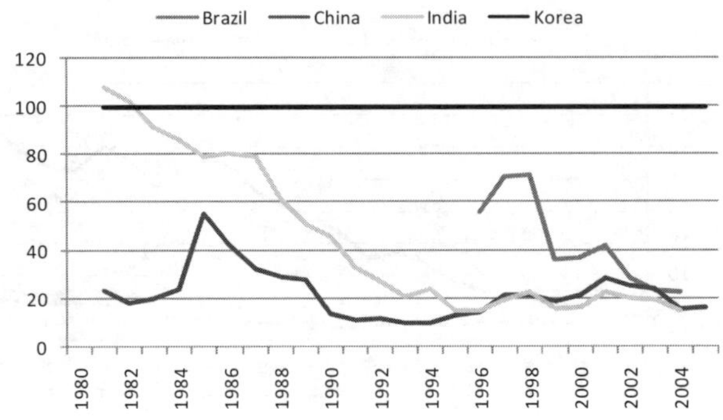

Value added per worker (current USD '000 per worker)

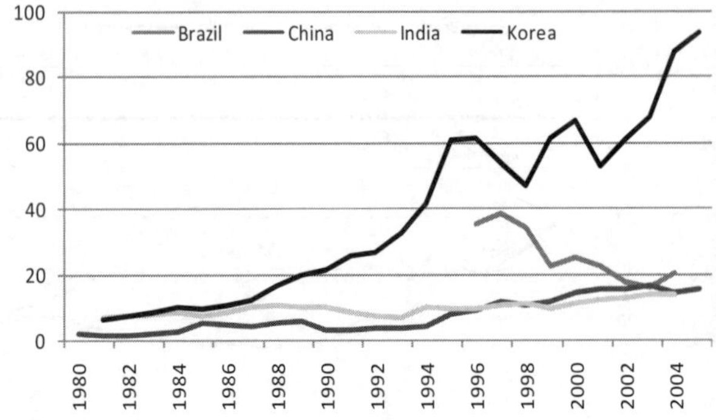

Product 351: Industrial chemicals

Value added per worker (Korea=100)

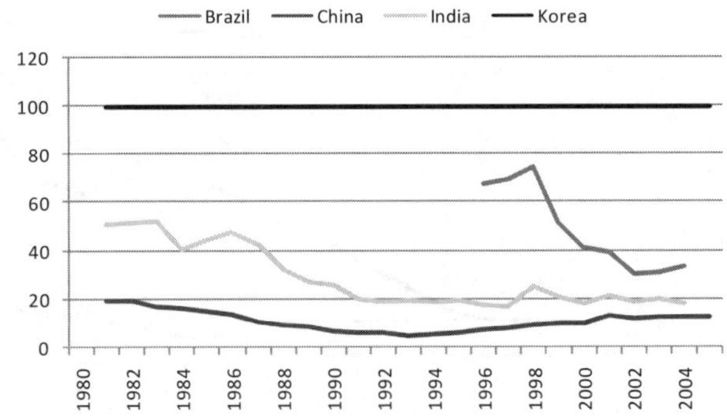

Value added per worker (current USD '000 per worker)

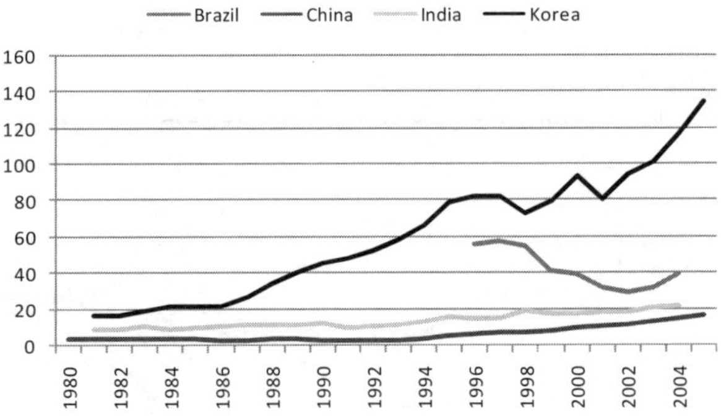

Product 371: Iron and steel

Value added per worker (Korea=100)

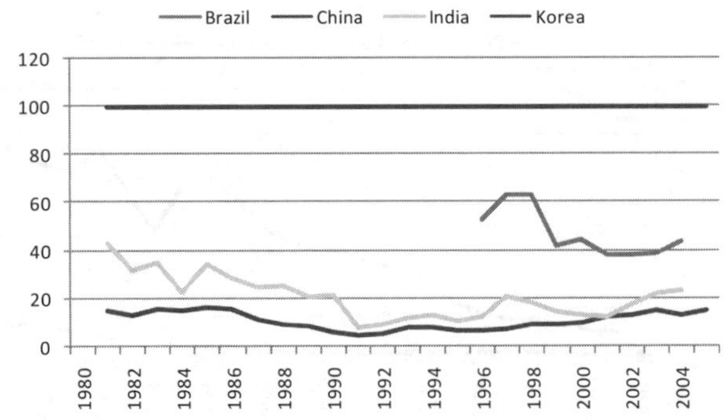

Value added per worker (current USD '000 per worker)

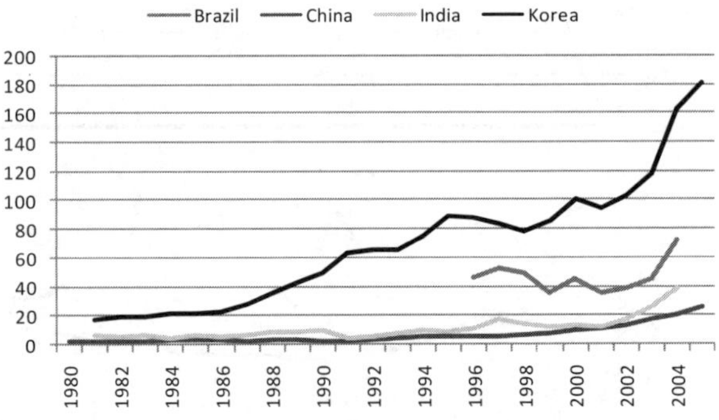

Figure 3.4. The BICs and Korea: distance of largest industries to the Korean technological frontier, 1980-2005 (cont'd.)

Product 322: Wearing apparel, except footwear

Value added per worker (Korea=100)

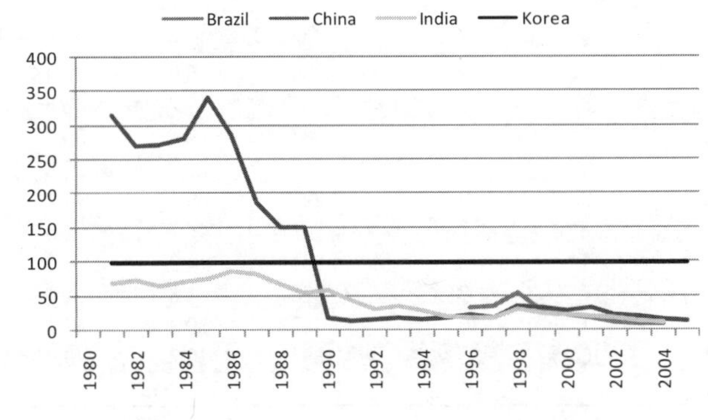

Value added per worker (current USD '000 per worker)

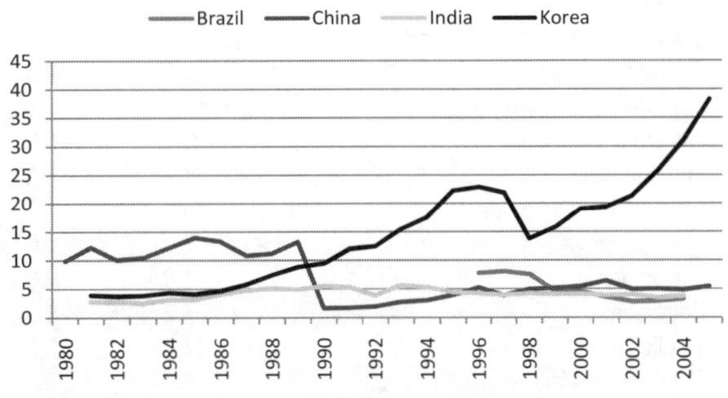

Product 384: Transport equipment

Value added per worker (Korea=100)

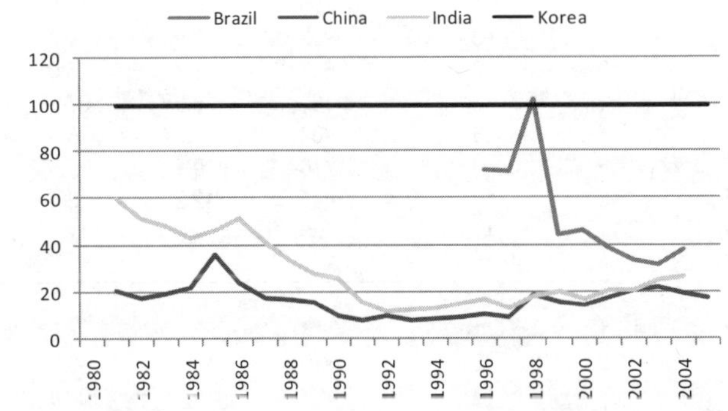

Value added per worker (current USD '000 per worker)

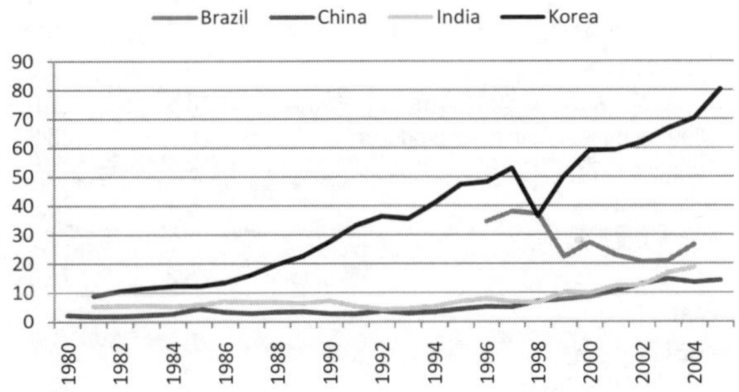

Source: Authors' estimates (col. 1) and calculations (col. 2) based on *INDSTAT Database*.

Another measure of Korean competitiveness is the technological content of its exports illustrated in Figure 3.7. Between the 1980s, when Korean exports were heavily concentrated in low-technology textiles and wearing apparel, and 2005-06 when its export basket was heavily concentrated in machinery for instance, there was a consistent shift in the technological content of exports from low-technology towards high- and medium-technology products. This trend is consistent with that of the OECD area (Lall *et al.*, 2006).

As Figures 3.4, 3.6 and 3.7 illustrate, Korean productivity took off at a phenomenal pace after 1985. So how did this happen? Table 3.1 provides statistical evidence of certain key characteristics of Korea's catch-up with the OECD area.

Table 3.1. Transformation of the manufacturing sector: change in manufacturing value added, leading industries

	Share in total manufacturing value added						Share of output exported (%)	
	1980-84	1995-99	2000-04	Rank in 1980-01	Rank in 2004-05	Growth 1980-04 (%)	1980-81	2000-04
Korea								
Machinery, electric	9.8	18.3	24.2	2	1	3 008	38.5	63.8
Transport equipment	8.0	12.3	13.7	4	2	2 004	5.0	32.9
Machinery, except electrical	3.9	9.6	9.7	8	3	3 085	19.7	45.8
Industrial chemicals	9.2	9.9	8.8	3	4	937	9.5	36.0
Iron and steel	7.2	5.2	6.5	5	5	906	27.9	22.8
Fabricated metal products	3.9	4.7	4.8	9	6	1 292	39.1	14.1
Food products	7.1	5.6	4.8	6	7	647	15.7	4.6
Plastic products	1.9	2.8	3.3	11	8	1 681	4.9	6.3
Petroleum	5.2	3.8	2.9	7	9	326	1.8	1.1
Textiles	11.5	5.3	2.9	1	10	147	30.3	43.4
Other non-metallic mineral products	3.5	3.3	2.4	10	11	657	13.2	3.9
% of total	*71.2*	*80.8*	*84.1*					
China								
Machinery, electric	3.6	11.6	16.2	9	1	3 047	3.2	33.7
Industrial chemicals	11.6	11.6	11.2	3	2	592	2.6	9.5
Iron and steel	7.3	7.0	10.1	4	3	694	1.6	7.2
Machinery, except electrical	14.6	8.3	8.1	1	4	256	0.7	67.5
Food products	4.6	6.6	6.8	8	5	942	7.9	10.3
Transport equipment	3.4	6.7	6.7	10	6	1 667	1.7	15.0
Textiles	13.6	7.0	5.7	2	7	161	8.6	37.2
Other non-metallic mineral products	5.1	5.8	3.8	7	8	462	1.3	5.8
Tobacco	5.2	5.3	3.6	6	9	716	0.1	1.5
Petroleum	5.4	3.8	3.5	5	10	441	12.2	3.1
Non-ferrous metals	2.0	2.2	3.4	11	11	788	2.9	12.1
% of total	*76.2*	*76.0*	*79.1*					
Brazil								
Food products		14.7	14.0	1	1	-14		28.9
Industrial chemicals		13.2	12.2	2	2	-16		8.9
Transport equipment		9.1	10.4	3	3	2		32.8
Petroleum		4.8	10.3	7	4	108		0.1
Iron and steel		4.1	7.6	8	5	64		29.4
Machinery, except electrical		7.4	6.7	4	6	-24		22.4
Machinery, electric		6.1	4.4	5	7	-38		18.6
Paper and products		3.8	4.1	10	8	-6		23.1
Fabricated metal products		3.9	3.5	9	9	-23		9.5
Printing and publishing		5.1	3.1	6	10	-46		1.7
Other non-metallic mineral products		3.2	2.9	11	11	-11		12.4
% of total		*75.6*	*79.3*					
India								
Industrial chemicals	14.8	20.6	16.6	2	1	301	0.8	13.9
Iron and steel	11.6	9.6	15.1	3	2	283	0.7	12.3
Petroleum	2.8	4.2	11.4	9	3	1 326	0.4	0.1
Transport equipment	8.9	8.9	9.7	4	4	268	6.0	7.4
Textiles	15.2	9.0	6.5	1	5	35	12.2	34.8
Food products	8.9	9.0	6.4	5	6	188	13.1	11.4
Machinery, except electrical	8.7	7.2	5.6	6	7	124	3.8	13.7
Machinery, electric	8.4	6.8	4.6	7	8	112	3.3	11.9
Other non-metallic mineral products	4.2	4.0	4.4	8	9	408	1.5	6.8
Non-ferrous metals	1.0	3.3	3.8	11	10	1 477	1.1	10.0
Fabricated metal products	2.8	2.5	2.3	10	11	163	12.4	26.9
% of total	*87.3*	*85.0*	*86.5*					

Source: Authors' calculations based on *INDSTAT Database*.

Structural transformation. Columns 5 and 6 in Table 3.1 indicate the dramatic transformation in the industrial structure of the Korean economy between 1980 and 2004. The rank of textiles, Korea's leading industry in 1980, fell to tenth place by 2004, underscoring the shift away from low-technology industries. Similarly, non-electrical machinery, which ranked eighth in 1980, ranked third in 2004. Overall the ten industries listed in Table 3.1 accounted for 84% of Korean MVA in 2004, up from 71% in 1980. The share of machinery and transport equipment had risen from 18% to 48% (columns 2-4). The growth rates in the MVA of these industries (column 7) also support this argument.

The remainder of this section argues that Korea's catch-up strategy involved a combination of interventions to promote export-led growth (even though import substitution efforts were initially pursued) and support for innovative industries. In short, the anti-export bias created by import substitution was offset by pro-export measures and the government closely monitored performance in the international market when prioritising areas for intervention.

Export-led. Growth in certain industries was powered by Korean exports. Between the early 1980s and 2004, the share of output exported increased from 38% to 64% in electrical machinery, which was Korea's leading industry, and from 5% to 33% in transport, its second largest industry (Table 3.1, columns 8-9). The overall share of exports in GDP increased from 23% to 43% between the 1970s and 2006. Export-led catch-up would not have been possible without the competitiveness of its leading industries; the trends in these industries' RCA validate this description of Korea's growth and catch-up strategy (Figure 3.6).

Trade liberalisation. After sharpening its technological capabilities, around 1984, Korea liberalised trade to level the playing field. Its tariffs dropped from about 18-20% in 1988 to less than 10% in the 1990s and are presently around 3-5% (Figure 3.5). A good example is the electronics industry which was nurtured with the aid of import substitution and domestically financed innovation, but evolved into a world-class industry through a strong export orientation. By 1993, Korea had already become a global power in the international market for electronics.

Industrial policy. Korea was committed to creating a strong economy based on domestic ownership, quite reminiscent of the "big push" type of industrialisation. Underpinning its innovation strategy were Korea's national champions or chaebol. Selective industrial policies targeted high-technology industries centred on chaebol that were protected from global competition and given open access to subsidised finance. The two key factors at this stage were government commitment and state-owned banks that bore the risk of technological innovation (Amsden, 1989).

In the first stage of development, the state adroitly used an industrial policy that was a blend of import-substituting trade policies and a dedicated technological adaptation strategy. Initially, the chaebol sought to master the use of imported technologies in selected low-technology export-oriented industries such as textiles, garments and footwear which became the leading industries (Kim and Nelson, 2000). Then they pursued a more sophisticated import-substitution strategy aimed at replacing imported inputs with domestically produced ones. There was always a careful balance between import substitution on the one hand and support to technological mastery to compensate for the implicit anti-export bias of the trade regime on the other.

Figure 3.5. Korea: trade liberalisation (reductions in weighted tariffs)

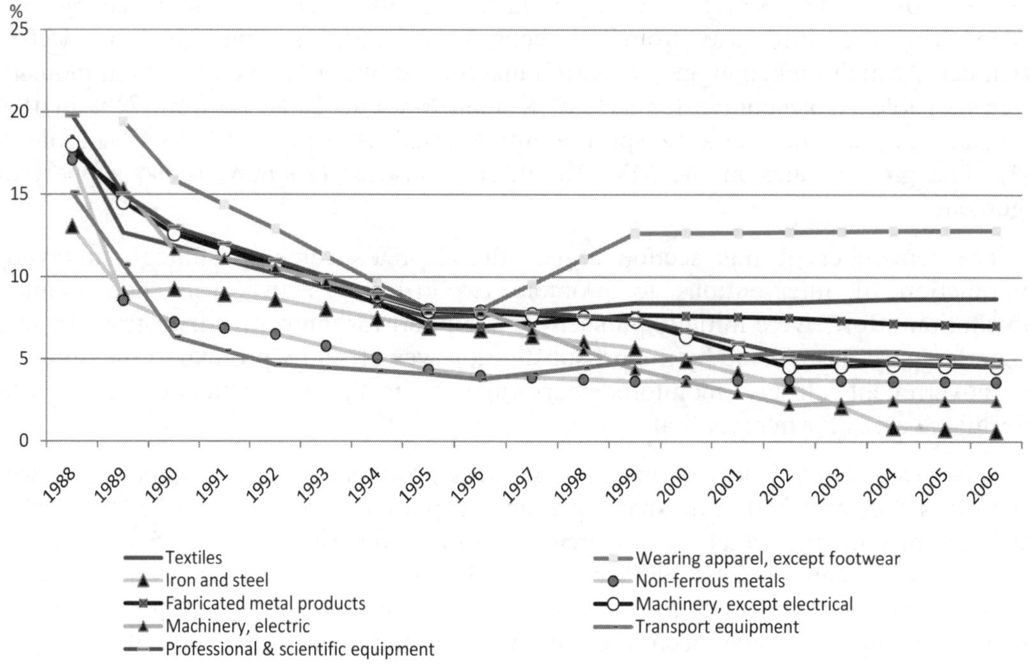

Source: Authors' calculations based on *TRAINS Database*.

The chaebol were the pillars of Korea's industrial development policy and the list of hand-picked industries was regularly revised. For example, during the 1970s, the focus was on heavy chemicals; in the 1980s on labour-intensive export industries; and from the 1990s onwards on high-technology information technology (IT) products and components (*e.g.* dynamic random memory chips, thin-film-transistor liquid crystal displays). In each case, there was a clear market test in terms of judging the success of the targeted industries by their performance in the global market.

The seeding and expansion of the chaebol created an oligopolistic market structure in major industries. For example, in the early 1980s, the top ten chaebol, including Hyundai, Samsung, Daewoo and LG, produced 20% of Korea's manufacturing output.[6] The chaebol operated as a highly leveraged Schumpeterian agent for change in which a centralised industrial structure favoured big firms seeking to pursue innovation based on scale economies (Wang, 2007). The targeted industries received long-term fiscal incentives and financial subsidies which included tariff exemption for importing intermediary goods, tax incentives, preferential access to capital, accelerated depreciation of imported equipment and subsidised prices for energy and transport. The government also intervened in the banking system, directing subsidised lines of credit to certain sectors, projects and firms.

6. The chaebol were also conceived as a mechanism to partly offset the disadvantage of limited entrepreneurial talent. Conglomerates like the Samsung Group and the Hyundai Group pioneered new industries (shipbuilding, automobiles and semiconductors). They sometimes co-operated with the government in economic planning and innovation but the government also encouraged competition among the chaebol in strategic areas to avoid excessive market power.

The Ministry of Trade and Industry used a two-phase policy towards foreign direct investment (FDI). In the first phase, it adopted a restrictive policy towards FDI and a lenient one towards contractual licensing. Government reserved the right to reject "undesired investment". Local content requirements and export quotas were imposed (Kim and Nelson, 2000). In the second stage, the Foreign Capital Inducement Act was revised to encourage FDI. There was special treatment for high-technology industries.

Innovation policy (distance-shortening strategies). The Korean government set up public R&D facilities, but the chaebol were the major source of assimilation and innovation of technology in the initial stages of the process. The technological content of production and exports was emphasised. The chaebol transformed themselves from technological followers into technological leaders by taking advantage of scale economies and receiving strong support from the state for their own R&D.

Korea invested roughly 0.5% of GDP in R&D in the mid-1960s with the government responsible for 80% of the investments. By 1995, gross domestic expenditure on R&D (GERD) already represented 2.3% of GDP with the private sector responsible for 80% of these expenditures. Korea's GERD was significantly larger than the OECD average (2.07% of GDP) but still behind that of the United States (2.51% of GDP). By 2006, Korea's GERD of 3.2% of GDP positioned it among the OECD leaders. The corresponding figures in 2005 were 2.6% for the United States and 2.25% for the OECD (for further details, see Chapter 7 and OECD, 2007). In parallel with the greater role played by the private sector in R&D, the government has continued to foster university-industry linkages through fiscal incentives and government procurement of advanced technologies. On almost all indicators of human resources for S&T, Korean scores exceed those for China by a large margin.

By the early 1990s, Korean firms had the technological capability to engage in innovation. According to the metric of the number of patents granted by the US Patent & Trademark Office, in 2003 Taiwan, China, became the eighth and Korea the ninth largest recipients, surpassed only by the United States, Japan, Germany, the United Kingdom, France, Canada and Italy, in that order (OECD, 2008a).

Other dimensions of the Korean model also contributed to an innovation-friendly environment: a high savings rate, broad macroeconomic stability (since the 1980s), and a high level of investment in education. The focus here, however, is on the distance-shortening policies pursued by Korea and discussed further below.

How much has China caught up?

In the early 1980s, China was very far from the Korean technological frontier[7] in most industries (Figure 3.4). It was at 80% in the food industry, industrial chemicals, and iron and steel industries, and about 60% in electrical machinery and transport. Textiles were China's and Korea's largest industry and China's distance from the Korean technological frontier in this industry was about 60% but in the wearing apparel industry, Chinese productivity was about 300% greater than Korea's.

7. Distance to the frontier is measured by manufacturing value added per worker.

Table 3.2. Industrial value added per worker

Average growth rates

		Brazil			China		
ISIC Code	Industry description	1980-96	1997-00	2001-05	1980-96	1997-00	2001-05
311	Food products	n.a.	-10.6	1.1	8.5	11.8	9.5
313	Beverages	n.a.	1.9	-9.2	6.4	13.2	7.5
314	Tobacco	n.a.	-4.0	-4.4	6.6	12.6	14.1
321	Textiles	n.a.	-6.9	-4.5	0.4	13.4	6.7
322	Wearing apparel, except footwear	n.a.	-10.8	-6.8	5.1	3.0	0.9
323	Leather products	n.a.	-8.7	4.4	7.9	0.6	-0.9
324	Footwear, except rubber or plastic	n.a.	-12.9	-4.2	n.a.	n.a.	n.a.
331	Wood products, except furniture	n.a.	-7.9	6.6	7.5	13.4	5.7
332	Furniture, except metal	n.a.	-8.1	-0.9	4.3	13.0	0.6
341	Paper and products	n.a.	1.7	-1.3	4.6	11.0	10.0
342	Printing and publishing	n.a.	-6.7	-3.5	5.8	21.2	5.2
351	Industrial chemicals	n.a.	-8.0	1.5	4.6	11.8	11.9
353	Petroleum refineries	n.a.	24.9	-1.2	1.6	15.5	7.6
355	Rubber products	n.a.	-6.0	2.7	0.4	7.8	8.8
356	Plastic products	n.a.	-10.6	-0.5	3.9	11.1	2.1
362	Glass and products	n.a.	-5.8	0.9	3.3	19.9	5.3
369	Other non-metallic mineral products	n.a.	-4.3	-0.1	3.7	14.6	12.1
371	Iron and steel	n.a.	1.1	15.6	5.4	14.2	22.1
372	Non-ferrous metals	n.a.	2.0	5.1	4.7	13.7	15.4
381	Fabricated metal products	n.a.	-9.8	0.6	1.3	12.0	6.1
382	Machinery, except electrical	n.a.	-7.8	-1.6	5.5	13.7	13.2
383	Machinery, electric	n.a.	-5.9	-3.6	16.5	13.6	1.4
384	Transport equipment	n.a.	-2.6	0.6	7.9	14.6	11.4
385	Professional & scientific equipment	n.a.	-3.0	-1.8	5.6	24.1	7.7
390	Other manufactured products	n.a.	-9.3	-6.0	7.9	5.7	1.0
		India			Korea		
311	Food products	9.0	2.6	1.3	11.4	3.9	5.3
313	Beverages	5.3	5.0	1.0	13.6	8.7	9.6
314	Tobacco	7.3	11.9	3.4	14.9	1.6	1.3
321	Textiles	3.1	6.5	1.2	12.1	2.5	5.3
322	Wearing apparel, except footwear	4.3	-0.1	-2.2	13.1	-1.5	15.3
323	Leather products	4.4	-5.8	5.8	11.7	2.2	6.8
324	Footwear, except rubber or plastic	3.6	8.0	-2.5	14.6	-2.6	9.9
331	Wood products, except furniture	6.3	-3.2	3.9	16.0	2.8	5.2
332	Furniture, except metal	1.8	36.4	4.3	14.1	3.1	7.4
341	Paper and products	4.2	19.3	-2.9	13.3	4.2	1.5
342	Printing and publishing	4.2	10.3	2.7	12.6	1.7	4.6
351	Industrial chemicals	4.1	5.2	5.4	11.4	3.9	8.2
353	Petroleum refineries	10.4	16.5	34.1	14.7	3.6	22.4
355	Rubber products	5.3	6.2	1.3	19.5	-0.5	9.4
356	Plastic products	2.9	11.6	5.5	13.8	0.5	6.7
362	Glass and products	3.5	16.4	12.5	16.6	11.8	4.3
369	Other non-metallic mineral products	5.7	11.3	7.1	11.4	5.3	6.2
371	Iron and steel	6.0	9.5	33.6	12.0	4.1	13.3
372	Non-ferrous metals	10.8	3.2	18.8	12.1	4.6	9.8
381	Fabricated metal products	1.8	4.5	5.1	12.8	-3.1	7.3
382	Machinery, except electrical	3.5	3.2	5.8	13.1	7.1	4.3
383	Machinery, electric	3.0	4.4	5.8	16.9	3.5	8.2
384	Transport equipment	3.8	8.4	17.9	12.2	8.6	6.4
385	Professional & scientific equipment	4.0	5.8	14.5	13.3	4.5	6.3
390	Other manufactured products	8.4	6.1	5.5	13.7	0.3	6.6

Source: Author's estimates based on *INDSTAT Database*. Data adapted to ISIC Rev. 2, 3-digit.

Throughout the 1980s and the early 1990s, China's distance to the frontier increased for two reasons. First, the Korean technological frontier was not static – in nearly every industry, it was a rapidly moving target, a reminder that catch-up is not about closing a fixed gap but requires a country to grow at a rate that outpaces productivity growth in the frontier country. Over 1980-96, China's leading industries had productivity growth rates significantly below those of Korea. As an example, Chinese productivity growth in industrial chemicals was 4.6% compared to 11.4% for Korea, only 5.4% in iron and steel compared to 12% for Korea, and so on (Table 3.2). Since 1997, however, Chinese productivity growth has been in the double digits while it has dropped to single digits in most Korean industries. The second reason why China's distance to the frontier widened throughout the 1980s is that the productivity dividend from its liberalisation reforms did not begin to accrue in large measure almost until the mid-1990s.

Compared to the other countries analysed, however, Chinese productivity growth was phenomenal during 1995-2000 (Figure 3.4). Even though Indian productivity levels were higher than China's in almost every industry prior to 1995, China's productivity growth outpaced India's and led to a narrowing of the productivity gap between China and India. The shrinking of the productivity gap between China and Brazil is even more remarkable. In 1995, the gap was significant in most industries but in the following decade, China's productivity consistently exceeded Brazil's. In short, while China's productivity is still far from the Korean technological frontier, its catch-up has far outpaced that of Brazil and India.

China's catch-up with Korea in terms of global competitiveness can be also tracked by its revealed comparative advantage in key industries (Figure 3.6). In the 1980s, China had the highest RCA levels among the BICs and Korea in textile exports. Even as China and Korea shifted away from textiles towards other industries, China maintained its edge over the Korean textiles industry. By 2002, Korea was no longer competitive in the industry (its RCA dropped below 1). In wearing apparel too, China's RCA was below Korea's in 1980, but had exceeded it by the mid-1990s, by which time Korea became uncompetitive in that industry.

Another remarkable aspect of the extent of China's catch-up with Korea is the transformation in the technological content of its exports. Lall (2000) observed that after World War II, countries whose export baskets moved from relatively low-technology products towards high-technology ones grew the fastest. In fact, Korea's catch-up with OECD countries is strongly correlated with its success in bolstering the technological content of its exports in the direction of OECD exports. China's success in rapidly transforming its weak technological capabilities and developing an RCA in more sophisticated and dynamic industries is reflected in Figure 3.7. By the 2000s, it exported a large proportion of relatively higher-technology products with a structure of exports that resembled Korea's far more than that of either Brazil or India.

Figure 3.6. China maintained its edge in textiles and rapidly closed the gap with Korea in wearing apparel

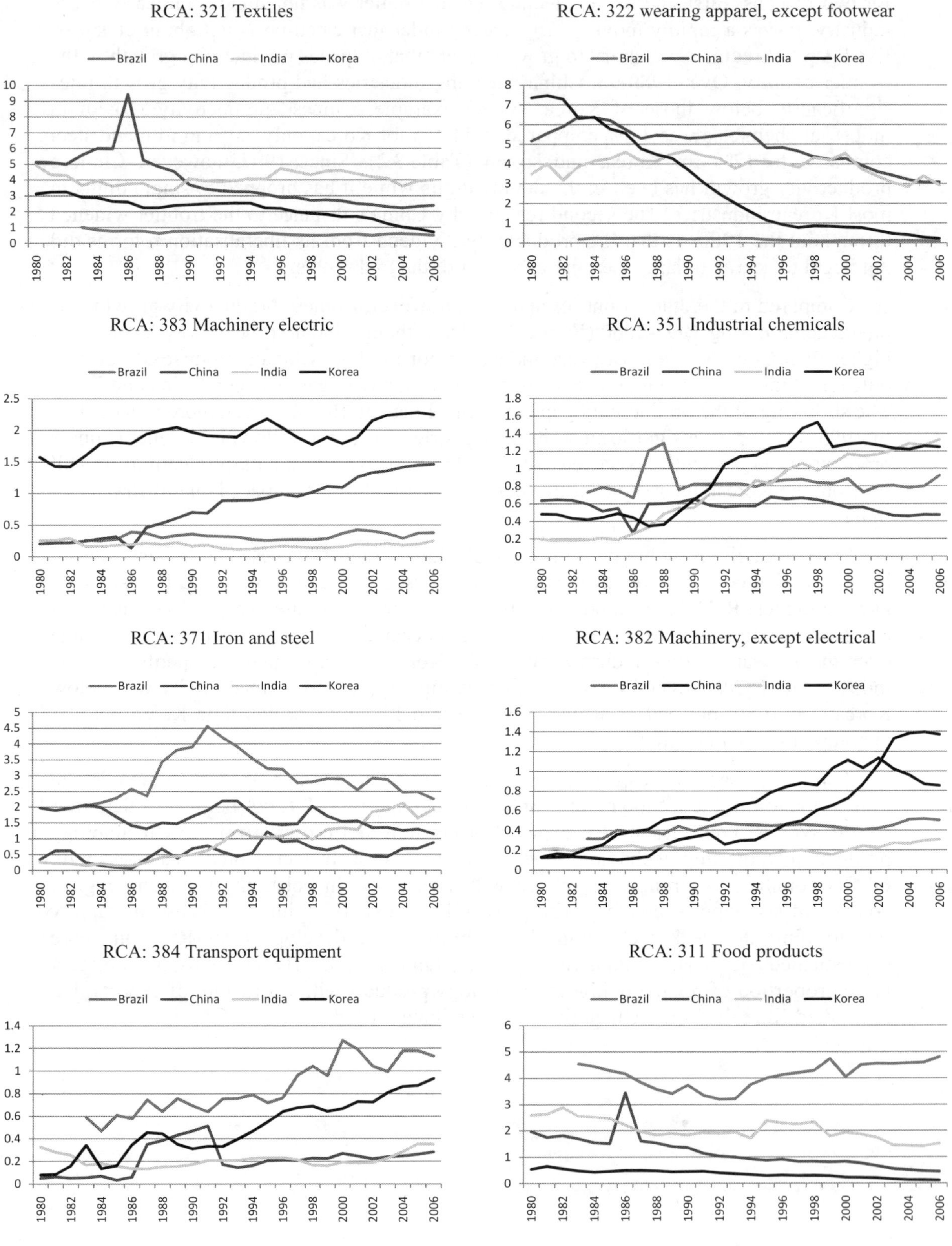

Source: Authors' calculations based on COMTRADE Database.

Figure 3.7. Export transformation and technological catch-up

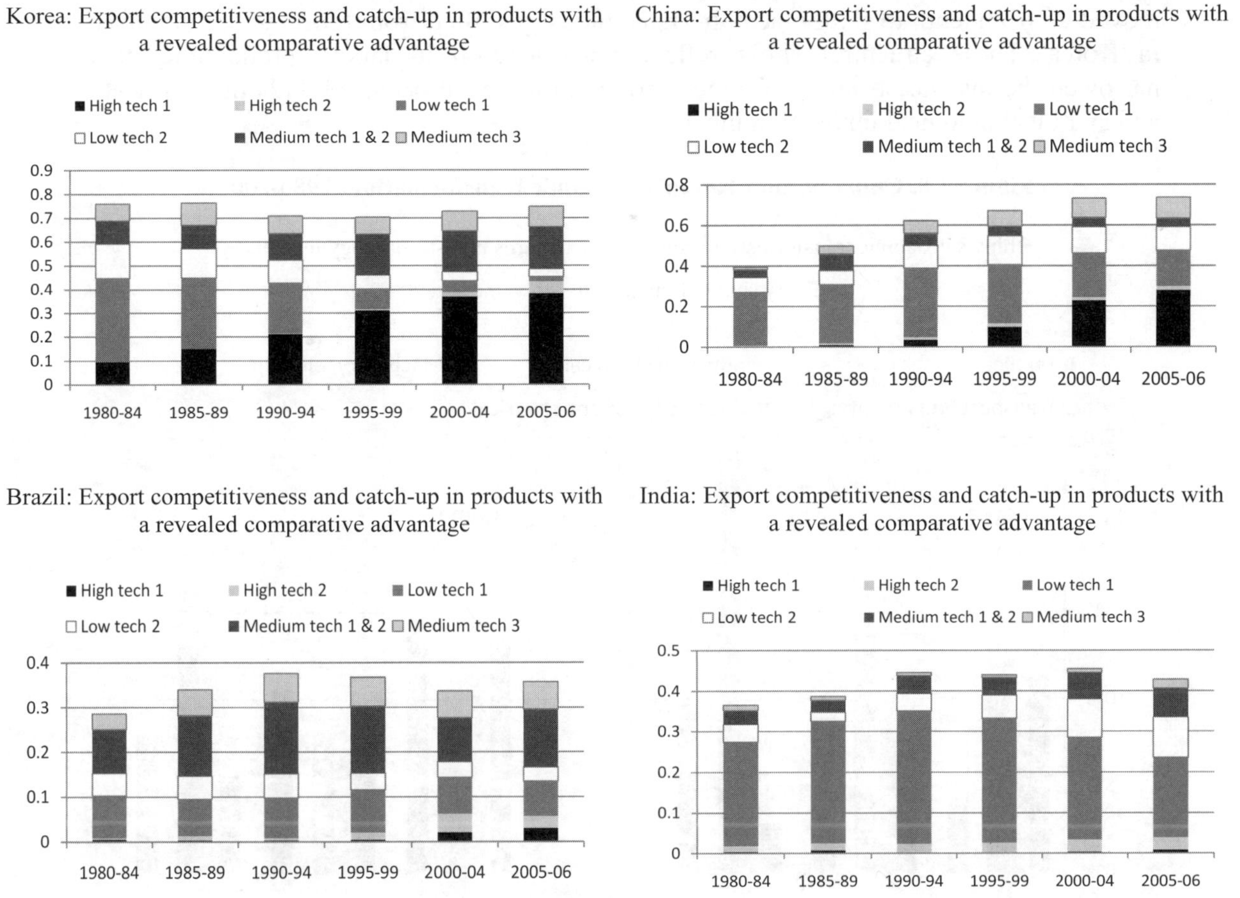

Relative to Korea, China's competitiveness has increased in two ways: the share of exports in which it has an RCA (greater than 1) is the same as Korea's, and the medium- and high-technology content of its exports has increased significantly more than in Brazil or India.

Source: Authors' calculations based on *COMTRADE Database*.

How did China catch up?

China's industrial transformation lagged that of Korea by a decade but its catch-up with Korea was no accident. It was the result of an industrial strategy designed to emulate the successful catch-up of Korea and other newly industrialising countries with OECD countries about two decades earlier. China shortened its distance to the technological frontier and grew faster than Brazil and India in at least three distinct ways. The first is selectivity. The sources of productivity growth were concentrated in a group of five or six industries whose individual shares were at least 5%. As their collective share in total manufacturing value added (MVA) increased from about 58% in 1980-81 to nearly 70% in 2004 these industries became the drivers of China's catch-up (Table 3.1).

Second, it shifted from industries with a lower technological content towards those with a higher one (Table 3.1). For example, between 1980 and 2005, the change in the rank of high-technology electrical machinery from ninth to first place and of low-technology textiles from second to seventh place was the outcome of technological innovation similar to Korea's where these two industries traded first and tenth places (Figure 3.8 and Table 3.1). In another example, the Chinese transport equipment industry moved up from tenth to sixth place; the same industry shifted from fourth to second place in Korea. These structural shifts reflected acceleration in labour productivity that narrowed the distance from the frontier and could not have occurred without significant and swift technological innovation in China.

Figure 3.8. China mimics Korea's economic transformation, 1980-2005

China's economic transformation from textiles towards high-technology machinery

Industry share of manufacturing value added

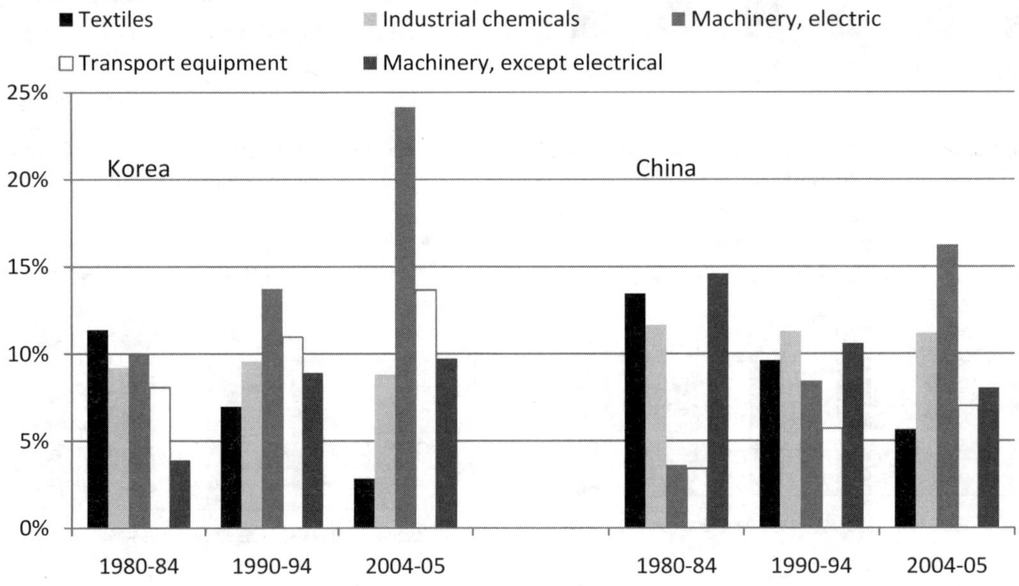

Source: Authors' calculations based on *COMTRADE Database* and *INDSTAT Database*.

The third characteristic of China's catch-up model indicates that, like the Korean export model, productivity growth is export-driven as is evident from the sharp increase in the export content of its leading industries (Table 3.1 and Figure 3.9). As an example, as much as 70% of the total production of non-electrical machinery was exported in 2004-05 compared to only 20% in 1990-94.

The Chinese catch-up model was founded on a strategy that gradually distanced the Chinese economy from its original inward-looking (socialist) model of self-reliance. As in Korea, China's export-led catch-up strategy relied on competition in the global market to serve as a screening mechanism to identify globally competitive industries. It was reinforced by policies that maintained a stable macroeconomic environment while attracting foreign investors.

Figure 3.9. Export led catch-up in Korea and China

Percentage of output exported

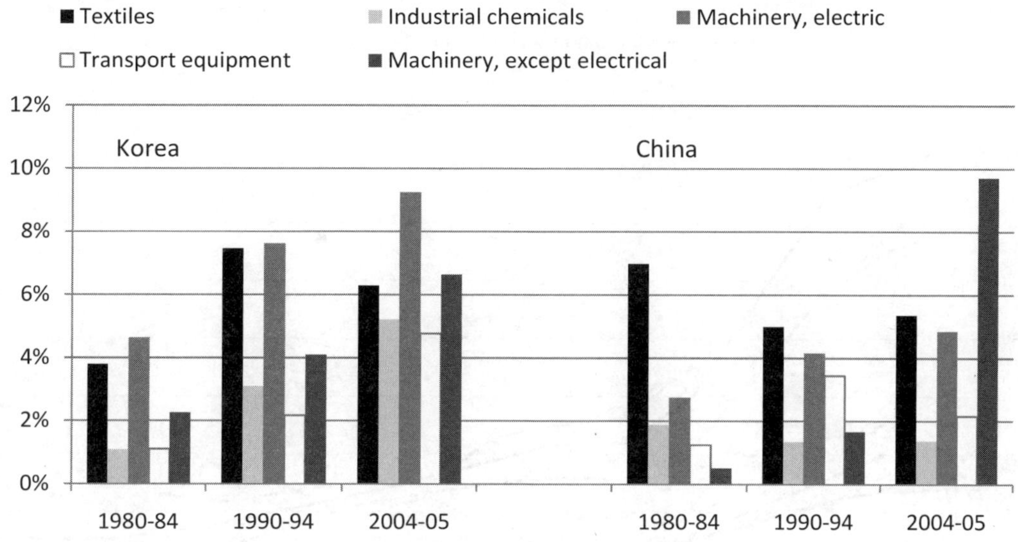

Source: Author's calculations based on *COMTRADE Database* and *INDSTAT Database*.

The timing and pace of foreign competition measured by tariff cuts sheds some light on the differences among the BICs in their pursuit of economic integration with the world economy. China cut its tariff rates drastically from about 32% in 1993 to 16% by 1996 and 2% by 2006 (Figure 3.10). In contrast, in the late 1980s, the tariffs on imports of electrical machinery were 81% in India and 46% in Brazil. The Chinese electrical machinery industry continues to grow following liberalisation but its relative position signals that the lowest rates of protection do not guarantee a competitive edge forever. Even though in absolute size it was larger than the same industry in Brazil, India or Korea, China's distance from the Korean productivity frontier was widening in 2005 and had begun to do so in 2000. This suggests that the pace of technological innovation in Korea is ahead of that of the BICs. A new wave of technological innovation is required if China wants to continue to catch up with Korea in this high-technology industry.

The Chinese brand of "distance shortening" along with industrial policy interventions seems to explain its success in moving closer to the frontier, significantly ahead of Brazil and India (Figure 3.4). China transformed itself from a high-technology original equipment manufacturing (OEM) contractor to a global competitor in high-technology exports. It did so by leveraging foreign technology through FDI, which became the means to access capital and introduce competition in the domestic economy. FDI was also used as a distance-shortening mechanism. Instead of waiting for its technological capabilities to mature, China used its "open door" policy to let the "market for technology" model co-ordinate FDI, foreign trade and technology transfer to modernise its economy (OECD, 2008b). Until the 1990s, China relied heavily on imported technology, especially from Japan, and help with the training of engineers at Toshiba, Changhong and Sanyo (Mathews, 2009). In recent years, its own technological capabilities have improved and "indigenous innovation" is being given priority (OECD, 2008b). This is distinct from

Korea's experience in which the country put the main onus of developing a pool of technological capabilities on itself.

Figure 3.10. China: trade liberalisation, 1992-2007

Reduction in weighted tariffs (%)

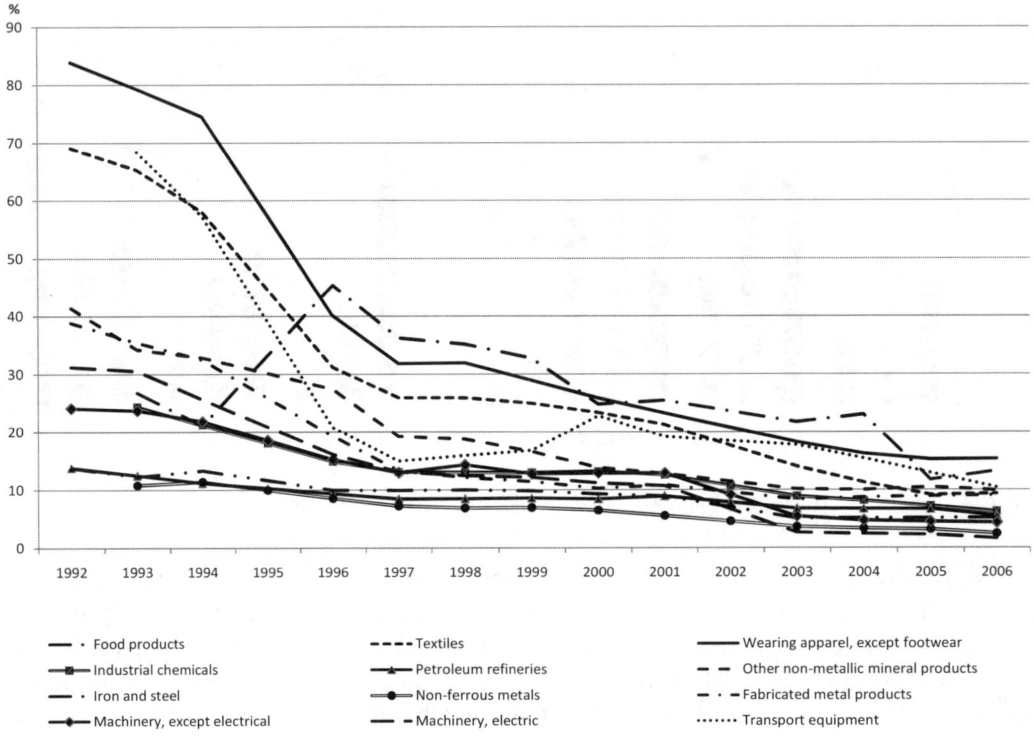

Source: Authors' calculations based on *TRAINS Database*.

In the past few years, the Chinese government has initiated a variety of measures to speed up convergence with OECD policies. These include enhanced public R&D and tax incentives, government-facilitated absorption of imported technology, active use of intellectual property rights (IPRs) and other international standards. Its GERD is 50% larger than Brazil's and its business enterprise expenditure on R&D (BERD) almost 100% larger. China is almost on a par with Korea in terms of the strength of industry-university linkages, proxied by the share of R&D financed by industry. The *National Guidelines for Science and Technology Development* offer tax incentives for business R&D, public procurement, industrial research alliances, and human resource development in S&T to ensure that its industry is not constrained by a paucity of skilled and technical workers (OECD, 2008a). In 2005, as a percentage of all new degrees, China had a larger proportion of science and engineering degrees than Korea and nearly four times as many as Brazil. The government is also offering special incentives to attract its diaspora: rewarding scientific talent, financing for basic research and support for technology innovation and commercialisation (OECD, 2008b). For this purpose, it has established 18 leading universities and 9 key research organisations. Its national innovation system shows that China has excelled in mobilising financial and human resources for S&T

development on a large scale and has made major progress in R&D indicators (OECD, 2008b).

Brazil: decline and recovery

From its position close to or on the Korean technological frontier in many industries in the 1980s, Brazil's productivity levels slid throughout most of the 1980s and 1990s, largely in response to an unstable macroeconomic environment and the absence of healthy competition. Around 2000, labour productivity growth started to pick up (Table 3.2). Even though recovery has been sustained, Brazil still has significant catching up to do.

Statistical evidence suggests that the distance of Brazil's leading industries from Korea's productivity frontier depends upon the timing of competition and the industry's distance from the frontier *prior* to reform.[8] In 2000, for example, the Brazilian food industry's distance from the frontier was about 25%, the same as China's (Figure 3.4). Even after recovering during 2000-05, this distance has not narrowed, mostly because the frontier has advanced faster. Similarly, the distance has widened in electrical machinery. Interestingly, in industrial chemicals, transport equipment and iron and steel, productivity levels are converging albeit gradually. Most notable is the relative productivity trend in the iron and steel industry in which Brazil was significantly close to the frontier in 2000. Moreover, in the petroleum refining industry, Brazil's technological performance exceeded Korean levels in 2000 (Figure 3.11).

Figure 3.11. Brazil's petroleum industry was close to the Korean technological frontier

Product 321: Petroleum refineries

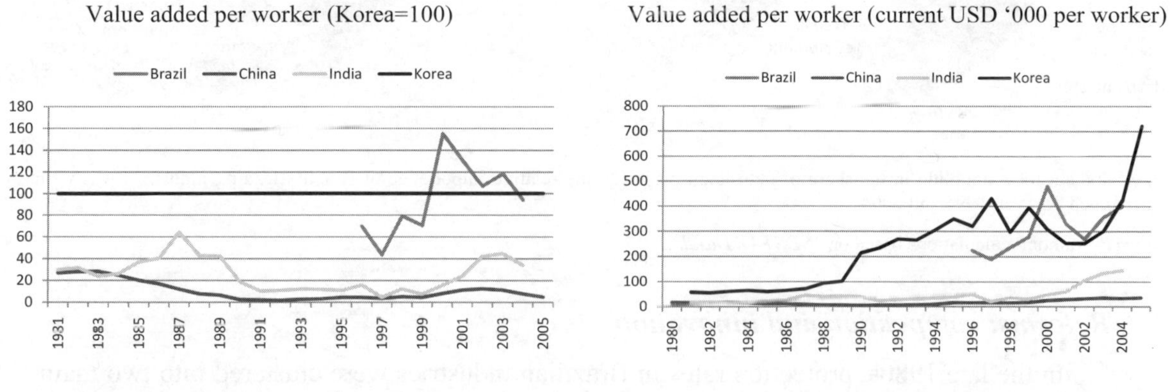

Source: Authors' calculations based on *INDSTAT Database*.

Korea's catching up with OECD countries and China's catching up with Korea were positively correlated with a rapid transformation of the industrial structure. This did not occur in Brazil. A widening of the distance from Korea's technological frontier in some of its largest industries and narrowing in others preserved on average Brazil's industrial structure: in 2005 it was not very different from what it was in the 1990s. This is discernible in the relatively unchanged shares of Brazil's medium- and high-technology

8. In 1996 and 2004, they were food products, industrial chemicals, transport equipment, iron and steel, petroleum, electrical machinery and non-electrical machinery – each had a share of 4% or more.

industries during this period (Figure 3.7). Interestingly, the share of medium-technology industries in Brazil was significantly higher than in Korea in 1985, but had not changed by 2005-06. And Brazil's high-technology industries remained relative laggards in comparison to the Korean technological leaders (Costa and Reis de Queiroz, 2002; Cruz *et al.*, 2006).

Unlike Korea or China, Brazil's catch-up strategy was neither selective not focused on leveraging innovation to catch up. It did not invest heavily in distance-shortening policies that would have pushed it closer to the frontier when it initiated trade liberalisation in the 1990s. Consequently, an inward-oriented economic orientation and a slower pace of reform limited industrial transformation even after Brazil opened up (Figure 3.12).

Figure 3.12. Brazil: limited structural transformation

Share of manufacturing value added

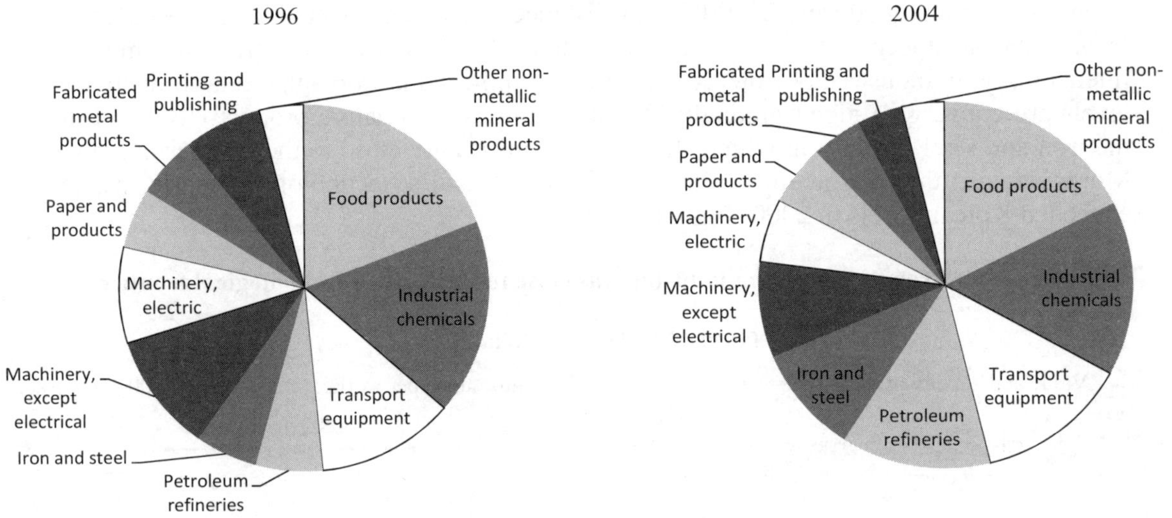

Note: Except for a doubling in the share of petroleum products, the relative importance of Brazil's leading industries was virtually unchanged between 1996 and 2004.

Source: Authors' calculations based on *INDSTAT Database*.

Reforms: competition and innovation

In the late 1980s, protection rates in Brazilian industries were clustered into two main sets (for a recent account of the evolution of Brazil's trade policies see Lattimore and Kowalski, 2008). The food, transport and machinery industries were farther from the frontier at nominal rates of 45-50%. The petroleum products, iron and steel, and industrial chemicals industries were closer to the frontier and had tariffs in the range of 25-35%. Tariff rates in the other industries were also in the lower range. Tariffs declined with liberalisation in the 1990s (Figure 3.13), but, with the exception of the food industry, the assignment of industries in the low or high tariff brackets did not change. In 2007, tariffs on most industries were between 8% and 17% compared with Korean tariffs that were below 5%.

Figure 3.13. Brazil: trade liberalisation

Weighted tariff rates (%)

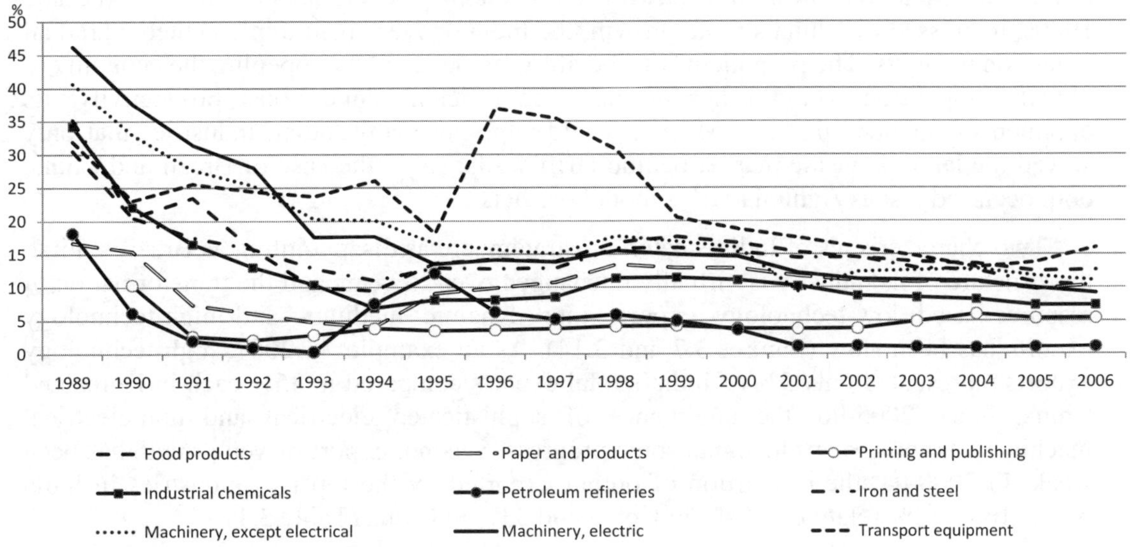

Source: Authors' calculations based on *TRAINS Database*.

Brazils' catch-up has suffered in part because of a lack of sustained effort to build the technological strength that underlies competitiveness. Undoubtedly, macro disturbances in the 1980s and 1990s also conspired against innovation. With one notable exception, the declining RCA in Brazilian industries illustrates the declining competitiveness of the manufacturing sector. A review of Brazil's innovation policy suggests that following reform, the only industry that evolved into a world class industry is food products (including ethanol related industries). It benefited disproportionately from S&T-oriented public investments in EMBRAPA, a government agricultural research organisation created in 1974. Similarly, public innovation funds channelled into Petrobras nurtured world-class petroleum-related activities.

While the Brazilian government has been spending more public funds to finance R&D (close to 1% of GDP), synergies between public-sector funded research institutes and private firms in the manufacturing sector remain limited. Compared with about 30% in the United States, about 55% of the total investment in technological innovation in Brazil is publicly funded (World Bank, 2007) but it has been less successful at energising technological innovation or patents that can be commercialised. Its import-substitution policies shielded its private sector, undermined incentives for private R&D and led to underperformance in innovation. Recent initiatives such as the Innovation Law and the Sector Funds, which are aimed at encouraging firms to invest in innovation, are attempts to correct these distortions (Katz, 2000).[9]

9. It has also been noted that overly "theoretical" research in publicly funded universities partially explains why the latter are of little use to entrepreneurial scientists and engineers who do not interact with universities. As an example, Brazil accounts for nearly 2% of articles published in internationally recognised research journals but only 0.18% of the patents (OECD, 2008a). Its average educational attainment is the next-to-lowest gross enrolment rate among the larger Latin American countries. Only 8% of the labour force has tertiary-level educational qualifications.

India: another latecomer

In terms of catch-up by the BICs, India is the indisputable laggard. In the early 1980s, India and China were neck and neck on the economic development plane. By the late 1980s, impressed by China's rapid growth, the Indian government implemented a gradual reduction in tariffs. The proponents of liberalisation believed that opening the economy to global competition would usher in increased efficiency and boost productivity. Its opponents contended that competition would wipe out many Indian industries that only served the large domestic market behind tariff walls. As in the case of Brazil and China, both occurred just as Aghion *et al.*'s model predicts.

Two characteristics of India's catch-up strategy are as disappointing as Brazil's. First, the structure of its industry is virtually unchanged over 25 years and the transformation of resource-based low-technology exports into dynamic medium- and high-technology exports has been slow (Figures 3.7 and 3.14). As an example, in 2007, high-technology exports represented only 5% of India's total exports compared to 35% each in Korea and China. Since 2005-06, the emergence of sophisticated electrical and non-electrical machinery exports seems to signal some progress. Second, export-driven growth has been weak. In 2000-04, the proportion of output exported by the top two industries in India was only 12-13% relative to 64% in Korea and 33% in China (Table 3.1).

Figure 3.14. India's leading export industries between 1980 and 2006

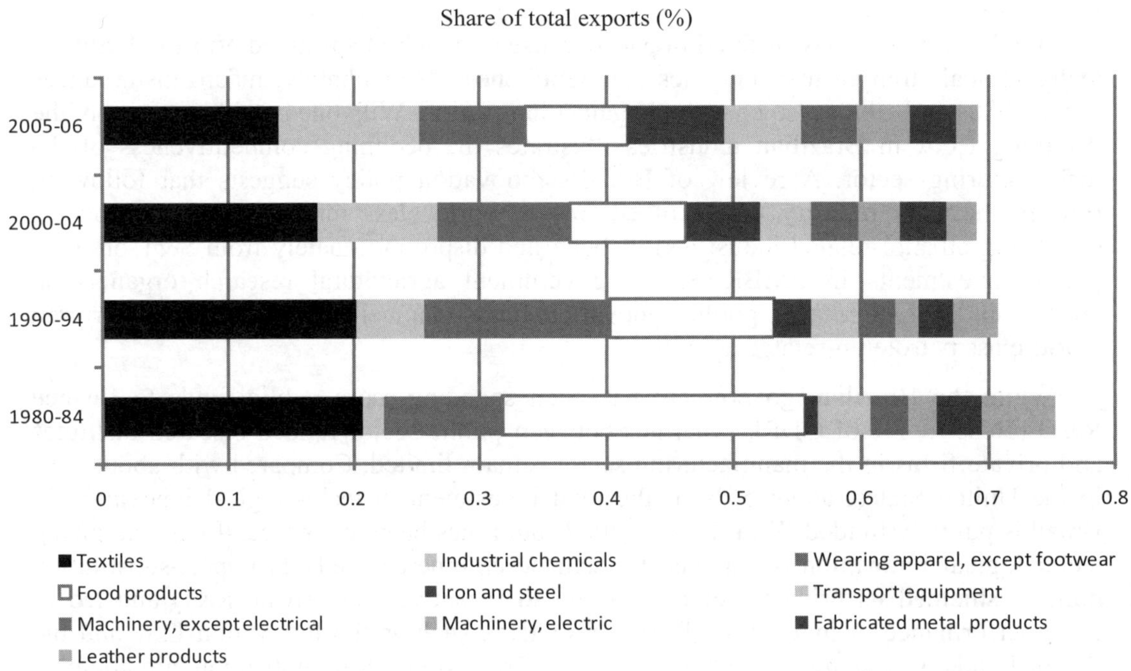

Source: Authors' calculations based on *COMTRADE Database*.

In the post-Independence period, the Indian government followed a national self-sufficiency policy that relied heavily on a complex web of government regulations and policies that were protectionist, highly interventionist, and marked by a plethora of industrial licensing requirements and strong government ownership of industry. India's trade regime was one of the most restrictive in Asia. The 1951 Industries Act adversely affected the technological sophistication of its industries for nearly four decades through industrial licensing that controlled the entry of new firms, expansion of existing ones and closure of inefficient ones. The policy regime was popularly known as the "licence raj" (Kochhar *et al.*, 2006; Topalova, 2004). In 1991, a balance of payments crisis triggered several structural reforms that disciplined the Indian economy and prepared it to join the club of emerging market countries.

In the early 1980s, in spite of the legacy of the "raj", several Indian industries were relatively close to the Korean technological frontier. Indian productivity levels were within 60% in textiles, 90% in wearing apparel, and 100% in the electrical and non-electrical machinery industries, *i.e.* these last were on the frontier (Figure 3.4). However, from the mid-1980s to the mid-1990s, productivity levels plummeted to about 20% and widened the distance of most Indian industries from the frontier. Progress in catch-up was mixed. Relative to China, India's distance to the frontier has barely lessened in the textiles and food industries. In electrical machinery, iron and steel, and transport equipment, India is neck and neck with China, and in non-electrical machinery and industrial chemicals, Indian industries are closer to the Korean frontier than China.

Trade liberalisation was very gradual. In the 1990s, when the government opened the economy to foreign competition, India was a decade too late in the catch-up game. From prohibitive rates of 75-100% in 1990, tariffs were cut slowly to 20-40% by 1996-97, and were between 10% and 20% in 2006, the highest among the BICs (Figure 3.15). Liberalisation also entailed other reforms. Between 1987 and 1995, the share of products subject to quantitative restrictions fell from 87% to 45%, import licensing was eliminated and export controls were relaxed. Only 33% of regulated industries were exempt from licensing in 1985; the remainder were de-licensed in 1991. Delayed and limited foreign competition was not without cost. It eroded Indian competitiveness especially in high-technology industries at a time when China's competitiveness in this area increased.

The Indian government implemented a blend of distance-shortening and industrial policies during the closed economy era of the licence raj. However, the absence of competition, especially foreign, and the lack of export orientation led India's industrial policies to have many unintended consequences. Specialisation in large skill-intensive industries seemed natural but government's commitment to a socialist pattern of development motivated policies that favoured small-scale enterprises (SMEs) and discouraged the emergence of large private industries. The government's bias against FDI led to the nationalisation of foreign firms in the late 1960s and 1970s. Consequently, Indian industry did not benefit from the financing, technology transfer and learning that come embodied in FDI. Large, private domestic firms were capital-constrained while the large state-owned manufacturing firms that catered to India's large domestic market benefited from state finance. In contrast, in Korea and China, the government provided cheap financing for its "hand-picked" industries to enable scaling and catching up.

India's distance-shortening policy was one of technological self-reliance and was implemented through high-calibre specialised public institutions such as the Council for Scientific and Industrial Research (CSIR) which developed extensive public R&D infrastructure for large state-owned enterprises (SOEs) and SMEs. Technological capabilities were sharpened by reverse engineering products and process technologies.

Figure 3.15. India: trade liberalisation

Weighted tariff rates (%)

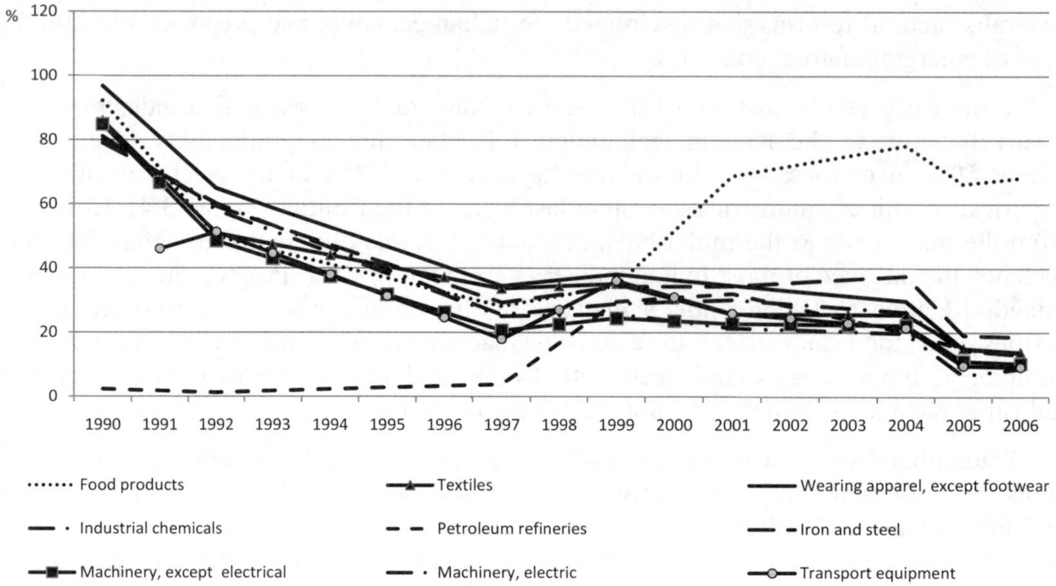

Source: Authors' calculations based on *TRAINS Database*.

The Indian government invested heavily in human resources via dedicated world-class institutions such as the Indian Institutes of Technology (IITs) and Management (IIMs) which sought excellence in high-technology engineering, science, and management skills (Dutz, 2007). In comparison to 11% and 12% in China, in 2000 the Indian government spent 86% of per capita GDP on each student in tertiary education compared to 14% in primary education (Kochhar *et al.*, 2006). Unfortunately, India's large private sector could not benefit from the vast pool of high-technology skills owing to the government's preference for SMEs.

When India opened up to foreign competition, its private sector was globally uncompetitive in most industries (Dutz, 2007), but past investments in technical skills were not completely wasted. The benefits of human capital created by Indian public institutions since the 1960s have accrued in large measure to India's booming software exports and budding medical tourism industries. Both industries are close to the global technological frontier. India's services exports are roughly 8% of GDP compared to only 5% in Korea. A further expansion of the Indian innovation system could potentially stoke a services exports-oriented catch-up strategy.

Role of policy: balance between competition and innovation

An important insight from the experiences of the four countries is not whether competition and innovation are necessary for sustained economic growth. Rather, it is about what kind of sequencing best maximises the gains from competition and innovation to spur catch-up. It is also about the balance between competition and innovation that policy makers can strike to accelerate growth. These issues are further explored by examining some statistics about competition and distance shortening and modelling the data using the Aghion *et al.* approach.

Table 3.3. Tariff and non tariff measures in BICs, total trade

	1990	1995	2000	2001	2002	2003	2004	2005	2006	2007
Weighted average tariffs in total trade										
Brazil	19.0	12.7	12.7	10.4	10.0	9.4	9.0	8.5	8.5	8.7
China	32.2	23.8	14.7	14.1	10.3	6.5	6.0	4.9	4.4	4.7
India	49.6	23.2	27.5	26.5	25.3	24.1	22.8	13.4	11.9	10.4
Korea	9.5	6.8	7.2	8.6	10.0	9.6	9.2	8.3	7.4	7.4
OTRI[1] (MFN applied tariffs plus non-tariff measures)										
Brazil				27.2				22.3	21.9	
China				21.1				11.2	10.1	
India				32.2				19.7	21.9	
Korea				10.0				-	10.0	

[1]OTRI (MFN applied tariff+NTMs). The OTRI is expressed as a tariff rate and applies to all goods. This index summarises the impact of each country's non-discriminatory trade policies on its aggregate imports. It is the uniform equivalent tariff that would maintain the country's aggregate import volume at its current level (given heterogeneous tariffs) including non-tariff measures. It captures the trade distortions that each country's MFN tariffs impose on its import bundle using estimated elasticities to calculate the impact of a tariff schedule on a country's imports. These measures are based on actual or current trade patterns and thus do not capture restrictions facing new or potential trade. They also do not take into account domestic subsidies or export taxes.

Source: World Bank, *World Integrated Trade Solutions* and *World Trade Indicators*.

Table 3.3 complements tariff information with a proxy for non-tariff measures (OTRI) and summarises the trends in trade liberalisation that increased competition in the BICs and Korea. In the early 1990s, Korean tariff and non-tariff barriers were already quite low at 10%; they have stabilised since at those levels. In 2006, the OTRI rates in Brazil and India were twice as high as those of Korea and China. The latter liberalised relatively rapidly while Brazil and India did so more slowly. If the industries of the three BICs were significantly and equally far from the Korean technological frontier, then Aghion *et al.*'s model would predict that liberalisation would hurt these industries. Slower and belated competition without export orientation in Brazil and India perpetuated uncompetitive and inefficient industries.

Table 3.4 summarises some of the outcome indicators of the main distance-shortening policies of the BICs and Korea. In the absence of a time series, recent statistics are used to track progress and identify several noteworthy issues highlighted in Table 3.4.

First, nearly every cell in Table 3.4 indicates that government-led innovation interventions were at work in all four countries. Second, the data suggest that governments value the strategic nature of distance-reducing policies and are willing to spend on them. However, the data do not indicate that there is a formula for combining the precise mix of public and private interventions to spur innovation. A few examples are noted below.

Consider the case of R&D financing. Consistently high scores of GERD and BERD in Korea and China point to the importance of R&D expenditures for distance-shortening policies. The university-industry linkages scores, which suggest the more efficient role of the private sector, are also significantly higher in these two countries than in Brazil and India. It should be noted however, that these data points do not indicate whether a higher level of government procurement of advanced technology or fiscal incentives for private R&D or both were necessary for encouraging private investments in R&D in Korea and

China. They also do not specify the optimal balance between these policy instruments. Is there a magic formula that Brazil and India have yet to discover?

Table 3.4. A comparison of distance-shortening with other policies in the BICs and Korea

	Korea	China	Brazil	India
Macro and trade liberalisation reforms				
Timing	1970s	1980s	1990s	1990s
Stabilisation	Yes	Yes	Yes	Yes
Degree of openness (trade as % of GDP: 1980s-2006)	67 to 84	27 to 71	18 to 26	14 to 46
Level of tariff and non-tariff barriers (text tables)	Very low	Very low	Medium low	Medium high
Industrial policy				
Preference for export-led growth (export/GDP (%); 1970s and 2005-06 (WDI)	23 to 43	5 to 40	7 to 14	5 to 21
Preference for hi- technology exports (% in total exports)[1]	35	35	8	5
Preference for natural resource-based exports *(% in total exports)*[2]	0.2	2.3	20.6	9.6
Intellectual property indicators				
Patent applications filed under PCT (2005)[1]	5 152	3 774	339	978
Triadic patent families[3] (2005)[1]	2811	356	58	113
Intellectual property protection 2007 (Scale 1-7) (KAM)	5.4	3.4	3.3	4.0
Total royalty payments and receipts (USD millions) 2006 (KAM)	6 497	5 478	1 506	446
Distance shortening in early stages	Mostly state-led	Mostly state-led	Mostly state-led	Mostly state-led
	Financing of R&D			
Gross domestic R&D expenditure/GDP (2006)[1]	3.23	1.43	1.02	0.71[4]
Business enterprise R&D expenditure /GDP (2006)[1]	2.49	1.02	0.49	
GERD per capita (current USD PPP)[1]	743	66	92	13
	University-industry linkages			
University-industry research collaboration in 2007 (Scale 1-7) (KAM)	5.4	4.1	3.4	3.5
University-industry linkages (indicator: share financed by industry)[4]	75%	69%	39%	16%
	Incentives for research			
Tax subsidies for USD 1 of R&D[2, 5]	0.18	0.339	0.254	0.266
Subsidies for firm level research (Scale1–7; 2003) (WEF)	3.52	1.86	2.03	2.87
Government procurement of advanced technology products (Scale 1-7; 2003) (WEF)	4.5	4.7	3.6	3.3
	Role of foreign innovators			
Prevalence of foreign technology licensing (Scale 1-7; 2003) (WEF)	5.32	3.79	4.65	3.27
Role of foreign innovators – (indicator: patents with foreign co-inventors (2002-04) [1]	4.60	27.87	28.42	
	Human resources for S&T			
Quality of S&T research (Scientific and technical journal articles per million population, 2005) (KAM)	339.5	31.89	52.9	13.35
Researchers/1 000 total employment (2006)[1]	8.65	1.60	1.48	0.3 (in 2000)
Science and engineering degrees as a % of all new degrees (2005)[1]	37.80	39.18	10.78	

KAM: The World Bank's Knowledge Assessment Methodology.

WDI: The World Bank's *World Development Indicators*.

WEF: World Economic Forum.

1. OECD (2008), *OECD Science, Technology and Industry Outlook*, OECD, Paris.
2. OECD (2007), *OECD Science, Technology and Industry Scoreboard*, OECD, Paris.
3. Patents applied for at the US Patent and Trademark Office, the European Patent Office and the Japanese Patent Office in 2005.
4. *OECD Factbook 2008*.
5. The corresponding tax subsidy for the United States is 0.66.

The role of foreign technology transfer through licensing underscores its importance for catch-up in all four countries, but the role of foreign co-inventors does not. Does Korea's low score suggest that foreign co-inventors are relatively unimportant or simply that China and Brazil are better at attracting foreign co-inventors?

Patent-related innovation indicators show that the BICs have gone beyond the technology adaptation stage and, like Korea, are playing catch-up. Consider, for example, the number of triadic patents, *i.e.* patents submitted to the European Patents Office, the US Patent and Trademark Office and the Japanese Patent Office in 2005. Korean patent records are firm evidence of its achievement in patenting. In 2005, even though it lagged the United States by a large margin (United States had 15 700 and Korea had 2 800), it had made phenomenal progress relative to 1985 when it had only 7 patents. Similarly in 2005, the figures for the BICs were impressive when compared to 104 for Singapore, 20 for Hong Kong, China, and 4 for Chile.

The main policy insight to be drawn from Table 3.4 is that each of the BICs and Korea had its own distance-shortening policies. Evidently, *in addition to* and *prior to* trade liberalisation, Korea and the BICs in varying degrees leveraged a package of distance-shortening policies to foster domestic innovation capabilities. This prepared them better to compete in the global market when they opened up. The issue was not a single indicator or two or even three, but a comprehensive package of support to innovation capabilities that could be measured by indicators such as those in Table 3.4. Among the policy instruments used were targeted fiscal incentives for innovation R&D, FDI for technology transfer, and heavy targeted investments in human resources for S&T.

Figure 3.16 further illustrates trade-offs between competition and innovation. It shows a diagram of the trade-off between competition and innovation. It is a graphical description of simulations using the parameters of the empirical model discussed below. It shows competition on the x-axis and growth in value added per worker which enables catch-up on the y-axis. The maximum growth rate of labour productivity is at point A which, in the context of the present analysis, represents the Korean technological frontier. Conventional neo-classical analysis assumes that when the rate of growth is slow, swift competition can help an industry to move speedily along the solid line towards the frontier (point A). However, as neo-Schumpterian theory argues, if distance matters, competition may induce the industry to move along a path with slower growth in value added. In fact, if the industry is too far from the frontier, opening up to competition can lead to negative growth in value added as shown by the trajectory CB (dotted line). For the same level of competition, industries closer to the frontier innovate and enjoy higher productivity growth. A sequencing of distance-shortening policies followed by competition can enable the industry to move closer to the frontier, but along a flatter slope, *i.e.* with relatively slow growth in value added until competition is introduced.

Consider the three arrows in Figure 3.16 used to illustrate the experiences of the BICs with distance-shortening and competition policies. In the early 1980s, most industries in China and India faced low levels of competition and were located far from the frontier, say at point C. Brazilian labour productivity levels, however, were much higher and placed Brazil closer to the Korean technological frontier. Rapid opening to foreign competition should have made industries in China and India uncompetitive and set them on a negative growth trajectory towards point B (or extinction). We know that this did not happen although the outcomes were very different. Distance-shortening policies were a precursor to China's opening to foreign competition and boosted its labour productivity growth, preparing it for global competition. In the late 1980s, when China liberalised swiftly, it benefited from accelerated growth in labour productivity. The Chinese model, like the Korean, balanced an innovation focus (via adoption of distance-shortening policies) with competition. Unsurprisingly, China's growth rate in labour productivity, as reflected in its pace of catch-up, is the fastest. In comparison, India did not significantly employ distance-shortening policies to boost the competitiveness of its manufacturing sectors. As a consequence, when it opened its economy, it was unable to compete in almost all manufacturing industries.

Figure 3.16. Tradeoffs between distance to the frontier and competition

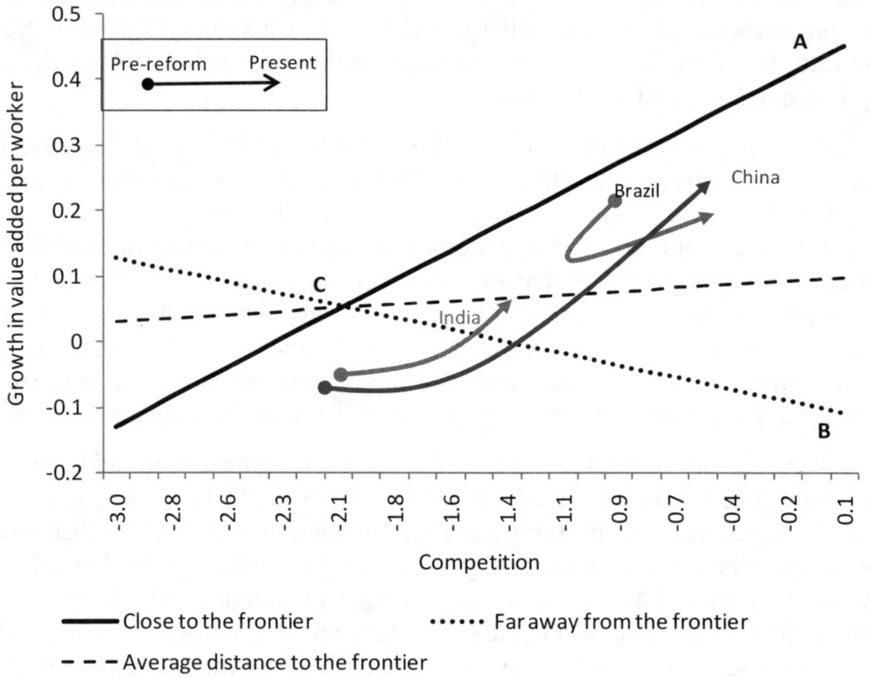

Note: Lines for Brazil, India, and China are provided as examples and do not necessarily represent the behaviour of all industries in these countries.

Source: Estimates from the empirical model discussed in the following section.

In the early 1980s, Brazil was closer to the frontier than China and India (Figure 3.4). This reflected the stronger initial position of the Brazilian economy at the end of its "miracle" years. During the debt crisis, the economy stalled and lost its competitiveness, as shown by lower growth in labour productivity. When the Brazilian economy displayed a productivity spurt in the 2000s, the high returns from China's distance-shortening policies and speedier rate of liberalisation had already made China the country with the fastest growth in value added per worker among the BICs (and indeed globally).

An analysis of how competition and distance-shortening policies have affected the *level* of competitiveness in the BICs is illustrated in Figure 3.17 with different variables on the axes. The x-axis displays distance to the frontier (from right to left) and the y-axis displays the level of value added per worker. Brazil's initial position was close to the frontier in the 1980s. In levels, its labour productivity is still higher than China's, as displayed in Figure 3.4. Even though China has achieved faster growth in labour productivity, its distance from the Korean technological frontier remains significant and can be measured on the y-axis. The following section tests the Aghion *et al.* model using data for Korea and the BICs.

Figure 3.17. Tradeoffs between distance to the frontier and competition

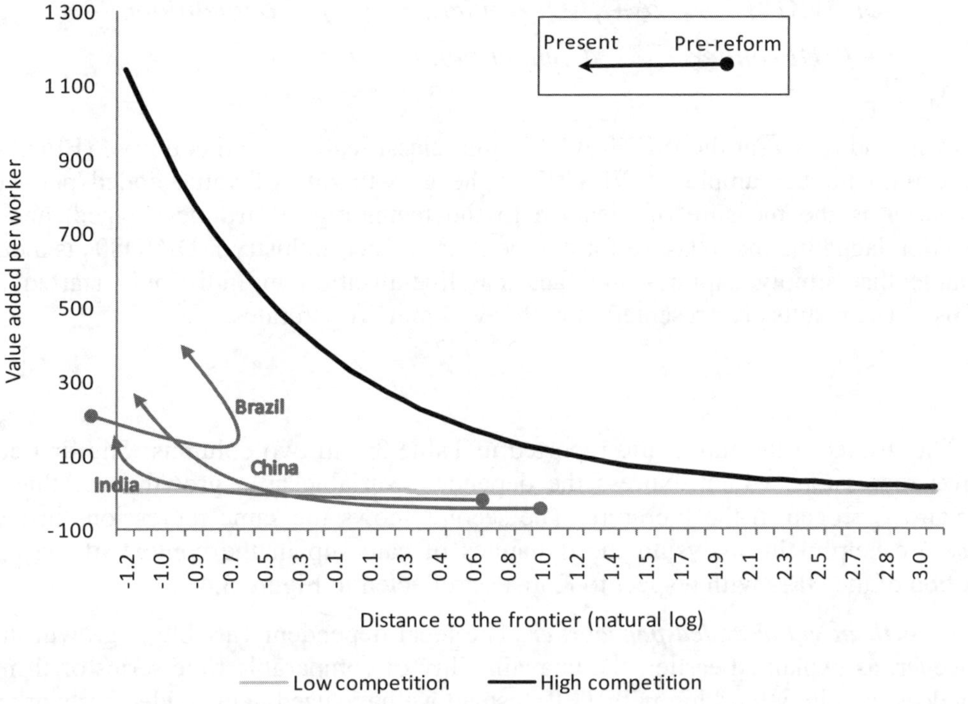

Note: Lines for Brazil, India and China are provided as examples and do not necessarily represent the behaviour of all industries in these countries.

Source: Estimates from the empirical model discussed below.

Empirical methodology

According to the Schumpeterian model, there is a positive relation between larger firms and innovation. Scherer's (1965) empirical research did not substantiate the Schumpeterian paradigm. Empirically, there is no *a priori* evidence to support or reject the view that competition adversely affects innovation. The economic environment, distance of industries from the technological frontier and other variables play an important role in affecting the impact of competition on innovation. Nickell (1996), for example, found a strong and positive relation between competition and innovation measured by firm-level TFP.

An investigation of the Schumpeterian paradigm needs to incorporate several factors. The first is a correction for the correlation between industry characteristics such as size and innovation. We deal with this by controlling for industry fixed effects. Second, the analysis must address reverse causality: whether growth in labour productivity (the dependent variable) affects the distance to the Korean technological frontier or whether a change in distance affects the growth and level of labour productivity. Panel data and an instrumental (exogenous) variable are used to correct for this problem. We use systems

GMM which uses lags of all exogenous variables to correct for endogeneity. Industry-specific weighted tariff rates represent competition.

We use the Aghion *et al.* (2006) model to test the effects of competition and innovation on catch-up. The model is specified as follows:

$$dLNVAPW_{i,j,t} = \alpha_1 + \beta_1 DFrontier_{i,j,t-1} + \beta_2 Competition_{i,j,t-1}$$
$$+ \beta_3 DFrontier_{i,j,t-1} * Competition_{i,j,t-1} + \varepsilon_{i,j,t}$$

for every industry *i* (at the ISIC Rev. 2 3-digit classification), and country *j* (Brazil, India and China) in the sample. *dLNVAPW* is the growth rate of value added per worker, *DFrontier* is the measure of distance to the technological frontier lagged, and tariff reduction lagged is the measure for competition in each industry *i*. DIND80s is a dummy variable that simply captures the fact that liberalisation in India only started in the 1990s.[10] The results are presented in both levels and growth rates.

Results

The results of the model are reported in Table 3.5 in two columns. The first column shows regressions which express the dependent variable as a growth rate; this is the standard approach in the literature. The second shows the same regressions in levels – these are helpful in analysing the dynamics of catch-up in the context of the relative position of the BICs with respect to Korea as reflected in Figure 3.4.

Growth in value added per worker. The ideal dependent variable is growth in TFP. However, as explained earlier, the unavailability of comparable time series of firm-level data does not allow working with TFP. Instead we have used value added per worker as a proxy for innovation – not an unreasonable substitution, as we confirmed by analysing investment climate assessments for the BICs at the level of large firms. In addition, use of the lagged values of the dependent variable allows us to control for the contribution of capital to output, *i.e.* the previous period's output levels already incorporate the effect of capital on labour productivity. The change in labour productivity can then be interpreted as the effect of a change in technological innovation.

Competition. The estimates show a positive and robust relationship between the competition and innovation variables. As expected, an increase in competition, indirectly measured by lowering tariffs, is positively associated with an increase in the growth of manufacturing value added per worker.

Distance to the frontier. This is measured by the ratio of labour productivity in industry *j* in Korea (benchmark) and labour productivity in industry *j* in one of the BICs. A negative and significant sign of this variable suggests that industries that are farther away from the technological frontier have fewer incentives to innovate and increase growth in value added per worker.

Interactive term. Suppose that two industries have the same level of competition. The industry that is farther away from the technological frontier will innovate less than the one that is closer to the frontier. The interactive term therefore evaluates the effects of competition on growth in labour productivity in a setting in which all industries are not

10. Brazil also only began a comprehensive trade liberalisation effort in the 1990s. The dataset, however, is limited to post-1990 data in this case.

equally close to the frontier. A negative sign suggests that competition is a necessary but not a sufficient condition to motivate an increase in an industry's innovative capacity. The impact of changes in the competition regime is stronger if industries are technologically more advanced or closer to the frontier but diminishes for industries located farther away from the technological frontier.

Table 3.5. GMM estimation results[1]

Dependent variable	Growth in value added per worker	Log of value added per worker
Distance to the frontier (lagged)	- 0.12552*	- 1.026538***
Competition	0.113523***	0.8717822***
Distance to the frontier*Competition	- 0.060857***	- 0.13327***
DIND80s	0.200927***	
Constant	0.286821***	5.717783***
Observations	1 045	1 855
Groups	73	98
Observations per group		
Minimum	6	8
Average	14.32	18.93
Maximum	20	24
Number of instruments	71	74
Tipping point in distance beyond which competition negatively affects innovative capacity	*1.865411*	*6.541214*

*** Significant at the 1% level.

GMM: System Generalized Method of Momentums using the Arellano-Bond Methodology and the xtabond2 command in STATA. The option collapse is used to avoid the excessive use of lagged instruments which diminishes the efficiency of the estimates.

DIND80s: A dummy for India in the 1980s.

1. The first estimates present the results of using a panel data estimation procedure with fixed effects for every combination of industry and country in the sample. To overcome the possibility of endogeneity in the specification of our model and heteroskedasticity, we use the system GMM estimator because it is well suited for dynamic "small-T, large-N" panels that may contain fixed effects and idiosyncratic errors that are heteroskedastic and correlated within, but not across, individuals. We use the two-step variant of the system GMM estimator because it is asymptotically more efficient, compared with the one-step variant. Nevertheless, two-step standard errors tend to be severely downward biased (Arellano and Bond, 1991; Blundell and Bond, 1998). To compensate for this, we use a finite-sample correction to the two-step estimation efficiency covariance matrix derived by Windmeijer (2005).

Source: Authors' estimates using ISIC Rev. 2 3-digit industrial data, COMTRADE data and TRAINS tariff data.

The results indicate that several industries in the BICs remain very far from the Korean frontier. The model can be used to calculate how far they are. If the number for distance is bigger than the tipping point of 1.86 (6.54 for the model in log levels), that industry's distance is too far from the frontier, *i.e.* its innovative capability was too weak relative to Korea's and it was adversely affected by trade liberalisation.[11] Each of the BICs has several such industries. Some have been more successful in narrowing the technological gap in the post-liberalisation period but the others are still below the tipping

11. Simulations are available from the authors upon request.

point. For example, the simulations show that with increased competition, Brazil has been able to innovate and narrow its distance from the Korean frontier in the food products industry but in China and India that industry was unable to withstand competition and has fallen farther behind. Distance can change when the frontier industry is moving faster than the incumbent (as in the case of Korea where most industries have grown faster; see Figure 3.4).

The positive sign on the competition variable and the negative sign on the interactive term in the model validate Aghion *et al.*'s results as well as the inverted-U hypothesis. In short, distance matters. The negative sign on the interactive term has led Aghion *et al.* to note that competition is desirable and should occur, but government policies should make it easier for the uncompetitive firms to shut down and let resources flow to more productive firms.

The results suggest that to benefit from the efficiency gains generated by competition, distance-shortening policies, such as those targeted at building domestic capabilities to innovate, become particularly relevant. By helping to shorten in advance the distance to the frontier, they can dampen the otherwise negative effect of the interaction between competition and distance. There is also a negative sign on the distance variable which indicates that, by itself, an increase in the distance to the frontier, caused for example by an increase in Korea's productivity levels, will not encourage the industry to innovate and grow. Overall, distance will have a negative effect on productivity growth. The farther the distance, the harder it will be for the country to catch up. This could apply to a poor country with an industry that is protected and too far from the frontier. The results also indicate that investment in distance-shortening policies alone may have low payoffs as the Indian experience shows.

The potential of distance-shortening policies for catch-up offers policy makers a variety of choices from which they can pick the one most suitable for their economy. It would, however, be a mistake to infer that Korean or Chinese distance-shortening policies are necessarily optimal and should be replicated by any country aspiring to catch up. Korea's programmes to forge industry-university linkages benefited from its strong endowment of human capital and its long-term commitment to education. Few developing countries can easily replicate the university-industry linkages that Korea enjoys today although going forward, there may some useful lessons. And China's ability to leverage innovation through conditional contracts with foreign investors was a function of the attractiveness (size) of the Chinese domestic market.

Conclusions

This chapter has analysed how three emerging market economies – Brazil, China and India – are catching up with Korea's technological frontier using Aghion *et al.*'s model across 30 industries. In general the analyses support Aghion *et al.*'s inverted-U hypothesis. When industries are closer to the technological frontier, they innovate to escape competition while a longer distance discourages them from innovating. For countries that are far from the frontier, catch-up is a greater challenge as the frontier itself is a moving target.

In addition to the primacy of competition policy, the results illustrate the role of distance-shortening policies. For a country which has industries that are technological laggards and are likely to be wiped out by foreign competitors, the model illustrates that distance-shortening polices may be a necessary complement to liberalisation for effective catch-up.

What can Brazil and India, and indeed other middle-income countries, learn from the Korean and Chinese models for catching up? Korea and China adopted a variety of distance-shortening policies (supporting R&D investments) in combination with financial subsidies to promote industries (with an emphasis on fostering high-technology industries) in parallel with an export-led growth strategy. They ensured that their targeted industries had mastered technological learning and were ready to innovate and keep competitors at bay (and/or had become natural partners for foreign companies) before pursuing trade liberalisation. Once the liberalisation effort was unleashed, however, both Korea and China did not hesitate to leverage swift competition to garner additional efficiency gains. Starting in the early 1980s, it took China less than two decades to transform its industrial structure away from resource-based primary industries and labour-intensive sectors towards the more high-technology industries that Korea had championed in its own catching-up experience.

Brazilian and Indian catch-up strategies have been less successful (see Chapter 7). The large anti-export bias of their import-substitution industrialisation strategies was not effectively counteracted by governmental policies, and insufficient attention to innovation-enhancing policies did not foster competitiveness at the international level. Upon opening up in the 1990s, they found themselves too far from the technological frontier and have been less successful in harvesting the gains from competition. Needless to say, there are world-class industries in both countries. But these often reflect the unintended consequences of inward-oriented policies combined with exogenous shocks rather than the sustained pursuit of innovation-driven policies. As an example, the impact of the ICT revolution on the tradability of services, coupled with investments in tertiary education, is at the heart of the success of India's software industry. Similarly, the pursuit of energy alternatives in Brazil, such as ethanol, was a response to the oil shocks of the 1970s, driven by the desire to minimise energy-dependency for security reasons.

It is important to recognise that a model of state-led innovation would be hard to replicate these days in view of the multilateral disciplines under the WTO that constrain discriminatory treatment in favour of national firms. Yet, both Korea and China have had significant success in their efforts to catch up with the OECD, and their experiences, as well as those of Brazil and India, provide useful insights on the interplay between innovation, competition and growth.

In sum, the experience of the BICs in terms of catch-up with Korea underscores the fact that there is no blueprint for distance-shortening policies. Certain outcomes of distance shortening policies seem to be common to the BICs and Korea and indeed in Korea's catching up with OECD countries. Striving to achieve the distance-shortening policy outcomes illustrated by the experience of Korea is a worthwhile goal. The task of identifying appropriate distance-shortening policies, for the BICs or other followers in the current global economic environment, remains, however, a formidable challenge.

References

Aghion, Philippe and Peter Howitt (1992), "A Model of Growth through Creative Destruction", *Econometrica, Econometric Society* 60(2) (March), pp. 323-351.

Acemoglu, D., P. Aghion and F. Zilibotti (2006), "Distance to Frontier, Selection, and Economic Growth", *Journal of the European Economic Association*, 4(1) pp 37-74.

Aghion, P., R. Blundell, R. Griffith, P. Howitt and S. Prantl (2004), "Entry and Productivity Growth: Evidence from Micro-Level Panel Data", *Journal of the European Economic Association* Vol. 2, pp. 265-276, April-May.

Aghion, P., N. Bloom, R. Blundell, R. Griffith and P. Howitt (2005), "Competition and Innovation: An Inverted U Relationship", *Quarterly Journal of Economics* May, Vol. 120, No. 2, pp. 701-728.

Aghion, P., R. Blundell, R.Griffith, P. Howitt and S. Prantl (2006) "The Effects of Entry on Incumbent Innovation and Productivity", National Bureau of Economic Research Working Paper No. 12027.

Aghion, P., R. Burgess, S. Redding and F. Zilibotti (2008), "The Unequal Effects of Liberalization: Evidence from Dismantling the License Raj in India", *American Economic Review,* Vol. 98 (4), pp. 1397-1412.

Aghion, P. and R. Griffith (2008), *Competition and Growth: Reconciling Theory and Evidence*, The MIT Press, Cambridge, MA.

Amsden, A.H. 1989. *Asia's Next Giant: South Korea and Late Industrialization*, Oxford University Press, New York.

Arellano, M. and S. Bond (1991), "Some Tests of Specification for Panel Data: Monte Carlo Evidence and an Application to Employment Equations", *Review of Economic Studies (UK)* 58, pp. 277-297.

Blundell, R. and S. Bond (1997), "Initial Conditions and Moment Restrictions in Dynamic Panel Data Models", Dept. of Economics, Discussion Papers in Economics, University College London.

Carlin, W., M.E. Schaffer and P. Seabright (2004), "A Minimum of Rivalry: Evidence from Transition Economies on the Importance of Competition for Innovation and Growth", *Contributions to Economic Analysis & Policy*, 3 (1), Berkeley Electronic Press.

Commission on Growth and Development (2008), *The Growth Report: Strategies for Sustained Growth and Inclusive Development*, The World Bank, Washington, DC.

Costa, I. and S. Robles Reis de Queiroz (2002), "Foreign Direct Investment and Technological Capabilities in Brazilian Industry", *Research Policy (Netherlands)* 31, Nos. 8-9, pp. 1431-1443.

Cruz, C., H. Brito and L. de Mello (2006), "Boosting Innovation Performance in Brazil", Economics Department Working Papers, OECD, Paris.

De Ferranti, D.M. *et al.* (2003), *Closing the Gap in Education and Technology, World Bank Latin American and Caribbean studies*, The World Bank, Washington, DC.

Dixit, A.K. and J.E. Stiglitz (1977), "Monopolistic Competition and Optimal Product Diversity", *American Economic Review* 67(3) (June), pp. 297-308.

Dutz, M. (ed.) (2007), *Unleashing India's Innovation: Towards Sustainable and Inclusive Growth*, The World Bank, Washington, DC.

Gorodnichenko, Y., J. Svejnar and K. Terrell (2008), Globalization and Innovation in Emerging Markets", IZA Working Paper, Institute for the Study of Labour, Bonn.

Grossman, G.M. and E. Helpman (1991), *Innovation and Growth in the Global Economy*, MIT Press, Cambridge, MA.

Katz, J. (2000), "The Dynamics of Technological Learning during the Import Substitution Period and Recent Structural changes in the Industrial Sector of Argentina, Brazil and Mexico", in *Technological Learning and Innovation,* Linsu Kim and R. Nelson (eds.), Cambridge University Press, Cambridge and New York.

Kim, Linsu and R. Nelson (2000), *Technology, Learning and Innovation: Experiences of Newly Industrializing Economies*, Cambridge University Press, Cambridge and New York.

Kochhar, K., U. Kumar, R. Rajan, A. Subramanian and I. Tokatlidis (2006), "India's Pattern of Development: What Happened, What Follows?", *Journal of Monetary Economics (Netherlands)* 53, No. 5, pp. 981-1026.

Lall, S. (2000), "Skills, Competitiveness and Policy in Developing Countries", Queen Elizabeth House Working Paper, Oxford University.

Lall, Sanjaya, John Weiss and Jinkang Zhang (2006), "The 'Sophistication' of Exports: A New Trade Measure", *World Development* (United Kingdom) 34, No. 2, pp. 222-37.

Lattimore, R. and P. Kowalski (2008), "Brazil: Selected Trade Issues," OECD Trade Policy Working Papers No. 71, OECD, Paris.

Loayza, N., P. Fajnzylber and C. Calderón (2005), *Economic Growth in Latin America and the Caribbean: Stylized Facts, Explanations, and Forecasts*, The World Bank, Washington, DC.

Mathews, J. (2009), "China, India and Brazil: Tiger Technologies, Dragon Multinationals and the Building of National Systems of Economic Learning", *Asian Business and Management*, Vol. 8, No. 1, pp. 5-32, Palgrave Macmillan.

Nickell, S.J. (1996), "Competition and Corporate Performance", *Journal of Political Economy* 104(4) (August), pp. 724-746.

OECD (2007), *OECD Science, Technology, and Industry Scoreboard*, OECD, Paris.

OECD (2008a), *Science, Technology and Industry Outlook*, OECD, Paris.

OECD (2008b), *Review of Innovation Policy: China*, OECD, Paris.

Ozawa, T. (2005), *Institutions, Industrial Upgrading, and Economic Performance in Japan: The 'Flying Geese' Paradigm of Catch-up Growth*, Edward Elgar, Northampton, MA.

Primo Braga, C.A. and G. Bannister (1994), "East Asian Investment and Trade: Prospects for Growing Regionalization in the 1990s", *Transnational Corporations* 3 (February), pp. 97-136.

Primo Braga, C.A., C. Fink and C.P. Sepulveda (2000), "Intellectual Property Rights and Economic Development," World Bank Discussion Paper No. 412, The World Bank, Washington, DC.

Romer, Paul M. (1990), "Endogenous Technological Change", *Journal of Political Economy* 98(5) (October), pp. S71-102.

Sabirianova, K., J. Svejnar and K.D. Terrell (2004), "Distance to the Efficiency Frontier and FDI Spillovers", Centre for Economic Policy Research, United Kingdom, Discussion Paper Series.

Sabirianova, K., J. Svejnar and K.D. Terrell (2005), "Foreign Investment, Corporate Ownership, and Development: Are Firms in Emerging Markets Catching Up to the World Standard?", Centre for Economic Policy Research, United Kingdom, Discussion Paper Series.

Scherer, F.M. (1965), "Firm Size, Market Structure, Opportunity, and the Output of Patented Inventions", *American Economic Review* 55 [5, Part 1 (December)], pp. 1097-1125.

Subramanian, A. (2008), *India's Turn: Understanding the Economic Transformation*. Oxford University Press, New Delhi.

Suh, J., H. Derek and C. Chen (2007), *Korea as a Knowledge Economy: Evolutionary Process and Lessons Learned*, World Bank Institute, Washington, DC.

Topalova, P. (2004), "Trade Liberalization and firm Productivity: The Case of India", IMF Working Paper, International Monetary Fund, Asia and Pacific Dept., Washington, DC.

United Nations Industrial Development Organization (UNIDO), Statistics and Information Networks Branch (2002), *Industrial Statistics Database. 3- and 4-digit Level of ISIC Code (Rev. 2)*. Statistics and Information Networks Branch, United Nations Industrial Development Organization, Vienna.

United Nations. Statistical Division (2006), *UN Commodity Trade Statistics Database*, COMTRADE, United Nations, Statistics Division, New York, *http://unstats.un.org/unsd/comtrade*.

University of Pennsylvania, Center for International Comparisons, Penn World Tables, *http://pwt.econ.upenn.edu/aboutpwt.html*.

Wang, J.-H. (2007), "From Technological Catch-up to Innovation-based Economic Growth: South Korea and Taiwan Compared", *Journal of Development Studies (UK)* 43, No. 6, pp. 1084-1104.

Windmeijer, F. (2005), "A Finite Sample Correction for the Variance of Linear Efficient Two-Step GMM Estimators", *Journal of Econometrics*, 126, pp. 25-51.

World Bank (1998), *Knowledge for Development: World Development Report 1998-99*, The World Bank, Washington, DC.

World Bank (2007), *Brazil: Knowledge and Innovation for Competitiveness*, Report No. 40011-BR, 2007, Brazil Country Management Unit, Human Development Unit, Latin America and the Caribbean Region.

World Bank, *World Development Indicators* (various years), The World Bank, Washington, DC.

World Bank, Development Data Group, Development Data Platform (DDP), The World Bank, Washington, DC, World Bank Group, *http://ddp.worldbank.org*.

World Bank, Development Data Group, and United Nations, Statistical Division, *World Integrated Trade Solution WITS*, The World Bank, Washington, DC, *http://library.worldbankimflib.org/ElectronicResources/eresourcelist.asp?catkey=728263;*
www-int.imf.org/depts/tgs/bts/bts-ie/cats/appcomtrade.asp?Active=0102;
wits.worldbank.org.

World Bank and European Bank for Reconstruction and Development (1999), *Business Environment & Enterprise Performance Surveys*, The World Bank, Washington, DC, *http://go.worldbank.org/RQQXYJ6210*.

Chapter 4

Priorities for growth in OECD economies

Sveinbjorn Blöndal and Sean Dougherty[1]

This chapter discusses the main results of work carried out at the OECD to benchmark economic performance and policy in its member countries, in order to make policy recommendations that may improve economic performance. It examines differences in policy recommendations that have been made for countries at various levels of development, characterises the main challenges each group of countries faces, and considers the main distinctions between these countries and five emerging non-members. Product market competition and human capital reforms are found to be especially important priorities in lower-income countries, which face substantial gaps in productivity to the frontier countries.

1 Sveinbjorn Blöndal is Head of the Macroeconomic Policy Division and Sean Dougherty is Senior Economist, Structural Surveillance Division, Economics Department, OECD.

Benchmarking structural policies

Differences in material living standards across OECD countries reflect in part different structural policy settings, as well as deeper institutional characteristics. Relatively low income per capita and a failure to converge towards the highest-income countries can therefore be signs of policies that are not as growth-friendly as they could be. Successive empirical studies by the OECD and others have sought to identify the policy levers that influence GDP per capita and its growth.[2] As part of these studies, various indicators have been developed to summarise performance on key components of GDP per capita and the stance of related policies in a consistent way across countries and over time.

The OECD's *Going for Growth* report seeks to help policy makers to achieve improved standards of living for their citizens. Drawing on knowledge of economic circumstances in individual countries, the exercise applies a systematic international benchmarking framework to analyse indicators of policy and performance. On this basis, it then identifies five policy priorities for each country that would help promote higher GDP per capita. These policy priorities are discussed and vetted by member countries, and the report itself is published annually under the responsibility of the OECD Secretariat. The report, which began in 2005, serves as a vehicle for the OECD to issue recommendations for reform across a range of policy areas, as part of its multilateral surveillance work on structural issues.[3] This systematic benchmarking relies primarily on objective policy indicators that have been linked econometrically to economic performance.

Empirical research linking policy with performance includes a long series of studies performed on a large number of OECD countries. These studies include the *OECD Growth Study* (2003), the *OECD Jobs Strategy* (1994) and its reappraisal (2006), and associated background studies, which took inspiration and drew extensively from the academic literature. These studies included estimates of short-run and long-run effects of product and labour market policies on GDP per capita, documented, for example, in Bassanini *et al.* (2001), as well as examinations of the effects of policies on employment and unemployment, as in Bassanini and Duval (2006).[4] Many of the OECD studies were

2. *Going for Growth* focuses on GDP per capita as its principal measure of material living standards. Welfare includes material standards of living as well as the value of leisure, inequality of income, use of non-renewable resources and environmental services. Many of these aspects of welfare are difficult to measure and are not available on a timely basis. Earlier *Going for Growth* editions have shown that there is a close relationship in most cases between GDP per capita and broader, though less timely, measures of economic well-being. This relationship is currently being re-examined by the Stiglitz-Sen-Fitoussi Commission.

3. *Going for Growth* is a core part of the mutual accountability and peer pressure that are central to the OECD's mission. This horizontal structural surveillance exercise supplements the country-specific surveillance that is reported in *Economic Surveys*, as well as thematic reviews in specific areas such as agriculture, education, environment, innovation, investment and regulatory policy.

4. Bassanini *et al.* (2001) and Boulhol *et al.* (2008) estimated growth equations using the pooled mean group estimator for a panel of 21 OECD economies, and examined the role of various types of structural policies on long-run economic growth. Bassanini and Duval (2006) estimated unemployment and group-specific employment rate equations for a panel of 24 OECD economies using various panel data estimators (including that of Arellano and Bond) to examine the role of policy settings and their interactions in affecting participation. These studies did not include most of the middle-income OECD countries, owing to overly short data availability, though there now exists a sufficient number of observations to incorporate them into some types of panel data analysis.

carried out for OECD committees, and documented in OECD Working Papers and various other publications. While much related work has been done outside of the OECD, the vast majority lacks a direct link to policy recommendations (OECD, 2009).

The broader literature on growth and development has offered a range of important insights into the role of policies, although it has focused more on deeper institutions, such as types of legal or political systems, which are usually the result of an accumulation of policy reforms over long periods (IMF, 2008). More broadly, institutions do seem to play an important role in economic development, and countries with higher levels of GDP per capita have institutions of much higher quality according to many measures (Kauffman *et al.*, 2008). However, the direction of causality is not always clear: "deep" institutions are also highly endogenous, and it is not at all easy to determine their causal role with respect to income levels or economic growth (Glaeser *et al.*, 2004; Acemoglu *et al.*, 2005). Moreover, the role of geographic factors and trade openness appears to be closely linked to institutions, making their identification difficult (Rodrick *et al.*, 2004; Boulhol *et al.*, 2008).

So far, the *Going for Growth* exercise has focused heavily on product and labour market policies (rather than deeper institutions) that can be shown to increase GDP per capita in OECD countries in a straightforward fashion. To do this, it has relied on studies using specific quantifiable policy indicators that have been vetted by policy makers and can be directly linked to policy actions. Given this, virtually all policy indicators that are currently used are produced by the OECD Secretariat. Though a large number of organisations produce various types of indicators that could potentially be relevant, these indicators usually lack a direct connection to policy levers (Furceri and Mourganne, 2009). Indicators of regulation of financial market competition have been lacking, although the OECD has proposed to expand the coverage of its indicators in this area.

It is important to emphasise that the recent debacle in financial markets does not call into question the beneficial effects of reforms of product and labour markets. A number of reforms throughout OECD countries in recent years have demonstrably shown their power to raise employment and productivity, and acting on the recommended reforms would measurably strengthen economic performance in the long term. In addition, more flexible product and labour markets are likely to strengthen countries' resilience and their capacity to weather future downturns with less disruption to output and employment.

This chapter first examines patterns of convergence (Figure 4.1), then discusses how indicators are used to benchmark countries' policy settings, and finally makes an assessment of the policy features that most distinguish middle-income OECD countries and some key emerging economies.

Are the OECD countries converging?

The academic literature has found some support for long-run convergence of growth rates across countries, conditional on institutional settings, although the evidence is weaker within the OECD, likely a result of the smaller set of countries (Mankiw *et al.*, 1992; Durlauf *et al.* 2005). Convergence in income appears to have spread in the OECD area in recent years, as 21 of the 30 member countries, as well as the European Union (EU19) as a whole, made progress in converging towards the United States over the decade to 2007, as shown in Figure 4.1. This record represents an increase compared with an earlier assessment in *Going for Growth*, which placed three fewer countries in the convergence category, with the European Union diverging slowly. The recent shift towards greater convergence of OECD countries was driven by an increase in labour utilisation in Europe, and a downward shift in productivity growth in the United States (OECD, 2009). While the fall in productivity growth in the United States appears to have been partly structural, it is still too early to tell whether the recent stabilisation in underlying trend productivity growth in Europe is durable.

Figure 4.1. GDP per capita levels and growth rates

Gap *vis-à-vis* the United States[1]

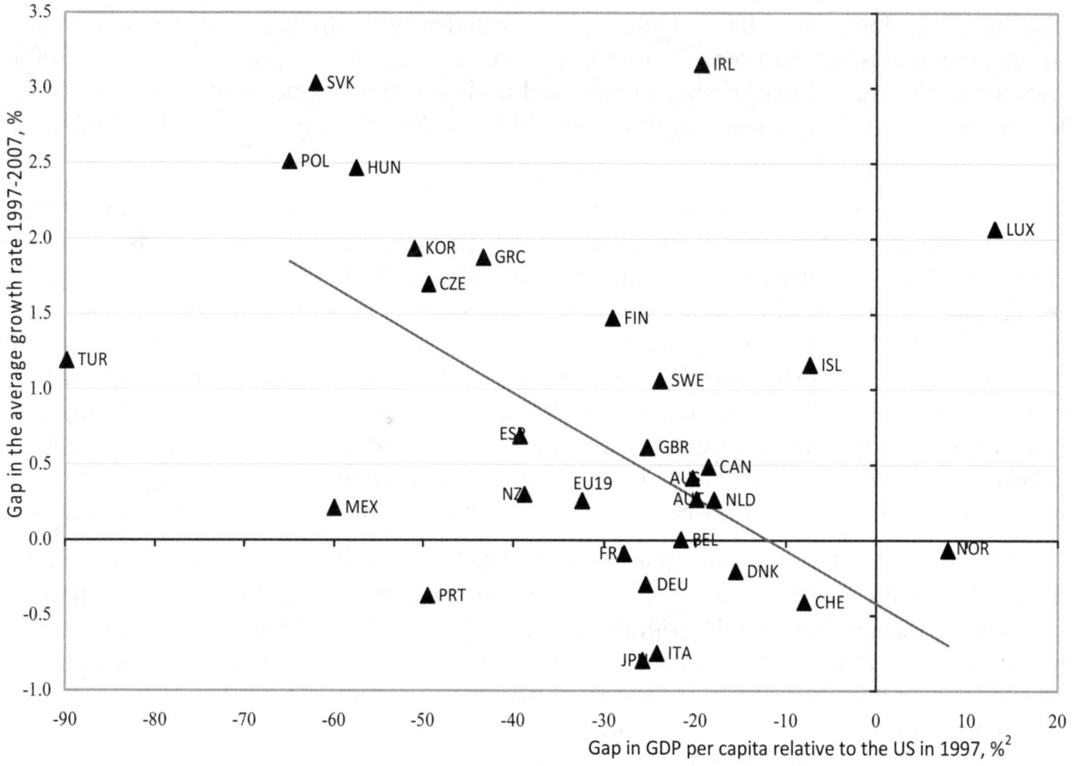

Notes: The average growth rate of GDP per capita is calculated on the basis of volume data from national accounts sources. The level of GDP per capita is calculated on the basis of current purchasing power parities (PPPs). Ireland, Luxembourg and Turkey are detected as statistical outliers using the method of Hadi (1994). The regression line is estimated on individual countries excluding these outliers.

For Luxembourg, the population is augmented by the number of cross-border workers in order to take into account their contribution to GDP.

Source: OECD, *National Accounts Database*; OECD, *Labour Force Statistics Database* and *OECD Economic Outlook, No. 84*.

Only a few countries were catching up rapidly: with one exception,[5] only countries with levels of GDP per capita that are less than one-half that of the United States – Hungary, Poland and the Slovak Republic – converged at a rate that exceeds 2% per year.

5. In Ireland, a severe cyclical downturn has put at least a temporary halt on convergence. Moreover, its catch-up is more evident for output than for income per capita. The distinction is largely due to the large repatriation of profits from foreign-owned companies and terms-of-trade losses due to falling prices of domestically produced computers and related equipment.

The two lowest-income OECD countries are converging more slowly: Mexico, which has stagnated in recent years, and Turkey, which is a statistical outlier in an estimate of a standard convergence equation. The ongoing financial crisis and its impact on activity may make it even harder to discern convergence patterns in the years to come.

The gaps in GDP per capita *vis-à-vis* the numéraire country can be broken down into contributions from labour productivity and labour utilisation, as shown in Figure 4.2. This breakdown (which is not dependent on the choice of the numéraire) shows that the countries can be divided into three groups, depending on their relative contributions:

- *Mostly a productivity gap*: The gap for the ten lowest-income OECD countries is accounted for primarily by the effect of low labour productivity, given their lower levels of physical and human capital per worker, although the five lowest-income countries also have a substantial gap in measured labour utilisation. Among countries with higher incomes, the income gap for Australia, Canada, Iceland, Japan and Switzerland *vis-à-vis* the comparison country primarily reflects productivity shortfalls.
- *Mostly a labour utilisation gap*: The income gaps of Belgium, France, Germany and the Netherlands can be mostly accounted for by low labour utilisation. This divergence reflects a range of factors, including relatively shorter working hours, lower participation rates for older workers and higher unemployment.
- *Both significant productivity and labour utilisation gaps*: For the European Union, as well as the Nordic EU countries, Spain and Italy, the income shortfall reflects gaps with the United States in both productivity and labour utilisation.

While the United States is used as the numéraire in the convergence figure and the breakdown analysis, this does not reflect any prior judgement about the strength of its policy model. In fact, the performance comparisons made in *Going for Growth* have been shown to be unaffected by the choice of numéraire, although use of the United States provides a straightforward way of summarising the data and gauging how well OECD countries are performing relative to its largest and generally best-performing country, absent special factors in the case of Norway and Luxembourg (see OECD, 2009).

Policy priority setting and results

The *Going for Growth* structural surveillance exercise seeks to identify five policy priorities for each OECD member country and the European Union, based on a systematic benchmarking approach. Three of these policy priorities are identified based on internationally comparable OECD indicators of policy settings and performance. The two additional priorities are often supported by indicator-based evidence, but may also draw on country-specific expertise. These priorities are meant to capture any potential policy imperatives in fields not covered by indicators. The policy indicators generally meet three main quality criteria: *i)* they can be tied to relevant performance indicators based on econometric evidence; *ii)* they relate to policies that are under the direct control of policy makers; and *iii)* they can be reliably measured with a sufficient degree of confidence to be credible to governments and the public.

Figure 4.2. The sources of real income differences

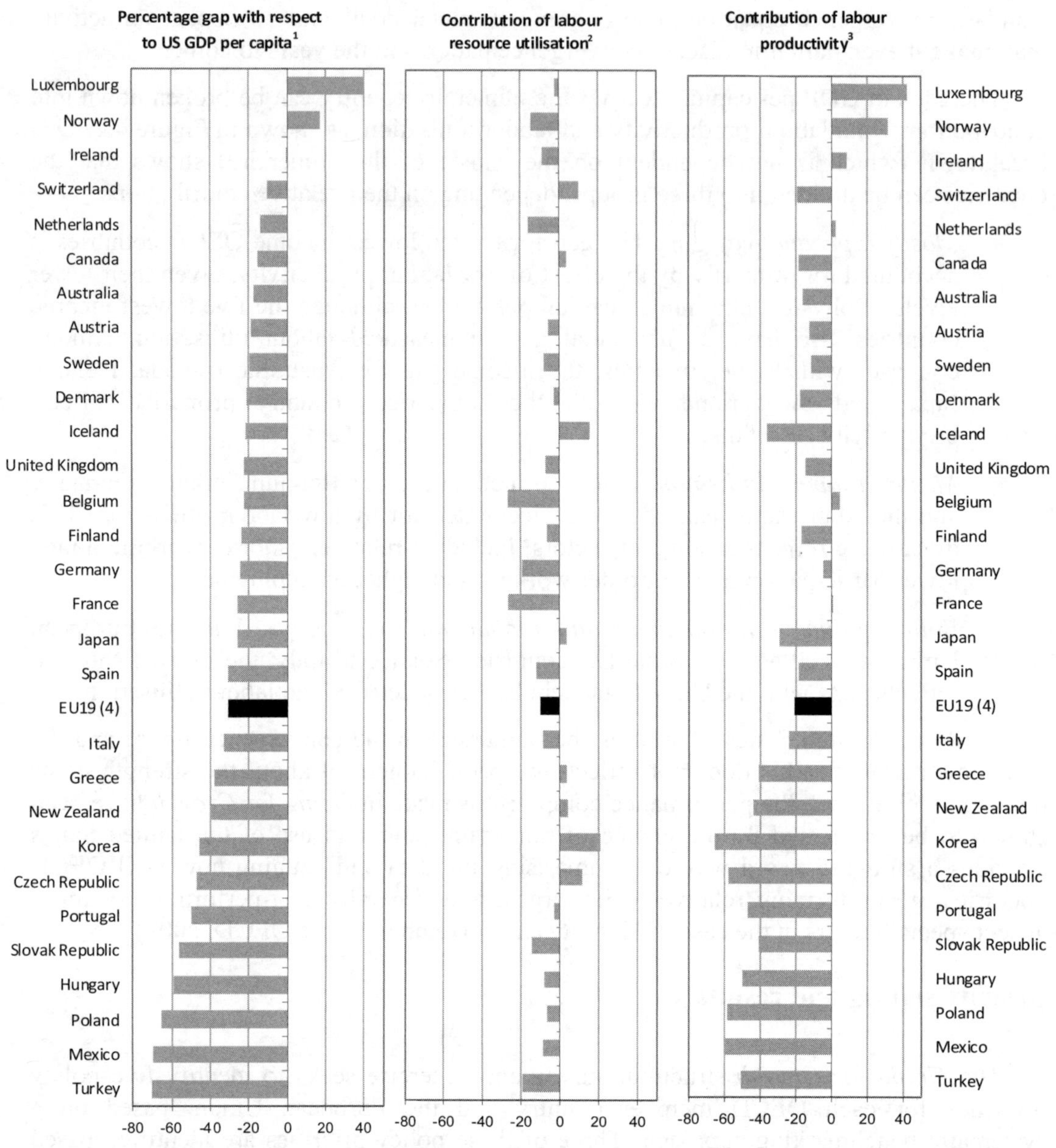

1. Based on 2007 PPPs. For Luxembourg, the population is augmented by the number of cross-border workers in order to take into account their contribution to GDP.

2. Labour resource utilisation is measured as total number of hours worked per capita.

3. Labour productivity is measured as GDP per hour worked.

4. The EU19 is an aggregate of countries that are members of both the European Union and the OECD. These are the EU15 countries plus the Czech Republic, Hungary, Poland and the Slovak Republic.

Source: OECD, *National Accounts Database*; OECD, *Economic Outlook 84 Database* and OECD (2008), *Employment Outlook*.

For the selection of the three indicator-based policy priorities, the starting point is a detailed examination of labour utilisation and productivity performance relative to the OECD average, so as to uncover specific areas of relative strength and weakness compared with other OECD countries. Each performance indicator is juxtaposed to the corresponding policy indicators, for which OECD empirical research has shown a robust link to performance, to determine where performance and policy weaknesses appear to be linked. This evaluation process is carried out for each of the approximately 50 areas for which OECD policy indicators provide coverage.

As an example, Figure 4.3 shows, for a sample country, a scatter plot of pairings of policy indicators (on the horizontal axis) with corresponding performance indicators (on the vertical axis). Since many of the approximately 50 policy indicators are associated with more than one performance area, there are potentially well over 100 pairings to be examined. The indicators of policy and performance are standardised by re-scaling them so that each has a mean of zero and a cross-county standard deviation of one, with positive numbers representing positions more growth-friendly than the OECD average. The scatter plot is thus divided into four quadrants, depending on whether a country's policy-performance pairing is below or above the average policy or performance score.

Figure 4.3. Example of selection of candidates for *Going for Growth* priorities

Diamonds represent policy-performance pairings

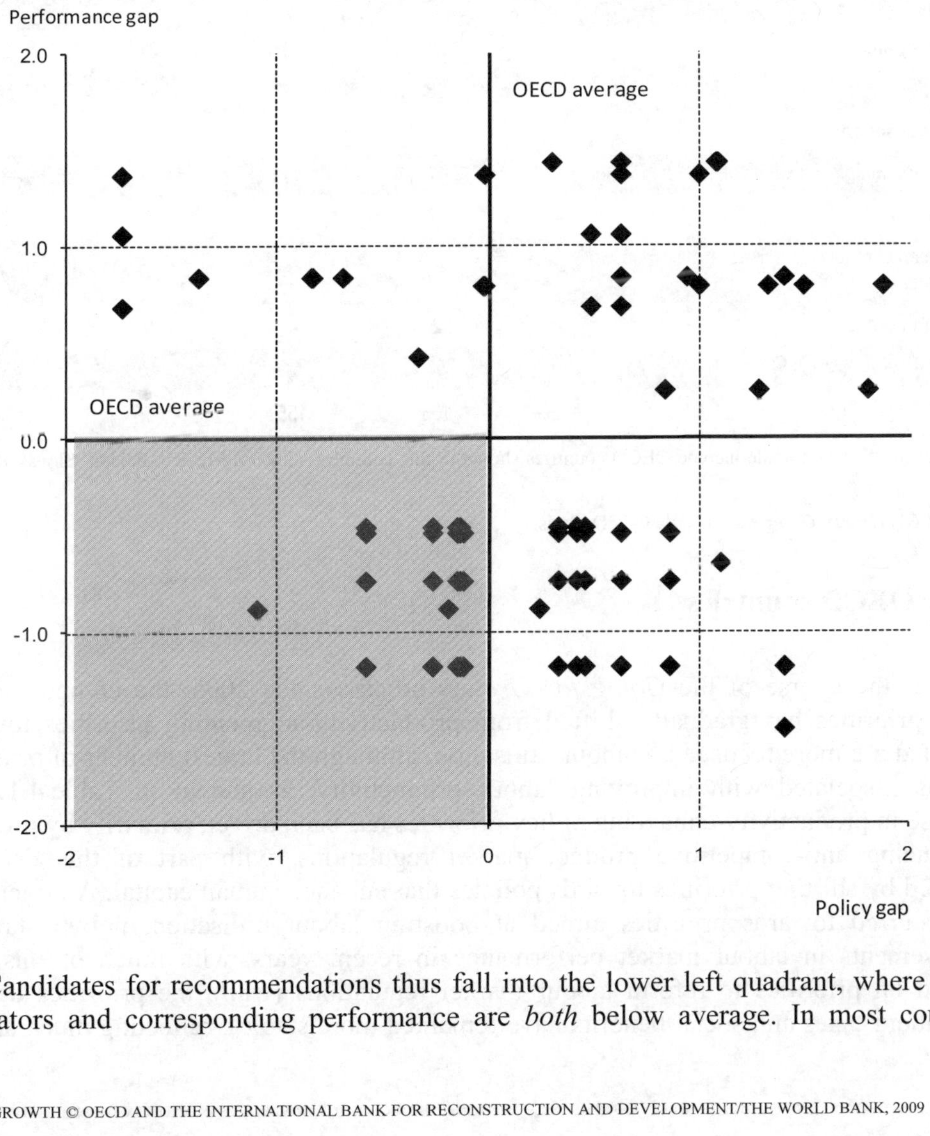

Candidates for recommendations thus fall into the lower left quadrant, where policy indicators and corresponding performance are *both* below average. In most countries there are more than three unique policy areas that qualify as potential priorities (for instance, Germany had 16 candidates in the 2009 exercise). When there are more than three candidate policy priorities, the list is narrowed using a combination of country expertise and the following criteria: *i)* the estimated quantitative impact of reforms in the policy area on GDP per capita as determined in previous OECD analysis; *ii)* the normalised distance of the policy stance from the benchmark (the OECD average), and *iii)* recent trends in policy and performance. The limit on the number of priorities means that for some countries, obvious policy imperatives may not be identified as priorities because other priorities are deemed more important.

Table 4.1. Distribution of policy priorities by year

30 OECD member countries and the EU

	2005	2007	2009	2009 MI-5[1]
Productivity				
Product market regulation	47	39	38	7
Agriculture	7	7	7	0
Human capital	16	22	24	5
Other policy areas	28	23	21	3
Total	*98*	*91*	*90*	*15*
Labour utilisation				
Average and marginal taxation on labour income	12	11	13	3
Social benefits	27	31	27	2
Labour market regulation and collective wage agreements	16	18	20	3
Other policy areas	2	4	5	2
Total	*57*	*64*	*65*	*10*
Overall	155	155	155	25

1. Priorities for the five middle-income OECD countries shown in this column (Hungary, Mexico, Poland, Slovak Republic, Turkey).

Source: OECD (2009), *Going for Growth*, OECD, Paris.

Priorities for OECD countries

Over the course of the *Going for Growth* process since 2005, the composition of policy priorities has gradually shifted from productivity-augmenting priorities towards those that are more focused on labour utilisation, although the largest number of priorities remains associated with improving labour productivity, as shown in Table 4.1. The decrease in productivity-enhancing policy priorities has been driven primarily by progress in reducing anti-competitive product market regulations, with part of this decrease balanced by shifting priorities towards policies that enhance human capital. A larger shift has occurred towards priorities aimed at boosting labour utilisation, notwithstanding improvements in labour market performance in recent years, with much of this shift focused on priorities to reform labour market regulations (while the priorities dealing with labour taxes and social benefits have remained more stable), reflecting much slower progress in this area: a concern highlighted in previous work on the political economy of structural reform (OECD, 2007). It should be kept in mind that labour-market related policies may have an impact on productivity as well as labour market performance. For example, overly stringent job protection legislation has been found to reduce productivity (Bassanini *et al.*, 2009).

The policy priorities for the five middle-income OECD countries (using the 2007 World Bank classification, and dubbed the MI-5 here) have been in a narrower range of policy areas that are more focused on productivity-related policies (Table 4.1, last column). This heavier weight on productivity-enhancing policies reflects these countries' relative weakness in these areas. Distributions of policy priorities by policy area are shown by country grouping in Table 4.2.

Table 4.2. Distribution of policy priorities by country grouping

Percent of total in 2009

	OECD+EU	EU only[1]	Lowest income 10[2]	MI-5[3]
Productivity				
Product market regulation	25	21	26	28
Agriculture	5	1	2	0
Human capital	15	16	20	20
Other policy areas	14	10	14	12
Total	*58*	*48*	*62*	*60*
Labour utilisation				
Average and marginal taxation on labour income	8	11	8	12
Social benefits	17	21	10	8
Labour market regulation and collective wage agreements	13	15	14	12
Other policy areas	3	5	6	8
Total	*42*	*52*	*38*	*40*
Overall	100	100	100	100

1. Priorities for the 19 EU countries that are OECD members, plus the European Union as a whole.

2. Priorities for the 10 lowest-income OECD countries (MI-5, plus the Czech Republic, Greece, Korea, New Zealand and Portugal).

3. Priorities for the five middle-income OECD countries (Hungary, Mexico, Poland, the Slovak Republic, Turkey).

Source: OECD (2009), *Going for Growth*, OECD, Paris.

Differences in the share of policy priorities by area are quite stark across country groupings, reflecting differences in weaknesses in performance for each group of countries. In particular, the EU countries have a greater share of recommendations in the labour utilisation area than the OECD as a whole, a sign of their relative weakness in this area. In contrast, the ten lower-income OECD countries have more recommendations for productivity-enhancing reform than the OECD as a whole, as do the five middle-income OECD countries. These recommendations are mostly concentrated in the areas of product market regulation and human capital policy. Such differences may not fully reflect the relative importance of reforms in each area, as the distance from best practice is probably larger for productivity-enhancing reforms among the lower-income OECD countries.

Product market regulation shows the large distance from best practice among the ten lower-income OECD countries. While reform has occurred in this area in these countries, their overall stance is still more restrictive than the OECD average, in some cases strikingly so. The overall score for this indicator is shown in Figure 4.4. The specific areas of policy weaknesses for these countries vary somewhat; entry barriers, public ownership and trade restrictiveness are particularly notable.

Figure 4.4. Restrictiveness of economy-wide product market regulation[1]

Scale of 0 (least restrictive) to 6 (most restrictive)

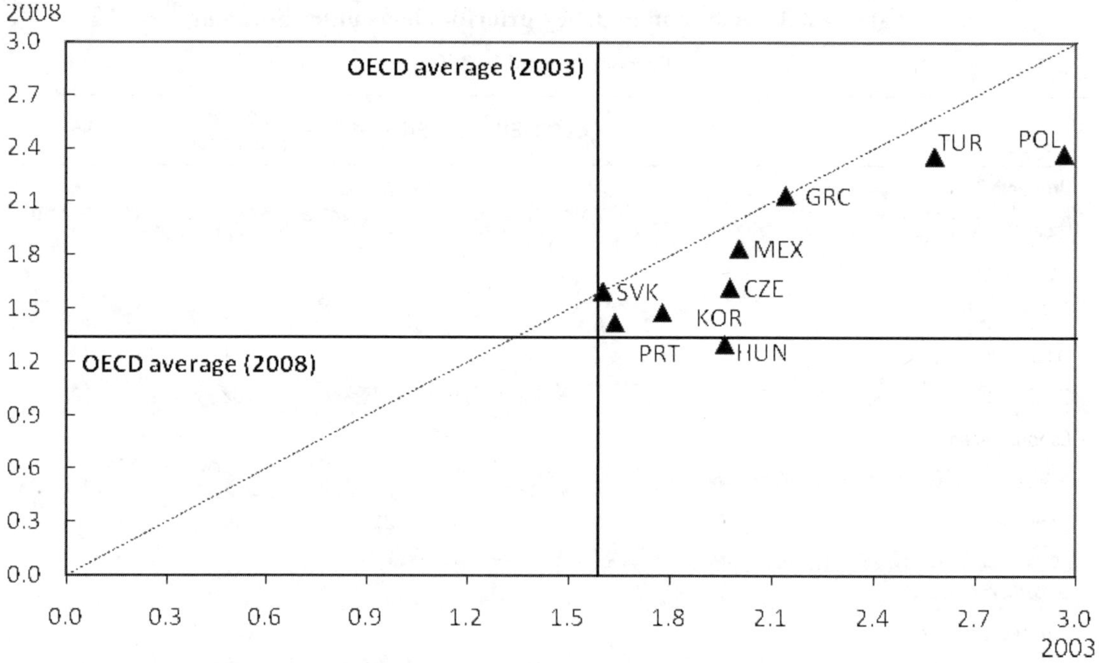

1. Because data for Greece and the Slovak Republic are not available for 2008, only 2003 data are shown.

Source: OECD, *Product Market Regulation Database*, *www.oecd.org/eco/pmr*.

Human capital is another area with large differences from best practice in policy settings among the lower-income OECD countries. While differences in secondary school achievement are not as great, achievement levels are still well below the OECD average for higher education, except in Korea and New Zealand (Figure 4.5). In Korea, very rapid progress has been made in raising achievement levels for the younger cohorts. Progress has been much slower in the lowest-income countries. This is disappointing, as the impact of changes in education policies can take many decades to be fully felt, even though it can be very large in the long run (OECD, 2009).

Policy reforms in the areas of product market regulation and human capital would often help to improve the convergence of the lower-income countries towards the frontier. However, it is important to realise that policies do not work in isolation; for instance, improvements in the intensity of competition may only enhance growth if they are accompanied by flexible labour market policies. It is important to consider policies as a package rather than in isolation, given interaction effects; moreover, policies may have different effects depending on a country's stage of development and the sophistication of its firms.

Figure 4.5. Tertiary education: achievement levels

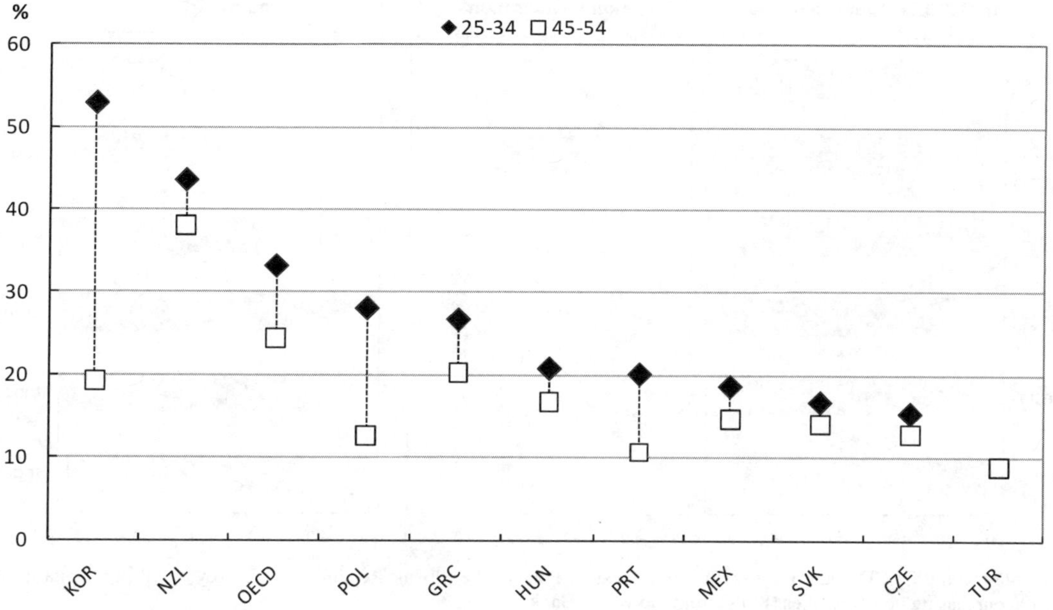

Source: OECD (2008), *OECD Education at a Glance 2008*, OECD, Paris.

One issue often discussed is the role of R&D and innovation support for countries that are far from the technological frontier. While innovation may play less of a role in the development of countries at earlier stages of development, R&D investment can play an important role in improving absorptive capacity, especially when it complements education policies. The effect of policies may not always be linear with respect to development, and there is some evidence of a possible hump-shaped relation of some policies with levels of development (Aghion *et al.*, 2009; see also Chapter 2 in this volume).

Higher levels of competition can also strengthen productivity growth through creative destruction. The entry of new businesses and the exit of laggards is a key mechanism in dynamic economies, bringing fresh ideas, business models and new products, and strengthening economic performance. These dynamics are particularly important for realising the reallocative efficiency gains that can arise from engaging in international trade (Dougherty, 2009), particularly as the global economy has become more and more integrated through trade and investment.

Priorities for emerging economies

The large emerging economies play an increasingly central role in the global economy, and the OECD has engaged with many of them. A special relationship of "enhanced engagement" has been offered to Brazil, China, India, Indonesia and South

Africa.[6] These countries' distinguishing characteristics relate to their large size and openness. They also have a lower level of GDP per capita than the OECD economies.

Figure 4.6. The sources of income differences for middle-income and emerging economies

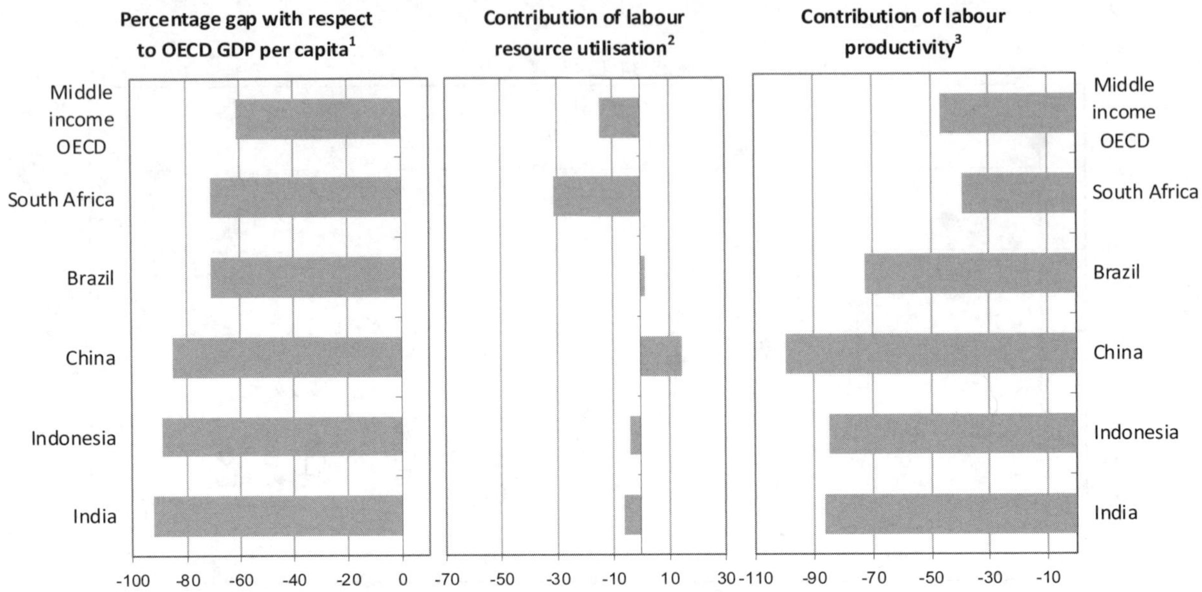

1. The middle-income OECD countries are Hungary, Mexico, Poland, the Slovak Republic and Turkey. GDP per capita is based on revised 2006 purchasing power parities (PPPs) from the World Bank.

2. Labour resource utilisation is measured as the employment rate, based on national labour force surveys, except for India where it is an OECD estimate based on the National Sample Survey.

3. Labour productivity is measured as GDP per employee.

Source: OECD and The World Bank.

Looking towards the future, the addition of new countries to the OECD *Going for Growth* exercise would raise a number of questions. For example: Do the same (linear) empirical relationships still hold; are the same policy indicators still relevant, for instance in countries where there are large informal sectors and weak property rights; and is the average performance still the relevant benchmark if new countries at an earlier stage of development are added?

The source of relative weaknesses for the large emerging countries – a shortfall in labour productivity – is quite similar to that of the middle-income OECD countries, as Figure 4.6 demonstrates (using a breakdown analogous to that of Figure 4.2, but not accounting for differences in working hours, since these are not available). With the exception of South Africa, the shortfall of these countries' level of GDP per capita relative to the OECD average is primarily in the labour productivity domain.

In terms of policies, some relevant policy indicators are already available for these large emerging countries, as a result of *OECD Economic Surveys* of non-members. This work suggests that, as for middle-income OECD countries, large weaknesses exist in the

6. The Organisation is also presently in the process of formally enlarging its membership. Five candidate countries are presently in the accession process: Chile, Estonia, Israel, the Russian Federation and Slovenia.

product market regulation and education domains. Moreover, country-specific analysis also suggests that policy recommendations are heavily weighted in this area as well. If such analogous differences hold for other policies, it may be relatively straightforward to extend the framework to the large emerging economies.

Annex 4.A1

The indicators used in *Going for Growth*

The current set of policy indicators used in the *Going for Growth* exercise is shown in Table 4.A1.1. The main types of indicators are shown with their source and most recent year. Most of the individual indicators are illustrated in a special chapter of the report. Forty-five detailed policy indicators are regularly used. All are produced within the OECD through its committees, and all are subject to vetting by country authorities. Most of the indicators are compiled directly by the Secretariat based on primary information, although some – such as the labour tax wedge – rely heavily on Eurostat or national authorities for compilation.

Table 4.A1.1. OECD structural policy indicators currently used in *Going for Growth*

Type of indicators (number)	Latest data
Minimum and median wages (1)	2007
Net unemployment benefit replacement rates (2)	2006
Average tax wedge on labour income (2)	2007
Marginal tax wedge on labour income (3)	2007
Social security contribution paid by employers (1)	2007
Implicit tax on continued work at older ages (2)	2007
Childcare/implicit tax on returning to work (2)	2004
Support for disability and sickness (2)	2006
Employment protection legislation (3)	2008
Collective bargaining and trade union density (1)	2003/04
Product market regulation (8 overall + 7 sectoral)	2008
Foreign direct investment restrictiveness (1)	2006
Educational attainment (2)	2006
PISA scores (1)	2006
Producer support estimates (1)	2006
Weighted trade/tariffs barriers (1)	2007
Health expenditure (1)	2005
Public investment (1)	2007
Subsidies to private R&D (3)	2006/07

A key feature of the policy indicators used is that the specific indicators are tied directly to performance outcomes in empirical studies, typically based on analysis of panel data. The coverage of the policy indicators thus closely follows work carried out by the OECD Secretariat in the context of its regular work programme. Areas that are well covered by this work include many policies in the areas of labour and product markets, as well as education and innovation.

The use of the OECD indicators is not always comprehensive: judgement about the quality, ongoing availability and reliability of indicators in some areas has limited their use, despite some empirical support. Some very recent OECD studies offer some room for expanding the indicator base relatively easily, including in the area of taxation (*e.g.* corporate tax rates), infrastructure (*e.g.* the regulatory independence indicator), and tertiary education (*e.g.* the structure of supply indicator). Studies in these areas have recently be completed, and preferred policy indicators from the analyses are being incorporated. Other projects currently under way which could be taken advantage of include those on health care and social mobility.

The performance indicators used in *Going for Growth* are principally based on a mechanical breakdown of GDP per capita, which includes employment, demographics, capital intensity, as well as productivity-related subcomponents of performance (Table 4.A1.1). These subcomponents are intended to measure key outcomes that relate to GDP per capita; contributions to productivity are somewhat more complex. For instance, trade and investment openness are treated as outcomes in this context (only *barriers* are included as policy indicators), since they play an important role in facilitating productivity gains through reallocation, as noted in the 2008 edition of *Going for Growth*.

Policy indicators

What criteria should be used to select the indicators? Following the original conception of *Going for Growth*, three main criteria have been used to date for selecting policy indicators: *i*) the extent of the empirical evidence; *ii*) the ability of policy makers to affect them; and *iii*) reliability of measurement. Each criterion is dealt with in turn.

First, *solid empirical evidence* based on theory is needed to tie policy indicators to relevant performance measures. A relatively high standard has been used for this determination: only indicators that were used in econometric analysis, and were demonstrated to be tied to improved economic performance, are included. This demonstration has come in the form of review by OECD committees, and has ensured considerable buy-in by member governments.

Second, policy indicators are usually under the *direct control* of policy makers. This criterion ensures that if weakness is found for a policy indicator, any resulting policy recommendations for a country can be traced back to specific reforms that need to be undertaken. Generally, this requirement has meant that only *fact-based*, and not *perception-based*, indicators would be considered, since perceptions may or may not be closely tied to actual policy settings. (At the same time, varying enforcement of objective policy settings may matter, and can be used as a useful complement to analysis; see Nicoletti and Pryor, 2006.) While indicators do not necessarily have to be based on primary data sources, a highly transparent and objective compilation methodology is essential. For instance, in order to measure product market regulation, synthetic indicators that measure regulatory settings in specific sub-domains are used, rather than the aggregates.

Third, policy indicators should be *reliably measured* with a sufficient degree of confidence to be credible to the government and the public. At a basic level this criterion requires that indicators measure what they aspire to measure ("construct validity"), are comparable across countries, and are recent (and can be regularly updated). But more fundamentally, in order to ensure the confidence of policy makers, policy indicators need to use objective compilation methods and be vetted by governments (*i.e.* through OECD committees), and made available to the public for review. In cases where secondary or commercial data sources are used (for components of the tariff indicator, for instance), governments need to have the opportunity to ensure the correctness of indicator values.

The criteria set out above are quite limiting in terms of the scope they offer for expanding the indicator base. The first criterion seems essential, although committee review may not be necessary if academic review is used; the second criterion could be relaxed to allow for some expert or perception-based indicators; however, this could make it hard to relate policy indicator underperformance to specific actionable policy recommendations, and could limit government buy-in of the results; the third criterion leaves little room for compromise given the role of committees at the OECD.

Performance indicators

For performance indicators, the criteria for their inclusion are simpler, and essentially statistical: relevance, comparability and timeliness. Given that the original motivation of the *Going for Growth* exercise is to improve material living standards, a mechanical breakdown of GDP per capita essentially defines the set of indicators that should be included. This limits the scope for broadening the set of performance indicators considerably, although there is some room for bringing in complementary indicators, especially those that relate to productivity. For instance, "entrepreneurship" or business demographics indicators such as firm entry and gross job creation rates (based on microdata) might be used as supplementary productivity performance indicators, if they are available on a regular basis, since firm dynamics were shown to be closely associated with productivity gains in the *OECD Growth Study* (2003) and in related research.

References

Acemoglu, D., S. Johnson, and J. Robinson (2005), "Institutions as a Fundamental Cause of Long-run Growth", in Aghion, P. and S.N. Durlauf (eds.), *Handbook of Economic Growth*, Volume 1A, North-Holland, Amsterdam.

Aghion, P., N. Bloom, R. Blundell, R. Griffith and P. Howitt (2005). "Competition and Innovation: An Inverted U-relationship", *Quarterly Journal of Economics*, Vol. 120, pp. 701-728.

Bassanini, A. and R. Duval (2006), "Employment Patterns in OECD Countries: Reassessing the Role of Policies and Institutions", OECD Economics Department Working Paper No. 486.

Bassanini, A., L. Nunziata and D. Venn (2009), "Job protection legislation and productivity growth in OECD countries", *Economic Policy*, Vol. 24, Issue 58, pp. 349-402.

Bassanini, Andrea, S. Scarpetta and P. Hemmings (2001), "Economic Growth: The Role of Policies and Institutions – Panel Data. Evidence from OECD Countries", OECD Economics Department Working Papers No. 283.

Boulhol, H., A. de Serres and M. Molnar (2008), "The Contribution of Economic Geography to GDP per Capita", OECD Economics Department Working Papers No. 602.

Dougherty, S. (2009), "Regulatory Costs and Trade-Related Reallocation in the OECD", paper presented at the Paris School of Economics GSIE Seminar, March.

Durlauf, S.N., P.A. Johnson and J.R.W. Temple (2005), "Growth Econometrics", in P. Aghion and S.N. Durlauf (eds.), *Handbook of Economic Growth*, Volume 1A, North-Holland, Amsterdam.

Furceri, Davide and Annabelle Mourougane (2009), "Taking Stock of Existing Structural Policy and Outcome Indicators", OECD Economics Department Working Papers, No. 676.

Glaeser, E., R. La Porta, F. Lopez-de-Silanes and A. Shleifer (2004), "Do Institutions Cause Growth?", *Journal of Economic Growth*, Vol. 9, pp. 271-303.

Hadi, A.S. (1994), "A Modification of a Method for the Detection of Outliers in Multivariate Samples", *Journal of the Royal Statistical Society*, Series B, Vol. 56.

IMF (2008), "Structural Reforms and Economic Performance in Advanced and Developing Countries", IMF Research Department Board Paper, *www.imf.org/external/np/res/docs/2008/pdf/061008.pdf*

Kauffman, D., A. Kraay and M. Mastruzzi (2008), "Governance Matters VII: Aggregate and Individual Governance Indicators 1996 – 2007", World Bank Working Paper No. 4654.

Mankiw, N. Gregory, D. Romer and D.N. Weil (1992), "A Contribution to the Empirics of Economic Growth", *The Quarterly Journal of Economics*, MIT Press, Vol. 107, No. 2, pp. 407-37.

Nicoletti, G. and F. Pryor (2006), "Subjective and Objective Measures of Governmental Regulations in OECD Nations", *Journal of Economic Behavior and Organization*, 59, pp. 433-449.

OECD (1994), *The OECD Jobs Study: Facts, Analysis, Strategies,* OECD, Paris.

OECD (2003), *The Sources of Economic Growth in OECD Countries*, OECD, Paris.

OECD (2006), *Employment Outlook: Boosting Jobs and Incomes*, OECD, Paris.

OECD (2007, 2008, 2009), *Economic Policy Reforms: Going for Growth,* OECD, Paris.

OECD (2008), *Education at a Glance*, OECD, Paris.

Rodrick, D., A. Subramanian and F. Trebbi (2004), "Institutions Rule: The Primacy of Institutions over Geography and Integration in Economic Development", *Journal of Economic Growth*, 9, pp. 131-165.

Chapter 5

The development of global innovation networks and the transfer of knowledge

Dirk Pilat, Koen De Backer, Ester Basri, Sarah Box and Mario Cervantes[1]

Confronted with increasing global competition and rising research and development (R&D) costs, companies can no longer survive on their own innovation efforts. Their innovation activities are increasingly international, and they are embracing more "open" approaches – collaborating with external partners, whether suppliers, customers or universities, to keep ahead of the game. Multinational enterprises (MNEs), in particular, have increasingly shifted R&D activities across borders within their global value chain and rely on outside innovation for new products and processes. Non-OECD economies play a growing role in this internationalisation process and accounted for 40% of the growth of global R&D in recent years. Moreover, migration of talent now plays an important role in shaping skilled labour forces throughout the OECD area. This chapter examines the globalisation of innovation, its drivers and impacts, and examines how policy can draw greater benefits from the globalisation process.

1. Dirk Pilat is Head, Structural Policy Division, Koen De Backer is Economist, Ester Basri is Senior Policy Analyst, Sarah Box is Economist, and Mario Cervantes is Senior Economist, all in the Directorate for Science, Technology and Industry, OECD. This paper draws on a number of OECD studies completed in 2008, cited as OECD (2008a) to (2008d).

Introduction

Innovation has become a key factor for success for OECD countries in the competitive struggle for market share and is a prerequisite for sustainable development in a globalised world. In a complex and highly competitive global market, companies have to innovate and develop commercially viable products and services faster than ever. To meet these new challenges, companies adopt new approaches to innovation which have implications for the appropriate policy environment for innovation.

Confronted with increasing global competition and rising research and development (R&D) costs, companies can no longer survive on their own R&D efforts but look for new, more open, modes of innovation. Their innovation activities are increasingly international, and they are embracing more "open" innovation approaches – collaborating with external partners, whether suppliers, customers or universities, to keep ahead of the game and get new products or services to market before their competitors. At the same time, innovation is broadening, as users of products and services, both firms and individual consumers, increasingly become involved in innovation themselves.

Multinational enterprises (MNEs), in particular, have increasingly shifted R&D activities across borders within their global value chain and rely on outside innovation for new products and processes. Non-OECD countries play a growing role in this internationalisation process. Companies are also more active in licensing and selling the results of their own innovations to third parties.

Alongside the growth in foreign direct investment (FDI), in trade, and in the internationalisation of R&D, mobility of human resources in science and technology (HRST) has also become a central aspect of globalisation. Migration of talent now plays an important role in shaping skilled labour forces throughout the OECD area, has important implications for innovation, and affects both OECD and non-OECD countries.

This chapter looks at key aspects of the globalisation of innovation, its drivers and impacts, and examines how policy can draw greater benefits from the globalisation process.

The global landscape for innovation

Globalisation is a major driver of innovation not only because it means more intense and global competition but also because it has created a more global landscape for innovation. A growing number of countries, including emerging economies, have developed important science and technology (S&T) capabilities and resources, and the internationalisation of R&D and of science, as well as the international mobility of researchers, has created an increasingly global supply of S&T (OECD, 2008a; OECD, 2008b).

While R&D investments remain mostly concentrated in the United States, the European Union and Japan, non-OECD economies account for a growing share of the world's R&D (Figure 5.1). China's gross domestic expenditure on R&D (GERD) reached USD 86.8 billion in 2006 after expanding at around 19% annually in real terms from 2001 to 2006. Investment in R&D in South Africa increased from USD 1.6 billion in 1997 to USD 3.7 billion in 2005. Russia's investment climbed from USD 9 billion in 1996 to USD 20 billion in 2006, and India's reached USD 23.7 billion in 2004. As a result, non-OECD economies account for a sharply growing share of the world's R&D –

18.4% in 2005, up from 11.7% in 1996. The growing weight of these countries in the global economy accounts for part of this shift, but so does the growing intensity of investment in R&D relative to GDP, notably in China.

Figure 5.1. Global R&D trends in major OECD regions and selected non-member economies

Evolution of global share of total R&D, 1996-2005

Percentage

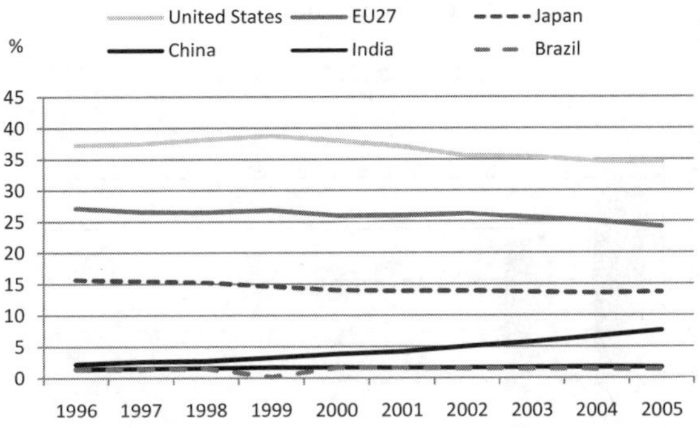

Change in global share of total R&D

Percentage

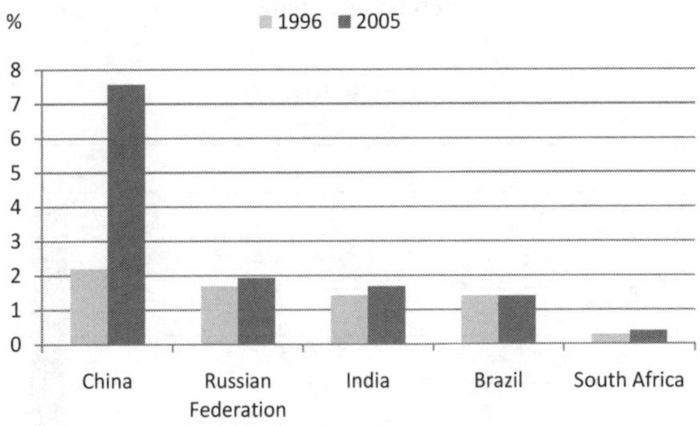

Source: Based on data for 79 non-OECD countries (UNESCO Institute for Statistics) and 30 OECD countries (*OECD Main Science and Technology indicators Database* 2008/1); see OECD (2008), *OECD Science, Technology and Industry Outlook 2008*, OECD, Paris.

Most countries have seen patents and scientific publishing increase in recent years. While the United States continues to account for the largest share of triadic patent families (patents filed in the United States, Japan and the EU to protect the same invention), its share has fallen, as has that of the EU25. At the same time, the share of patent families from Asian economies increased markedly between 1995 and 2005, albeit from a low level. Publication of scientific articles has also increased, but remains highly concentrated in a few countries, with the OECD area overall accounting for over 81% of global production. Nevertheless, scientific capabilities are growing strongly in some emerging economies (Figure 5.2).

In parallel to this global supply of S&T resources, innovation strategies increasingly depend on global sourcing in order to sense new market and technology trends worldwide. International sourcing of technology and knowledge has become an important reason for MNEs to internationalise their R&D activities. As markets have opened up, MNEs have become more mobile and increasingly shift activities in their global value chains (OECD, 2007), including R&D, across borders in reaction to differences in location factors (including costs of innovation). Recent empirical evidence shows that the top 700 R&D spending MNEs[2] increasingly invest in R&D outside their home country in line with the growth in the global supply of S&T resources (OECD, 2008b). A survey of the largest R&D investors, undertaken by UNCTAD from November 2004 to March

2. More than 95% of the 700 firms with the largest R&D expenditure are MNEs and they account for close to half of the world's total R&D expenditure and more than two-thirds of the world's business R&D (UNCTAD, 2005). The top R&D-performing MNEs often spend more on R&D than many countries and their presence is felt not only through activities in their home countries but also increasingly abroad.

2005, suggests that the pace of internationalisation in R&D may be accelerating (UNCTAD, 2005): as many as 69% of responding firms stated that their share of foreign R&D is set to increase (only 2% indicated a decline and the remaining 29% expected no change). The average firm in the UNCTAD survey spent 28% of its R&D budget abroad in 2003, including in-house expenditure by foreign affiliates and extramural spending on R&D contracted to other countries.

Figure 5.2. Global shares of patenting and scientific publications (selected countries)

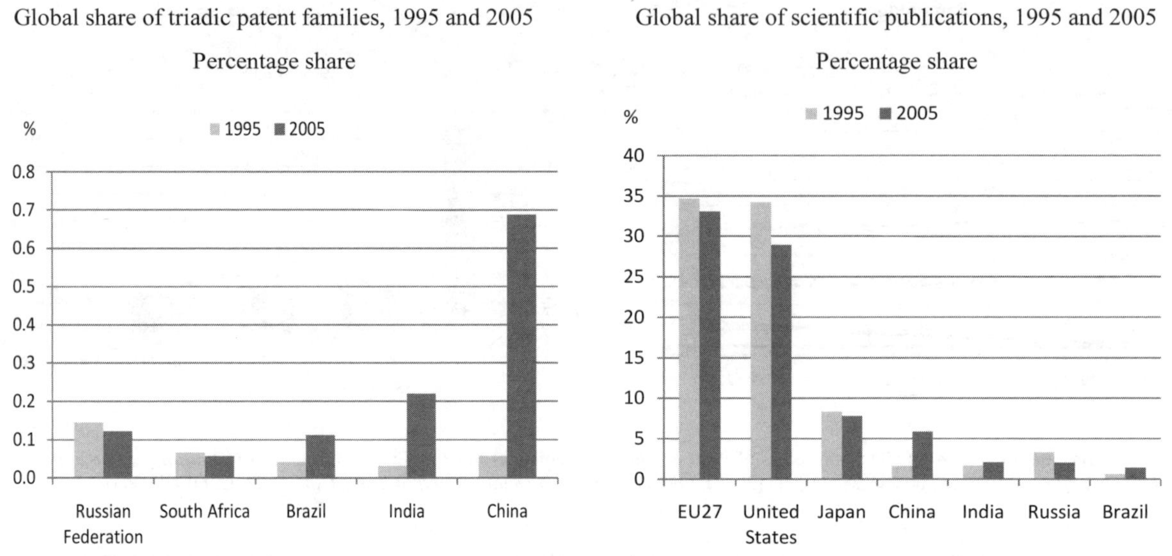

Source: OECD Patent Database, 2008.

Source: National Science Foundation, *Science and Engineering Indicators 2008*.

Technology sourcing has become a major consideration for locating R&D abroad, and the geographic dispersion of MNEs is increasingly a means of knowledge creation rather than knowledge diffusion. Their decentralised R&D activities have been defined as "home-base augmenting" (Kuemmerle, 1997) or "asset-seeking" (Dunning and Narula, 1995).

Location decisions for R&D facilities that augment those of the home base are typically supply-oriented, based not only on the host country's technological infrastructure, but also on the presence of other firms and institutions from which investing firms can benefit: spillovers from other R&D units, access to trained personnel, links with universities or government institutions, the existence of an appropriate infrastructure for specific kinds of research, etc. The R&D of these affiliates is more innovative and/or aimed at technology monitoring, and is largely determined by the quality of the components of the regional or national innovation systems. The features of a host country that attract such innovative R&D vary, depending on the industry and the activity.

This new motivation complements the traditional demand-oriented reasons for R&D abroad: market proximity to "lead users" and adaptation of products and processes to local conditions. R&D activities have also been undertaken in affiliates abroad to support an MNE's local manufacturing operations and often follow in the wake of FDI in manufacturing. In this case, technological knowledge tends to flow from the parent firm's laboratory to the foreign-based facility so that the technological advantages of the affiliate primarily reflect those of the home country (where the core of innovation activities is concentrated) and foreign R&D units exploit the parent company's technology.

While "home-base augmenting" activities are increasing, "home-base exploiting" motivations remain important. The empirical evidence showing that companies offshore R&D activities in which they are strong at home suggests that asset-exploiting activities are mostly undertaken abroad (see also below). Moreover, the distinction between adaptive and innovative R&D centres may seem clear in theory, but it is less so in the real world. Criscuolo *et al.* (2005) found that although most FDI in R&D still falls into the home-base exploiting category, it most often tends to be simultaneous with home-base augmenting R&D.

Most internationalisation of R&D by MNEs still takes place within the main OECD regions. However, with the increasingly global supply of S&T resources, emerging countries are attracting more R&D (OECD, 2008a). An UNCTAD survey on future R&D investments found that China was the location mentioned most often (Figure 5.3), followed by the United States. India was in third place, and Russia was also among the top ten target locations. Other emerging economies named were Singapore, Chinese Taipei and Thailand.

Figure 5.3. Most attractive foreign R&D locations

Percentage of responses

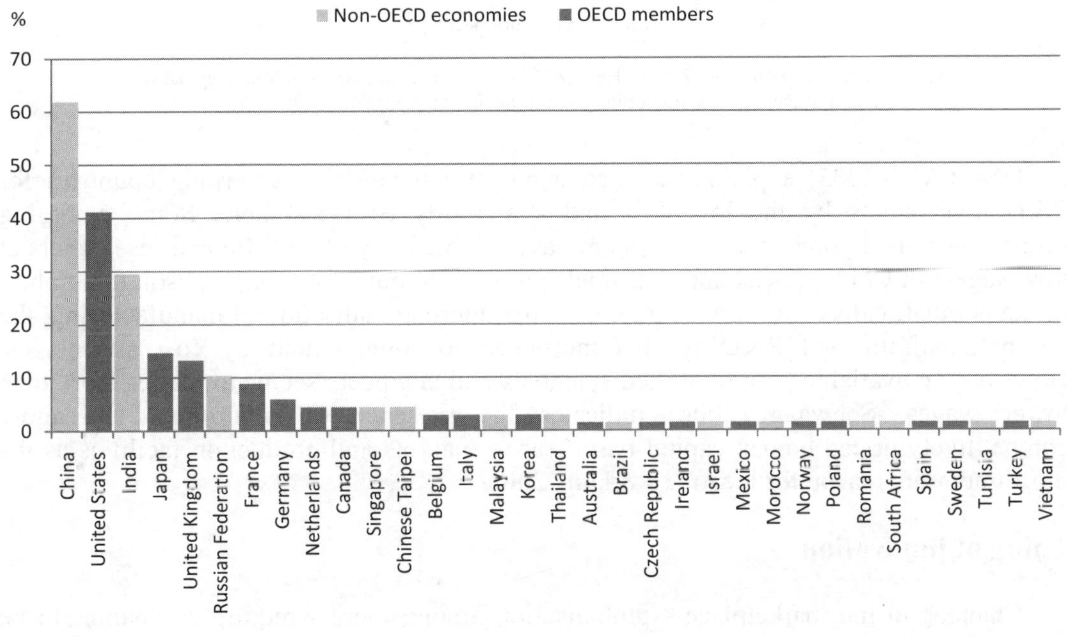

Source: UNCTAD (2005), *World Investment Report. Transnational Corporations and the Internationalization of R&D*, United Nations, New York and Geneva, in OECD (2008), *The Internationalisation of Business R&D: Evidence, Impacts and Implications*, OECD, Paris.

Thursby and Thursby (2006) surveyed US and EU MNEs on their plans to increase/decrease technical employment (*i.e.* employees who conduct or support R&D: researchers, research assistants, lab technicians and engineers involved in R&D) over the next three years (survey conducted in April 2005). They found that emerging countries such as India and China were expected to continue to be major beneficiaries of future R&D expansion, while prospects for Western Europe were rather pessimistic (Figure 5.4). More than half of the US respondents reported that they had either recently expanded or planned to locate R&D facilities in China and India; the percentage was somewhat lower for western European MNEs.

Figure 5.4. Anticipated change in R&D employment by MNEs over the next three years

Percentage, base year 2005

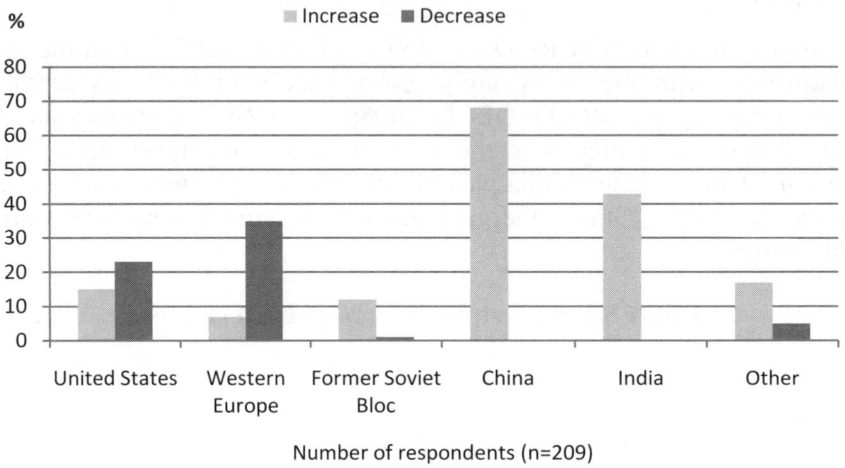

Source: Thursby and Thursby (2006), "Here or There? A Survey of Factors in Multinational R&D Location and IP Protection", Marion Ewing Kauffman Foundation, Washington, DC.

UNCTAD (2005) explains the increasing attractiveness of emerging countries for R&D investments by the low cost and availability of researchers. Some emerging economies with a good education system have a large body of well-trained researchers at low wages. In China, for example, a small proportion, but a very large absolute number, of the population has a tertiary degree. Like the internationalisation of manufacturing, the internationalisation of R&D is also motivated to some extent by cost advantages. However, an available pool of skilled scientists and engineers seems to matter more than lower wages. Schwaag (2006) indicates the presence of a stronger and more competitively priced human capital base near to markets and production facilities as the most important reason for locating R&D in China.

The opening of innovation

Changes in the marketplace – globalisation among them – require companies to be open to external ideas that supplement internal R&D in order to remain competitive. Owing to more intense and global competition and technological progress, product life cycles have been drastically shortened, forcing companies to innovate more quickly and develop products and services more efficiently. Moreover, the growing integration of different technologies has made innovation more costly and riskier. The greater the need for interdisciplinary cross-border and cross-sector research, the less a single company has the capability to innovate successfully. Companies therefore increasingly look for partners with complementary expertise to obtain access to different technologies and knowledge.

Open innovation models have become an integral part of the innovation strategies and business models of companies in recent years. Innovation is increasingly based on knowledge assets beyond the boundaries of the company and co-operation has become an important way of tapping into knowledge resources outside in order to generate new ideas and bring them quickly to the market (the "outside-in" approach). At the same time companies may spin out into a separate business those technologies and intellectual property which they have developed internally but are outside their core business and thus better developed and commercialised by others (the "inside-out" approach).

The most important benefit of open innovation to companies is that it provides a larger base of ideas and technologies. Companies look at open innovation as close collaboration with external partners, such as customers, consumers, researchers or other people that may have an input to the future of their company. The main motives for joining forces between companies is to seize new business opportunities, to share risks, to pool complementary resources and to realise synergies. Companies recognise open innovation as a strategic tool to explore new growth opportunities at lower risk. Open technology sourcing offers companies more flexibility and responsiveness without necessarily incurring huge costs (Figure 5.5).

Figure 5.5. The changed business environment: closed versus open innovation

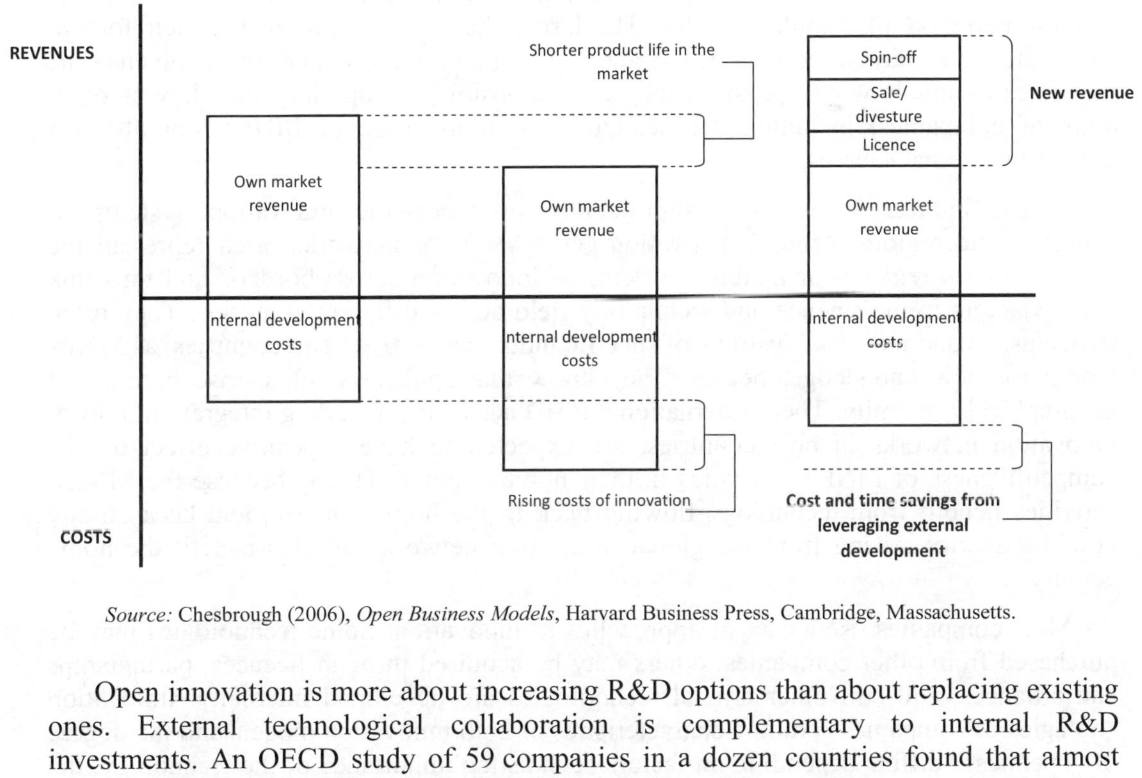

Source: Chesbrough (2006), *Open Business Models*, Harvard Business Press, Cambridge, Massachusetts.

Open innovation is more about increasing R&D options than about replacing existing ones. External technological collaboration is complementary to internal R&D investments. An OECD study of 59 companies in a dozen countries found that almost three-quarters of them devoted the bulk of their R&D budget – 80% or more – to in-house R&D activities (OECD, 2008c). At the same time most companies are actively involved in open innovation practices: 51% of the companies allocate up to 5% of their R&D budgets to research in other companies, while 31% allocate more than 10% outside.

The trend towards more openness in innovation is not totally new. The emphasis of open innovation primarily reflects a greater awareness of the organisation of innovative activities (technological as well as non-technological) across firm boundaries with more equal importance to internal and external sources of innovation. Recently, globalisation has significantly altered the scope for outsourcing and open innovation as it has broadened the choice of potential partners, giving rise to the development of global innovation networks.

The term "open innovation" does not refer to free knowledge or technology, but the collaborative methods applied, and may still imply the (significant) payment of licence fees for intellectual property.

The development of global innovation networks

In order to match the growing demand for innovation from customers, suppliers, etc., with the worldwide supply of science and technology, (large) companies increasingly develop innovation "eco-systems" across countries. In these innovation networks, companies link up with people, institutions (universities, government agencies, etc.) and other companies in different countries to solve problems and tap into new ideas.

In this global innovation climate, it is becoming increasingly important for companies to be involved in both external and intra-firm networks. Thus global innovation networks include a company's own R&D facilities abroad as well as collaboration with external partners and suppliers (see Box 5.1). Depending on their expertise, the different partners in these networks play multiple roles. The larger the role companies and their foreign R&D facilities play in global eco-systems, the more intense and more diverse the transfers of know-how will be, since they are responsible for sourcing know-how in other units of companies, including the headquarters in the case of MNEs, but also for accessing external sources.

Global innovation networks significantly influence the innovation systems of countries and regions. MNEs' innovation eco-systems or networks often represent the nodes between regional or national systems of innovation across borders, and thus link various actors in the science and technology field across different countries. They often span clusters and industrial districts of specific industries over several countries, as MNEs search for new knowledge, because they know that spillovers often arise because of geographical proximity. These international R&D activities, including integration in local innovation networks in host countries, are expected to have a positive effect on the competitiveness of MNEs' activities in their home country. This is because the MNEs' activities benefit from technology flowing back to the home country, and because any new discoveries arising from the global innovation network can also benefit the home country.

Most companies use a mix of approaches to innovation. Some technologies may be purchased from other companies, others may be acquired through licences, partnerships and alliances, and still other critical technologies are developed internally. Innovation strategies of companies combine characteristics of both innovation models and the degree of "openness" differs depending on factors such as the importance of the technology, the strategy of the firm, the characteristics of the industry, etc.

Box 5.1. Global open innovation in Novartis

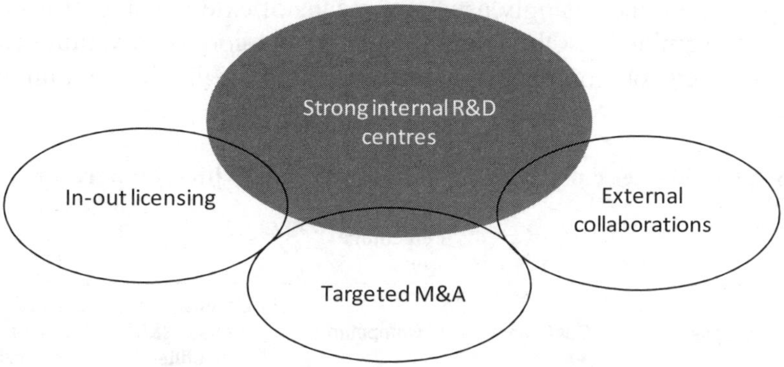

The innovation strategy of Novartis (one of the 59 company case studies included in the OECD project) is built around strong internal R&D centres complemented by in-out licensing, targeted mergers and acquisitions (M&A) and external collaborations. The R&D centres of Novartis are concentrated in Switzerland, the United Kingdom, France, the United States, Japan and India; overall, Novartis has more than 8 000 associate R&D personnel in 59 countries worldwide. In biotechnology, Novartis has set up external collaborations with 120 companies and 280 academic centres; these represent more than 30% of the R&D budget. In addition, Novartis has financed 150 entrepreneurial ventures over the last five years.

The attractiveness of (global) open innovation depends on the technological and industrial context in which companies are operating. The open innovation model is perhaps most evident in the information and communication technology (ICT) and pharmaceuticals-biotechnology sectors, but it is also increasingly important in the automotive and aerospace industries. Four key factors determine the potential for open innovation on a global scale for a company or industry. The first is how easy it is to innovate, either radically or incrementally – the "opportunity conditions". The second is how easy it is to protect innovation – the "appropriability conditions". The third is "cumulativeness" the degree to which the innovations of today are the basis for innovations of tomorrow. The final element is the degree of multidisciplinarity and cross-functional complexity of knowledge.

In industries characterised by rather short technology life cycles (*e.g.* the ICT, electronics and telecommunications industries), companies have sought external partners in order to keep up with new developments in and around their industry. In industries characterised by rather long technology life cycles and strong protection of intellectual property rights (IPR) (*e.g.* pharmaceutical, chemical and materials industries), companies mainly look outside the firm to keep up with research. In industries in which patents are important but can be more easily circumvented (*e.g.* the transport equipment industry and the fast-moving consumer goods industry), companies set up collaborations to keep pace with new developments. They seek technologies or products that have proven their market potential, which they can improve, scale up and commercialise.

Companies use different methods to source external knowledge. The two traditional models are: partnerships with external parties through alliances, joint ventures and joint development; and acquisition or sale of knowledge through contract R&D, purchasing or licensing. Open innovation, however, is increasingly realised through "corporate venturing": equity investments in university spin-offs or in venture capital investment funds. Companies also use venturing to look for external partners to commercialise their innovations which they do not use internally (divestments, spinning out, spinning off). Spin-off companies are increasingly used as a means of externalising projects. Some case study companies examined in OECD work had set up a corporate venture capital fund to develop new projects or companies based on ideas originating within the company (OECD, 2008c).

Table 5.1. Companies collaborating in innovation activities, by partner, 2002-04[1]

Percentage

	Suppliers	Customers	Competitors	Consultants and private R&D institutes	Universities and other higher education	Government and public research
Belgium	73	59	27	42	37	26
Czech Republic	80	68	40	39	34	19
Denmark	66	65	35	44	32	16
Germany	44	51	27	18	53	26
Ireland	72	78	19	31	31	18
Greece	46	32	47	27	27	10
Spain	52	23	17	23	26	28
France	65	50	36	32	26	18
Italy	56	39	37	50	36	11
Luxembourg	79	73	49	36	33	27
Hungary	71	53	37	34	37	14
Netherlands	75	55	31	38	31	24
Austria	43	45	22	42	58	30
Poland	67	39	20	19	15	21
Portugal	71	60	35	45	39	25
Slovenia	79	70	43	42	41	28
Slovak Republic	84	80	56	49	39	30
Finland	92	93	77	74	75	59
Sweden	75	65	25	46	41	15
United Kingdom	74	73	36	41	33	25
Iceland	68	68	48	23	17	45
Norway	70	67	36	61	45	49

1. Or nearest available years.

Source: OECD (2008), *OECD Science, Technology and Industry Outlook 2008*, OECD, Paris.

Companies traditionally seek to retain their core capabilities, directly determining what to outsource or with whom to collaborate. Companies' core competencies (in technology and markets) are developed as much as possible internally. In contrast, open innovation may be a faster and less risky way to diversify, in technology, markets or both, than internal development. The innovation model is relatively more open for non-core technologies and markets. Since not all required knowledge is available in house, companies actively search for all the expertise necessary to bring the new technology or product to the market and gain a competitive advantage.

The diffusion of global innovation networks can be explored through existing data on R&D, innovation, patenting and licensing. Companies collaborate on innovation most frequently with suppliers and customers, while co-operation with competitors and private R&D labs and consultants seems to be somewhat less frequent. While universities and government research institutes are generally considered to be an important source of knowledge transfer for the innovation activities of companies, collaboration with public research organisations (higher education or government research institutes) is less frequent (Table 5.1). This may be because public research focuses more on upstream research and exploration activities that may have only a modest impact on innovation.

Large companies are four times more likely to collaborate on innovation than small and medium-sized enterprises (SMEs). Large companies are also much more active in public research, although there is more cross-country variation for large firms than for SMEs. Nevertheless, it should be kept in mind that the data only reveal the existence of some sort of collaboration, not its type or intensity (Figure 5.6).

Figure 5.6. Companies collaborating on innovation activities, by size,[1] 2002-04[2]

As a percentage of all companies

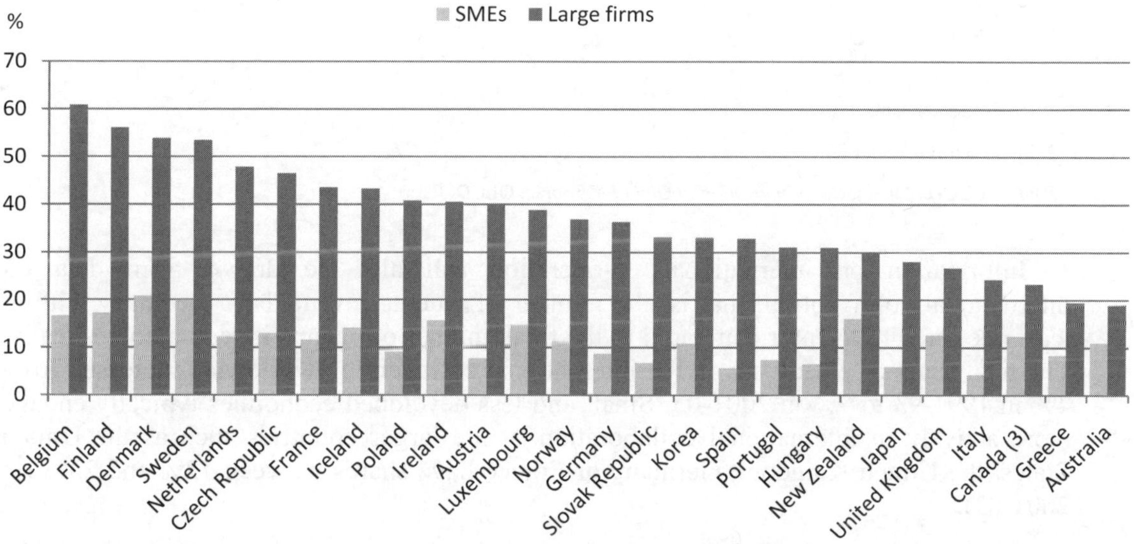

1. SMEs: 10-249 employees for European countries, Australia and Japan (persons employed); 10-99 for New Zealand, 10-299 for Korea, 20-249 for Canada.

2. Or nearest available years.

3. Manufacturing sector only.

Source: OECD (2008), *Open Innovation in Global Networks*, OECD, Paris.

While international technology collaboration is found to play a prominent role in the innovation process of companies (Figure 5.7), firms seem to favour innovation partners that are geographically close. Again, no quantitative information on international collaborations is available, but companies seem to enter such arrangements only if they are strongly motivated, since international partnerships are more costly and difficult to manage. The company case studies indicated that technology partners in the same geographical area generate fewer confidentiality problems (OECD, 2008c).

Figure 5.7. Companies with foreign co-operation on innovation, 2002-04[1]

As a percentage of all companies

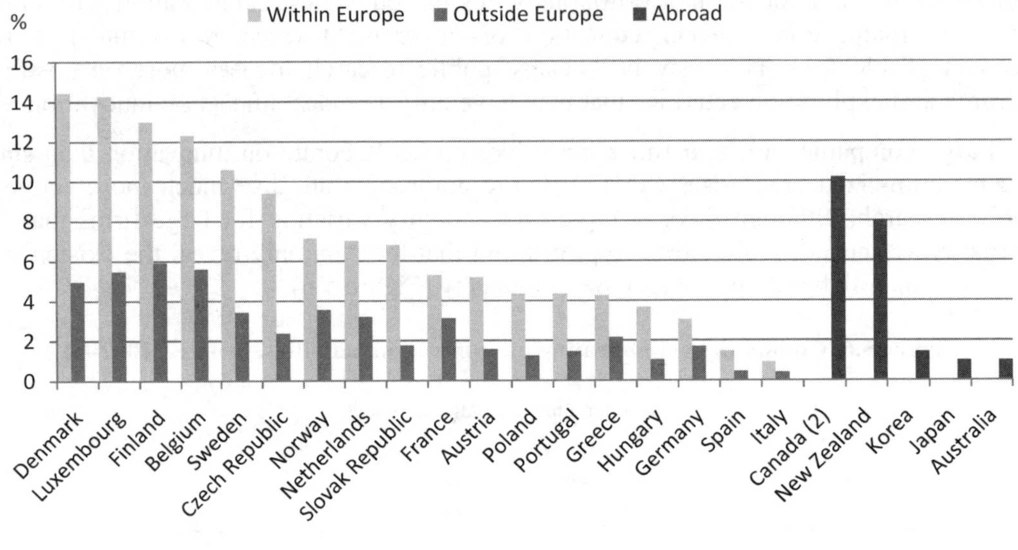

1. Or nearest available years.

2. Manufacturing sector only.

Source: OECD (2008c), *Open Innovation in Global Networks*, OECD, Paris.

Information on international co-operation can also be drawn from data on international co-invention, that is, the number of patents invented by a country with at least one foreign inventor compared to the total number of patents invented domestically. The total share of patents involving international co-invention worldwide increased from 4% in 1991-93 to 7% in 2001-03. Small and less developed economies typically engage more actively in international collaboration, while larger countries such as the United States, the United Kingdom, Germany or France have shares between 12% and 23% (in 2001-03).

The breakdown of collaboration by main partner country confirms to some extent the importance of geographical proximity in international co-invention. EU countries collaborate essentially with other EU countries, while countries such as Canada and Mexico collaborate more frequently with the United States. For instance, more than 20% of inventions in Canada and Mexico involved collaboration with a US inventor (OECD, 2008c). China, India, Israel, Japan and Korea also appear to co-operate significantly with the United States.

The empirical evidence also shows that markets for technology licensing are large and growing. Separate studies have demonstrated that licensing is growing rapidly in the United States and Europe, as well as in Japan, notwithstanding some regional differences. International licensing in particular appears to be on the rise, although much of this reflects transactions among affiliated businesses.

Since innovation is increasingly the basis for the competitive advantage of companies, the growing number of interactions with external parties such as customers, suppliers, universities, etc., has important repercussions for the protection and safeguarding of intellectual assets and intellectual property: patents, trademarks, trade secrets, etc. Open innovation may increase the risk of leakage of proprietary knowledge and involuntary spillovers. Other potential disadvantages are the extra costs of managing co-operation with external partners, the loss of control, the adverse impact on the flexibility of the company, the dependence and possible over-dependence on external parties and the potentially opportunistic behaviour of partners.

Theft of intellectual property is typically identified as the most important risk in global innovation networks. Unique knowledge may be revealed to external partners who may later become competitors or make better use of the fruits of the venture or the know-how. Working closely with external partners can create uncertainty about how the benefits of technology collaboration are to be appropriated. SMEs may face greater risks in collaborating with larger companies because of their typically smaller resources and limited expertise in intellectual property rights (IPR) issues.

The effective management of intellectual property (IP) is therefore crucial, not only to identify useful external knowledge but especially to capture the value of a firm's own IP rights. In the past, IP management was often relatively closed, since intellectual property was mainly created and used internally, and intellectual property protection was used to prevent or block off competitive moves. Patents were especially important for protecting companies' inventions from imitation. The sometimes low utilisation rate of IP assets in the commercialisation of products and services was a direct consequence of this. Most patents did not directly generate revenue for patent owners via their incorporation into products, processes and services or through licensing revenues.

Companies engaged in open innovation practices now often organise licensing activities and strategic alliances for a proactive intellectual property strategy that aims at sharing technologies rather than keeping IP as a defence mechanism. Patent licensing has been found to generate significant financial benefits for patent holders. Furthermore, companies increasingly create cross-licensing agreements and other collaborative mechanisms in order to facilitate technology collaboration.

Successful open innovation also depends on the open character of the business model. As knowledge has become companies' key resource, open innovation needs to be embedded in an overall business strategy that explicitly acknowledges the potential use of external ideas, knowledge and technology in value creation. Owing to the integration of different technologies, industry borders are shifting or even disappearing, necessitating new business models and organisational structures, including the effective management of human capital (open culture, diversity, etc.).

International mobility of the highly skilled

Alongside sustained growth in foreign direct investment, in trade and in the internationalisation of R&D, mobility of human resources for science and technology has become a central aspect of globalisation. Migration of talent now plays an important role in shaping skilled labour forces throughout the OECD area.

The importance of mobility stems from its contribution to the creation and diffusion of knowledge. Not only does it aid in the production and dissemination of codified knowledge, it is also an important means of transmitting tacit knowledge. In the broadest sense, tacit knowledge is any knowledge that cannot be codified and transmitted as information through documentation, academic papers, lectures, conferences or other communication channels. Such knowledge is more effectively transferred among individuals with a common social context and physical proximity.

Various factors contribute to the flows of the highly skilled. In addition to economic incentives, such as opportunities for better pay and career advancement and access to better research funding, mobile talent also seeks higher quality research infrastructure, the opportunity to work with "star" scientists and more freedom to debate. Less amenable to potential government policy, but still important, are family or personal ties that draw talent to certain locations.

Once in another country, people diffuse their knowledge. In the workplace, knowledge spreads to colleagues, especially those in close contact. Knowledge also spills over to geographically proximate individuals and organisations and can contribute to the emergence of local concentrations of activity. Mobile HRST also act as a vital complement to the transfer of knowledge through flows of goods and capital across borders.

For receiving countries, the inflow of talent has positive effects relating to knowledge flows, including the possibility of increased R&D and economic activity owing to the availability of additional skilled workers, improved knowledge flows and collaboration with sending countries, increased enrolments in graduate programmes, and potential firm and job creation by immigrant entrepreneurs. Mobility can help to link domestic firms to foreign knowledge and to stimulate spillovers from foreign R&D to local R&D units and the economy at large. At the same time, receiving countries must ensure that inflows of scientists and researchers do not delay reforms to policies that may be limiting the domestic supply of HRST.

For sending countries, work on the effects of emigration has often focused on migrant remittances and brain drain, with particular emphasis on the impact on developing countries. Remittances are an important source of income for many low- and middle-income households in developing countries. The main concerns about brain drain centre on the loss of productive labour and its associated output, the fiscal cost of educating workers who then move abroad, and the potential impact on much-needed institutional development and structural change. However, these concerns must be balanced against the question of whether these researchers and scientists could have found productive employment at home.

Emigration of skilled workers, such as researchers and scientists, can also be beneficial for the creation and diffusion of knowledge in their country of origin. In particular, emigration possibilities may encourage the development of skills. In addition, when skilled individuals move to larger and "denser" economies they can benefit the sending country by producing "better" knowledge than they could at home, accumulating human capital faster and improving their productivity, thereby increasing the potential return flows of knowledge. This can increase the global stock of knowledge.

"Brain circulation" can stimulate knowledge transfer to sending countries. This may mean the return of skilled migrants to their home country after a period abroad, or a pattern of temporary and circular migration between home and abroad. Professionals diffuse the knowledge they acquire to their home country and maintain networks, thereby facilitating continuing knowledge exchange. To make the most of brain circulation, the home country needs to have sufficient absorptive capacity and returning talent needs to be able to re-enter local labour markets at a level that is appropriate for their skills and knowledge.

The existence of a diaspora further enhances the transfer of knowledge. A stock of skilled HRST abroad can act as a conduit for flows of knowledge and information back to the home country, and social and other links increase the probability that knowledge will continue to flow along this conduit even after individuals move back or move away. In some emerging economies, diaspora networks play a vital role in developing science and technology capacity.

Taken together, these effects suggest that knowledge flows associated with the emigration of researchers and scientists can provide benefits to sending countries. The mobility of researchers therefore is not necessarily a zero-sum game in which receiving countries gain and sending countries lose.

Figure 5.8. Foreign-born highly skilled people in selected OECD countries, 2001

By country of birth and country of residence, thousands

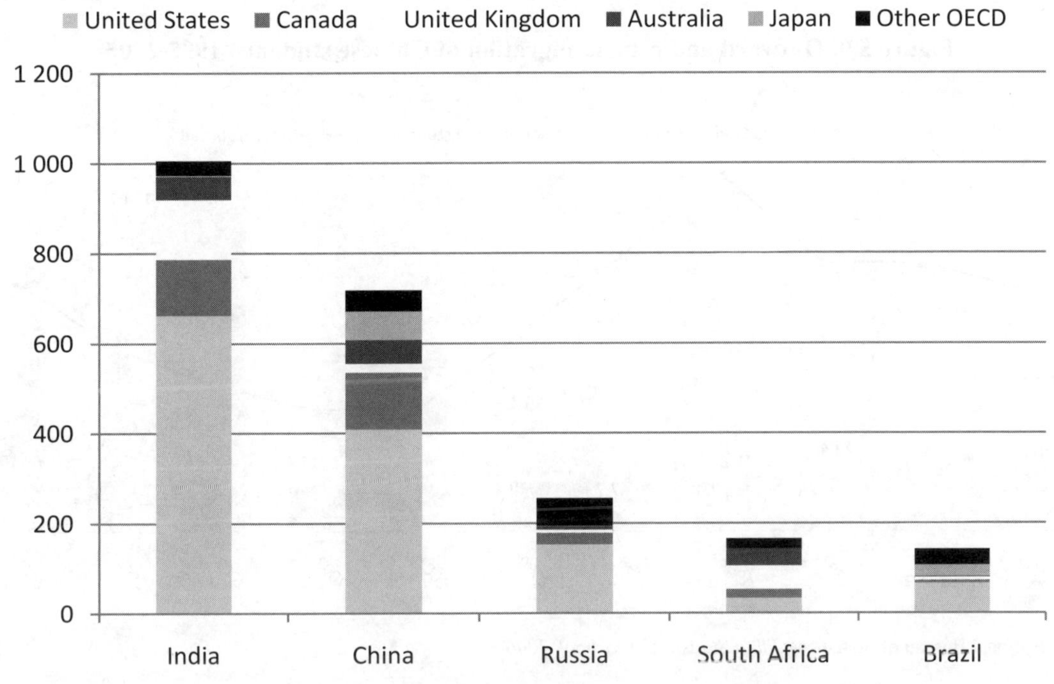

Source: *OECD Database on Immigrants and Expatriates*.

Data on international mobility of HRST show that most OECD countries are net beneficiaries, with inflows exceeding outflows. Australia, Canada, France and the United States, in particular, have experienced strongly positive net inflows of tertiary-educated migrants. However, a more detailed picture reveals that, in relative terms, Ireland and New Zealand have experienced large outflows. In absolute terms, Germany and the United Kingdom have the highest number of skilled expatriates, while Luxembourg, Norway and the Slovak Republic have the least. For some countries, intra-OECD flows add substantially to the stock of highly skilled individuals. Non-OECD migrants play an important role in several OECD countries, and the main sources are Asian, led by China and India (Figure 5.8).

The international mobility of students is a further aspect of the internationalisation of HRST. OECD countries benefit from the inflow of talented students and scholars, and many now actively seek to attract foreign students. Benefits also occur when domestic students study abroad and gain knowledge and experience in another country. Data show that the numbers of students enrolled outside their country of citizenship has risen sharply since 1995.

Return and circular flows of migrants add to the mobility picture. Data show a tendency for many "permanent" or long-term migrants to return to their country of origin (Figure 5.9; OECD, 2008e). Return rates appear to be higher for skilled workers and for those from countries at a greater cultural, economic and geographic distance from the host country. This trend is consistent with the notion of a globalising labour market in which the mobility of skilled workers is affected by changes in relative labour market conditions. The decision to return is driven strongly by lifestyle and family considerations and the availability of attractive employment opportunities at home. For students, the considerations are similar.

Figure 5.9. Outward and return migration of Chinese students, 1995-2005

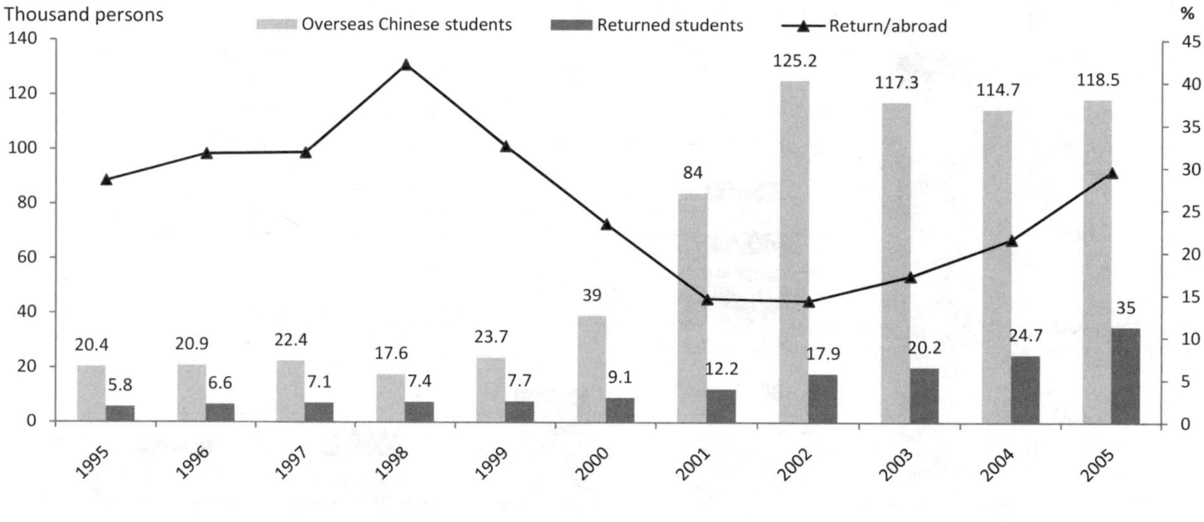

Source: National Bureau of Statistics, *China Statistical Yearbook 2006*.

Quantitative evidence on the impact of mobility patterns is not readily available. Many variables and factors influence science and technology outcomes and are hard to disentangle. Nevertheless, data and information can be used to build a picture and to see some links between mobility and broader science and innovation outcomes. A clear effect of the mobility of highly skilled workers is the increasing internationalisation of the labour market for the highly skilled. Both in private industry and academia, foreign staff are sought for their specific knowledge or abilities, their language skills, and their knowledge of foreign markets.

The links between mobility and innovation are less clear, although some evidence suggests that immigrants contribute strongly to patent applications and the creation of technology firms. Studies from several countries highlight a trend towards more international co-authorship of academic articles. Some work suggests that the impact of collaborative work, as measured by citations, is higher than the average impact of national work.

In the broader context of R&D and innovation activity, many countries have greatly improved their ability to exploit and perform research and innovation over the past decade. This is changing the geographical spread and intensity of research and scientific activity. The growing sums spent on R&D in non-OECD countries and their human capital resources, coupled with the increasingly internationalised activities of technology firms, all suggest that the opportunities for mobile talent will continue to grow.

Implications for policy

Benefiting from global innovation networks

While OECD countries are still the most active investors in and performers of R&D, globalisation has reduced barriers to entry and created opportunities for new players in emerging economies – and also at home – to tap into global networks. At the same time, the emergence of global players such as China and India, as new markets but also as platforms for research and talent, has raised concerns in OECD countries about the offshoring of R&D and related high-skill jobs and/or erosion of existing national R&D infrastructure and capacity. For smaller and catching-up OECD economies, the emergence of global players increases competition for R&D-related FDI and for research talent, which may make the catching-up process more difficult. The development of global innovation networks provides opportunities for developing countries to access research and innovation networks to accelerate their own development, but also creates a risk that national resources may be shifted away from country needs to meet the short-term objectives of foreign-based platforms.

The emergence of open innovation also raises policy issues. While open innovation is essentially business-driven, it has implications for a range of policies. Insofar as open innovation is about "open" business models for innovation, countries' framework conditions (*i.e.* product and labour markets, IPR and competition policies, a strong public research base, etc.) are extremely important policy levers. At the same time, because open innovation involves going beyond firms' and nations' boundaries, it may create issues for government research and innovation policies. Most OECD countries' S&T policies are predominately national in scope, but it is becoming clear that policies designed for geographically circumscribed knowledge-based activities or for vertically integrated value chains of firms need to be reviewed.

The globalisation of R&D and the emergence of open innovation strategies in firms clearly raise intellectual property issues. While strong IP protection can attract R&D-related FDI, excessively strong protection can act as a barrier to open innovation strategies that rely on knowledge sharing and access. Access to IP allows innovators to create new IP which can in turn be made available to other users.

Traditional policies and instruments for stimulating research and innovation are under pressure to adapt to the global context for innovation. For example, the globalisation of R&D implies that the leverage effect of public instruments may become less effective if national firms can readily shift R&D or expand it in offshore markets with greater growth potential. Another possible implication is the need for greater coherence in policy making across government ministries and departments to increase the leverage of existing mechanisms.

At the same time, while it is clear that national innovation policies must look beyond geographically circumscribed knowledge-based activities and vertically integrated value chains, it is not altogether clear that the effects are the same for large countries with large internal markets for R&D and innovation as for small countries that are more dependent on international flows of knowledge and capital. For larger countries, nationally focused innovation policies may still matter and they may not necessarily be at odds with globalisation. Indeed, for larger countries it may be more important to ensure that regional and local initiatives have a global dimension. Similarly, "policy coherence" is becoming more important, but it is arguably more so – and easier – for smaller countries. It is therefore not surprising that smaller economies have taken the lead in opening up national programmes for R&D to firms abroad irrespective of the location of their production capacities.

Ultimately, however, open innovation and the globalisation of R&D are, in the first place, business strategies of firms in response to market opportunities and challenges raised by the globalisation process itself and by technological change. Globalisation creates new market opportunities which imply new innovation strategies in response to (new or unmet) market needs. The response builds on technological progress and international trade, including global value chains, generating further acceleration of the internationalisation of R&D and globalisation more generally.

The challenge for governments is to help firms adjust their innovation strategies to a changing environment. As mentioned, framework conditions are a clear area for policy action, but there are other areas in which policies have a role to play. However, if they are isolated from other measures, they may not succeed in encouraging innovative behaviour. For example, extending networking policies or improving industry-science relations alone may not be enough to attract foreign R&D or to promote more collaborative research and open innovation. However, if such initiatives are integrated with other policy measures, such as those to promote entrepreneurship and new firm creation, they may stimulate capacity in the overall innovation system. The OECD project has highlighted the importance of "building trust" in the marketplace as well as at the interface between the public and private sectors. Stable framework conditions and government policies therefore seem to be important.

A question on the minds of policy makers is whether new or radically different policies are needed to meet these challenges. As countries have different economic structures and resource endowments, policies will have to be differentiated according to the national context. Small countries with a weak research base will need to focus on strengthening their knowledge capabilities in order to contribute to and participate in global innovation networks. Other countries may need to focus on improving policy coherence. In sectors in which global value chains are extremely fragmented, the ICT infrastructure may be especially important. And in countries where multinationals play a large role, the focus may need to be on designing policies to help local firms capture spillovers from global innovation networks.

For developing countries, global innovation networks may offer new opportunities to draw benefits from the globalisation of innovation. Thus far, only a few emerging countries, notably China and India, have been able to attract FDI for R&D. Moreover, the spillovers from FDI for R&D on emerging economies have sometimes been relatively limited, perhaps owing to a lack of absorptive capacity or a mismatch between the types of R&D performed via FDI and the market needs of these economies. The development of global innovation networks may enable a wider range of emerging countries to benefit from the globalisation of innovation, as these networks explicitly seek to access and draw on global knowledge, including in developing countries. Nevertheless, if countries are to participate, they will have to make an effort to enhance their attractiveness.

Attracting talent

Policies to attract and retain talent are also gaining in importance in many OECD countries, though with a wide range of "intensity". Most OECD countries regard such policies as important and have policies to encourage and assist mobility. These run from economic incentives to encourage inflows, to immigration-oriented assistance, to procedures for recognising foreign qualifications, to social and cultural support, to support for research abroad. Some countries focus on just a few policy mechanisms, while others offer "something for everyone".

Only a few countries' policy approaches are part of an explicit mobility strategy. For those in which policies are not part of such a strategy, there is a greater risk of incoherence among policies on inflows, outflows and the diaspora. Ideally, mobility policies should be part of a wider mobility strategy that contributes to the country's economic and social objectives and sets out the rationale for intervention. There is generally more support for inflows of researchers and other HRST than for outflows, perhaps because countries judge outward mobility to be adequate or because they are reluctant to encourage outward mobility, despite arguments about the benefits of brain circulation.

National policies appear generally to target the same population, with little orientation towards national scientific and technological interests. Since many countries offer support for mobility, as opposed to permanent migration, researchers may use these policies to work in a number of countries. In most cases, national policies do not place restrictions on the country of origin (inward mobility) or of destination (outward mobility). In theory then, mobility policies often have a global focus.

OECD countries already have a wide selection of policy tools at their disposal, which they use more or less intensively to promote HRST mobility. The question then is, what is the role for international mobility policy in the future, given what is known about mobility and knowledge flows and about current mobility, R&D and innovation patterns? In designing future mobility policies, a key first step is to identify a rationale for intervention and clear objectives. For mobility, the main rationale may be the potential positive externalities from knowledge spillovers and information asymmetry issues. The obstacles to mobility commonly cited include legal and administrative barriers, lack of funding, personal issues and language.

Few policies have been evaluated, so it is difficult to point to best practices. However, some lessons can be drawn from evaluation material provided by countries in response to an OECD questionnaire, including the importance of setting appropriate funding levels and programme durations for the target population. More work on evaluation would be valuable.

Given differences among countries, it is not possible to identify a "recipe" for what governments should do more of, what they should do less of, and what should stay the same. One promising avenue, however, is removal of barriers to short-term and circular mobility. Shorter (and potentially repeated) periods abroad may avoid some of the obstacles that currently hinder mobility, and would support knowledge flows associated with brain circulation and the diaspora.

While a uniform recipe for specific policies may be inappropriate, it is clear that policy coherence is important for all countries. The first task is to ensure the co-ordination and coherence of various mobility policies. This includes considering the consequences of mobility for development in sending countries that are the target of development and aid policies. Linking policy design and implementation in these areas can better achieve the goals of both mobility and development and contribute to more effective management of migration. The second task is to ensure that mobility policies fit within the broader policy environment for innovation and to create a sound base for innovation and scientific endeavour. In particular, to improve innovation outcomes, it is not sufficient simply to increase the number of HRST; these people must operate in a system that enables them to use, create and disseminate knowledge.

Finally, an important message from the OECD's work on mobility of talent is that the global competition for skilled people is growing. Many OECD countries and a growing range of non-member economies aim to attract the same pool of highly skilled researchers and scientists. Relying extensively on international flows and mobility policies to fill existing or future gaps in supply may therefore entail risks. Policy will also need to focus on addressing shortcomings in national policies that may limit the supply of HRST.

References

Chesbrough, H. (2006), *Open Business Models*, Harvard Business Press, Cambridge, Massachusetts.

Criscuolo, P., R. Narula and B. Verspagen (2005). "Role of Home and Host Country Innovation Systems in R&D Internationalisation: A Patent Citation Analysis", *Economics of Innovation & New Technology*, Vol. 14, pp. 417-433.

Dunning, J. and R. Narula (1995), "The R&D Activities of Foreign Firms in the US", *International Studies of Management and Organisation*, 25, pp. 39-73.

Kuemmerle, W. (1997), "Building Effective R&D Capabilities Abroad", *Harvard Business Review*, March-April 1997, pp. 61-70.

OECD (2007), *Staying Competitive in the Global Economy: Moving Up the Value Chain*, OECD, Paris.

OECD (2008a), *The Internationalisation of Business R&D: Evidence, Impacts and Implications*, OECD, Paris.

OECD (2008b), *OECD Science, Technology and Industry Outlook 2008*, OECD, Paris.

OECD (2008c), *Open Innovation in Global Networks*, OECD, Paris.

OECD (2008d), *The Global Competition for Talent – International Mobility of the Highly Skilled*, OECD, Paris.

OECD (2008e), *OECD Reviews of Innovation Policy: China*, OECD, Paris.

OECD Policy Brief: "Research and Development Going Global", *www.oecd.org/publications/policybriefs*.

Schwaag, S. (2006), "China, from Shop Floor to Knowledge Factory", pp. 227-266 in M. Karlsson (ed.), *The Internationalization of Corporate R&D*, Swedish Institute for Growth Policy Studies, Sweden.

Thursby, J. and M. Thursby (2006). "Here or There? A Survey of Factors in Multinational R&D Location and IP Protection", Marion Ewing Kauffman Foundation, Washington, DC.

UNCTAD (2005), *World Investment Report. Transnational Corporations and the Internationalization of R&D*, United Nations, New York and Geneva.

Chapter 6

Innovation strategies for growth: insights from OECD countries

Jean Guinet, Gernot Hutschenreiter and Michael Keenan[1]

Monitoring of broad innovation policy trends and in-depth country-specific reviews reveal how governments from countries at different levels of economic development and with different industrial specialisation seek to strengthen innovation as an engine of growth. This chapter demonstrates that while there is some convergence in terms of broad objectives and policy principles, the government of each country has to custom-design multi-dimensional innovation policies which need to be continually adapted to meet new challenges in the global market. In addition to a synthesis, the chapter presents insights from three country reviews to illustrate common elements and country-specific features of government strategies to promote innovation-fuelled growth.

1 Jean Guinet is Head, Gernot Hutschenreiter is Deputy Head and Michael Keenan is Policy Analyst, Country Review Unit, Directorate for Science, Technology and Industry, OECD.

Introduction

In the last decade a series of major OECD projects,[2] drawing on an expanding body of literature, have contributed, together with relevant work of other international organisations,[3] to an emerging consensus among policy makers about the critical importance of innovation among the factors determining economic performance (Box 6.1).

Box 6.1. Innovation has become the key driver of economic growth

Most of the rise in living standards since the Industrial Revolution has been the result of new and improved products, processes and services. However, innovation has now become even more important for a wider spectrum of economic and social activities. Globalisation is forcing all countries to move their economic activity further up the value chain to continue to compete and prosper. Continued leadership, but also the capability to catch up, will therefore come from staying a step ahead of the competition in higher value-added elements of the economic process. Economic research provides new empirical evidence of this tightening relationship between innovation capability and economic success at both the macro (aggregate) and micro (firm) level:

- *At the macro level*, about half of the cross-country differences in per capita income and growth is due to differences in total factor productivity (TFP), which in turn is mainly driven by technological development and innovation, with a strong influence of R&D (World Bank, 2007). Recent empirical research (Coe *et al*., 2008) confirms the role of both domestic and foreign R&D capital as significant determinants of TFP. Human capital and institutional factors, notably those that condition the efficiency of national innovation systems (NIS), also have a significant impact on TFP. Moreover, countries in which doing business is facilitated and the quality of tertiary education is high tend to derive more benefits from domestic R&D, from R&D spillovers from abroad and from human capital formation.

- *At the micro level*, it has been demonstrated that in all sectors of activity, from high-technology to the more traditional resource-based industries, innovative firms exhibit better performance and create more and better jobs. For example, recent OECD analysis of innovation at the firm level (OECD, 2008b) shows that product innovation increases business firms' labour productivity.

From microeconomic to macroeconomic performance. For business innovation to translate into better macroeconomic performance, structural change is required to shift resources from non-innovative towards innovative firms irrespective of the industry. In successful countries the government facilitates such processes by providing favourable framework conditions and specific support to induce more companies to enter the "innovation game" in the first place and to reward the efforts of already innovative companies. The OECD study shows that firms that receive financial support from government or engage in co-operation (with other firms and/or public research institutes) invest more in innovation (OECD, 2008b).

2. Such as the Technology and Economy Programme (1991); the Jobs Strategy (1994), notably its component on Technology, Productivity and Job Creation (see OECD, 1998); and the Growth Study (2003).

3. Such as the Knowledge for Development programme (K4D) of the World Bank.

In May 2007, the OECD Ministerial Council Meeting, recognising the growing importance of this policy area, mandated the Organisation to develop an Innovation Strategy (IS) to help member countries to devise more efficient policies to strengthen innovation as an engine of growth. In-depth country reviews of innovation policy have been undertaken in order to base the IS on a rich set of concrete national experiences with the evolving nature of innovation processes and the ability of public policy to boost the performance of innovation systems. These reviews are conducted for both OECD member and selected non-member economies. The wide variety of countries reviewed so far[4] – in terms of level of economic development, size, institutional features, etc. – offers fertile ground for identifying commonalities and differences in policy challenges and responses to the growing importance and evolving nature of innovation.

A central goal of this chapter therefore is to explore the dynamics and impacts of some of the new trends and drivers in innovation processes and policies in a variety of contextual settings. In a first part, the chapter summarises the main lessons learned from all completed reviews and other relevant OECD work on national responses to the changing dynamics of innovation, with a special focus on those concerning innovation policy governance, mix and instruments (Figure 6.1). In a second part, the analysis of selected issues (*i.e.* economic diversification, internationalisation and policy co-ordination) is further deepened through a comparison of three small open economies: Chile, Norway and Switzerland. A final section draws some general conclusions and pointers for further work.

Policies to promote innovation-led growth: broad lessons from OECD country-specific work

Strategic goals: some convergence, but a widening "innovation divide" and enduring country specificities

The agenda for high- and middle-income countries is converging. Most advanced countries have now adopted more articulated and ambitious innovation strategies (the European Union: the Lisbon Strategy; the United States: the Competitiveness Agenda; Japan: Innovation 25; as well as Australia; Finland; France; Germany; the United Kingdom; etc.). But the most striking new development is the recognition of the importance of innovation by the most dynamic catching-up economies. Innovation policy in countries as diverse as Chile, China, Mexico and South Africa reflects a change in the understanding of the role of and interplay between the creation and diffusion of technology, with a rejection of the idea that countries need to "exhaust" their potential for catching up before entering on their "own" innovation and R&D activities.

The shared determination of a growing number of countries at different levels of development to promote more innovation-driven growth creates *frictions of convergence* in several areas: intellectual property rights (IPRs), competition for talents, "forced technology transfer", standards, etc.

At the same time the marginalisation of low-income countries and of low skills in high-income countries is a risk for several reasons. Increasing returns on investment in

4. Chile (OECD, 2007d), China (OECD, 2008c), Greece (OECD, forthcoming), Hungary (OECD, 2008d), Korea (OECD, 2009a), Luxembourg (OECD, 2007a), Mexico (OECD, 2009b), Norway (OECD, 2008a), New Zealand (OECD, 2007b), South Africa (OECD, 2007c) and Switzerland (OECD, 2006).

knowledge lead to geographical concentration of innovative activities; the youngest populations are located in areas with lower education and training capacities; the demand for low-level skills falls while its global supply increases; the growth strategy of rich countries/individuals contrasts with the survival strategy of poor countries/individuals.

Figure 6.1. New trends in innovation processes and policies

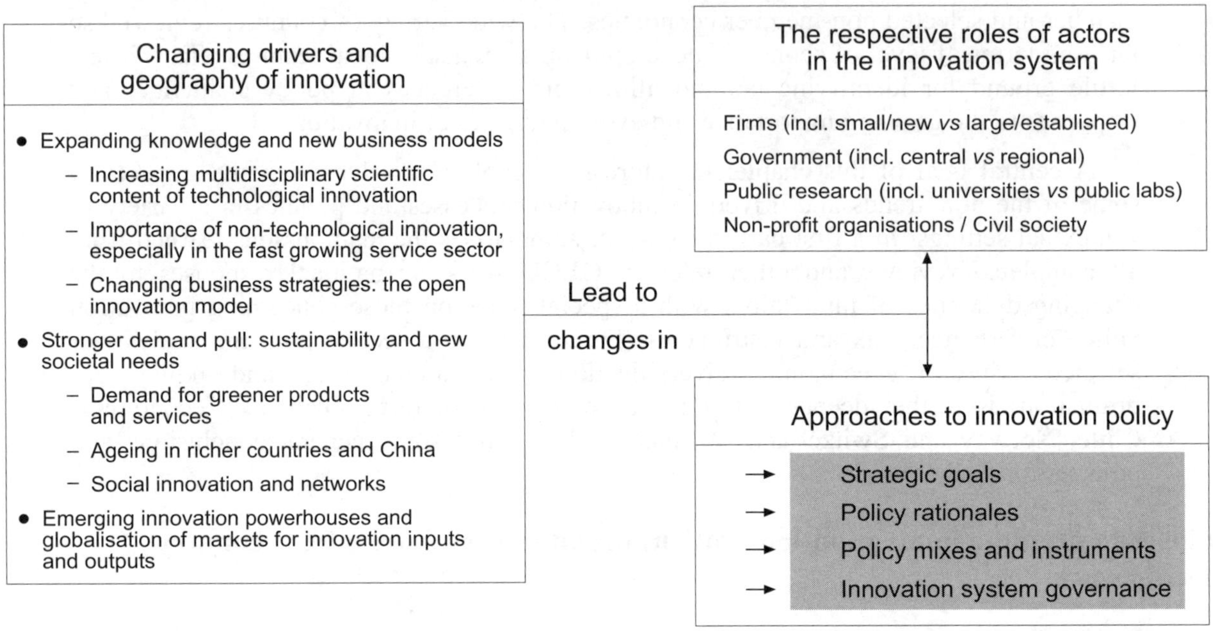

This creates *conflicts of divergence* within and among countries: immigration pressures, social unrest, insecurity, environmental damage (*e.g.* deforestation), counterfeiting and piracy, etc. Issues such as innovation and development, the social impacts of innovation, links between growth-enhancing innovation and purely welfare-augmenting innovation should therefore receive more attention.

One main lesson from the OECD innovation policy reviews is that no single configuration of the national innovation system is appropriate for all economies. Each country possesses a unique inheritance which conditions its ability to exploit successfully the opportunities offered by increasing globalisation, economic growth, social change, and developments in science, engineering and technology. A successful innovation system is one which, given the wider economic and technological environment, enables a country to build successfully on its inherited strengths and to remedy, offset or work around its inherited weaknesses in order to exploit to the maximum extent possible its potential for future sustainable economic growth and social well-being.

Rationales for innovation policy revisited

Broadening the conceptualisation of innovation

Discussions of the core scientific and technological functions of national innovation systems often jump quickly from "science and technology" to "research and development". Consequently, maps of the R&D system easily become viewed as maps of the innovation system. This tendency is reinforced by heavy reliance on data on R&D

inputs and outputs, the only internationally comparable available indicators of the main features of innovation systems. Yet, this is a serious distortion, since it leaves out many other kinds of S&T – and non-S&T – activity that play central roles in innovation. The OECD innovation policy reviews have sought to avoid this bias as far as possible by paying closer attention to the following characteristics of contemporary innovation:

1. *Innovation systems are not solely concerned with the types of innovation that are globally novel.* The *Oslo Manual* (OECD, 2005) acknowledges that important forms of innovative activity involve changes that are new to particular industries or individual firms, something that is confirmed by results from a growing number of innovation surveys. In this respect, technology diffusion is an important source of innovation. Cross-border diffusion of technological knowledge is clearly very important for small countries and especially for those that trail behind the technological frontier (Gerschenkron, 1962; Abramovitz, 1986; Fagerberg, 1994). However, even for large, technologically advanced economies such as the United States or the European Union, cross-border knowledge diffusion is of key importance for economic performance in the longer term (Hollenstein and Hutschenreiter, 2001).[5]

2. *Design, engineering and management play key roles in innovation systems.* The core activity at the heart of almost all innovation is the creation of a set of specifications (or "designs") of the change that is to be made. It tends to draw upon existing knowledge without any direct input of new knowledge from R&D. Unforeseen problems often emerge, however, and these may require R&D for their solution.

3. *Technological knowledge, even if publicly available, is in many cases not unconditionally appropriable.* Rather, potential innovators require certain capabilities, referred to as "learning" or "absorptive capacities" (Cohen and Levinthal, 1989) if they are to adopt and make efficient use of existing knowledge.

4. *Innovation encompasses not only "hard" technological innovations, but also softer forms such as organisational arrangements and procedures.* This is especially truc in the services sector. Recent work confirms that services are more innovative than previously thought, and that in some areas, they are more innovative than the average manufacturing industry. However, although they are not well understood, most innovation in services appears to be non-technical and to result from small, incremental changes that do not require much formal R&D (Tamura *et al.*, 2005).

5. *Demand is a central driver of innovation.* A major obstacle to innovation in many countries participating in the OECD innovation reviews is the lack of sophisticated demand for innovative products and services.

6. *The complexity and scope of innovation processes requires a broad set of complementary skills.* The skills necessary for successful innovation go well

5. The importance of international knowledge diffusion was illustrated by Eaton and Kortum (1996) who found that more than 50% of the productivity growth in each of the 19 OECD countries in their sample could be attributed to innovations from just three countries (the United States, Germany and Japan). Only these three countries, together with France and the United Kingdom, derive more than 10% of their growth from domestic research. On this issue also see Eaton and Kortum (1999) and a survey by Gong and Keller (2003).

beyond those associated with the conduct of research. Other types of technical (*e.g.* design) and "soft" (*e.g.* knowledge management, entrepreneurial, etc.) skills need to be widely distributed among the workforce, acquired in a number of places (*e.g.* not just in the university, but also in the workplace) and in a number of ways, and continually updated (*e.g.* through lifelong learning).

7. *Innovation systems are increasingly internationalised.* Many maps of national innovation systems place heavy, sometimes exclusive, emphasis on national activities and interactions. This too easily obscures from policy attention international elements that can be critically important in influencing how all aspects of the system function. Yet, these international components of the system are very diverse and growing in importance. They include inward flows of technology embodied in final consumer goods and services; collaboration along global value chains in creating, transferring and implementing innovation in local production for export; collaboration with foreign partners in scientific research or technological development; inward and outward flows of foreign direct investment by multinational enterprises; and the emigration, return and original immigration of all sorts of qualified scientific and technological human resources.

8. *Innovation is an increasingly distributed phenomenon.* The increasing complexity and the costs and risks involved in innovation processes enhance the value of collaboration between actors with complementary assets. In this environment, transaction-cost advantages can be an important factor for innovation performance. Social capital in the form of trust and networking can reduce transaction costs.

Governments need to do more than correct market failures

The idea that "market failure" leads to under-investment in research has been the principal rationale for state funding of R&D since the early 1960s. In the innovation systems perspective, the presence of bottlenecks or other failures that impede the operation of the innovation system can be crucial obstacles to growth and development. Given the broader conceptualisation of innovation outlined above, the scope for "failures" is considerable. Arnold (2004) has identified the following four types:

- *Capability failures*: Innovation capabilities may be lacking, for example, owing to managerial deficits, lack of technological understanding, learning ability or "absorptive capacity" to make use of externally generated technology.
- *Failures in institutions*: Failure to (re)configure institutions so that they work effectively with the innovation system.
- *Network failures*: These refer to problems in the interaction among actors in the innovation system and relate to phenomena such as weak links between system actors, missing complementary assets in clusters, etc.
- *Framework failures*: Deficiencies in regulatory frameworks, as well as other background conditions, such as the sophistication of demand, culture and social values can have a negative effect on innovation and economic performance.

A fifth type of "failure" identified in OECD innovation policy reviews – which the reviews have helped to address – is political failure. This refers to the relative lack of political attention to innovation, a situation that must often be overcome before some of the other failures can be effectively tackled.

These failures justify state interventions to ensure that innovation systems perform as a whole provided that the state is actually capable of reducing failure. State intervention can play a number of roles to support innovation and research. These include the development of absorptive capacity, particularly in firms; the funding of strategic and basic research; the fostering of networks and other system linkages; and the provision of strategic intelligence to inform not only policy makers, but also other actors in the innovation system. Some of these types of intervention are examined below.

Figure 6.2. The scope of innovation policy

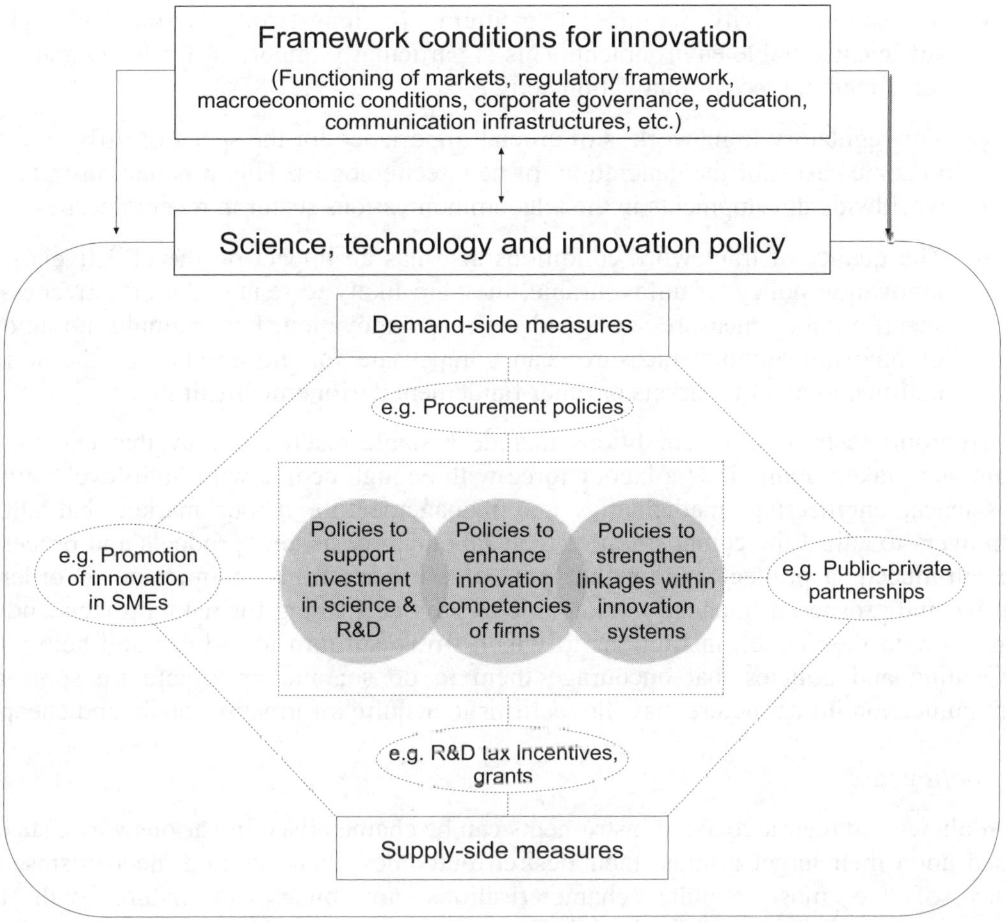

Source: The authors.

Policy mix and instruments

In the broadest sense, innovation policy should cover a wide range of macroeconomic and microeconomic policies and some aspects of social policy. Of course, many of these policies have objectives that are not directly related to innovation, yet it is important for the various areas of policy that affect innovation to reinforce one another so that innovation performance is not weakened by conflicts. It should be recognised that new innovation policy rationales do not invalidate instruments and goals associated with earlier rationales. Instead, the overall effect of broadening concepts and rationales is an added layer of policy instruments, which further increases policy complexity. A mix of intervention rationales across different departments and agencies is mirrored by a mix of policy tools and instruments. This mix makes it difficult to achieve co-ordination and coherence across agencies, something that is further discussed below.

Getting the framework conditions right

Science, technology and innovation (STI) policies that specifically seek to boost innovative activity are less effective if the surrounding framework conditions are not suitable. There are several reasons why framework conditions are a key prerequisite for strong innovation performance:

- Innovation activity requires a medium- or long-term horizon and thus a sufficiently stable environment. This is particularly important for R&D and more fundamental types of innovation activity.
- The regulatory framework is of crucial importance for the speed of diffusion, and in some cases for the generation, of new technologies. This was demonstrated by worldwide developments in the telecommunications sector in recent decades.
- The quality of framework conditions also has an impact on the effectiveness of innovation policy. If unfavourable, they are likely to reduce the effectiveness of specific policy measures designed to foster innovation. For example, no amount of innovation policy measures can compensate for the absence or the serious malfunctioning of markets or other fundamental economic institutions.

Appropriate framework conditions include a stable macroeconomy that encourages business to take a long view; a labour force with enough people with high-level training in science, engineering, mathematics and management; a labour market that allows employers to adjust the composition of their labour force as new products and processes are introduced; a product market that puts pressure on firms to innovate in order to survive and prosper; a financial market able to provide funding for risky projects and for firms new to the market; institutions to which firms can turn for advice and help when innovating and policies that encourage them to do so; and up-to-date transport and communication infrastructure that allows firms to acquire information easily and cheaply.

STI policy mix

Policies and their associated instruments can be characterised in various ways, such as based upon their target groups, their desired outcomes, their funding mechanisms, etc. Many of the most popular characterisations are binary in nature with two counterbalancing variants, and a key challenge is to strike the appropriate balance. These include:

- *Supply-side/demand-side policies*. Innovation policy has traditionally been more supply-side oriented, *i.e.* it has tended to focus upon the supply of new technical knowledge, owing in part to its R&D bias. Yet, demand is a major potential source of innovation, although its critical role as a driver of innovation has still to be recognised in government policy. This is beginning to change, and there is growing policy interest in orienting public demand towards innovative solutions and products. This has the potential to improve delivery of public policy and services and generate innovative dynamics and benefits from the associated spillovers. In particular, public procurement has emerged as a potentially powerful instrument to drive research and innovation by providing "lead markets" for new technologies. Firms are attracted to invest in research owing to the lower risks associated with the presence of an informed customer waiting for the resulting innovations. Moreover, innovations developed in this way may then be deployed in private-sector markets.

- *Direct/indirect support measures*. Historically, direct funding of business R&D and innovation activities by the public sector has been more popular than fiscal incentives, *e.g.* tax relief. However, by 2008, 21 OECD countries offered tax relief for business R&D, up from 12 in 1995 and 18 in 2004. Most have tended to make it more generous over the years (OECD, 2008b). The appeal of R&D tax credits stems from their non-discriminatory nature in terms of research and technology fields or industrial sectors. Ideally, the two types of measures should be used as complements to make the best use of their respective advantages.
- *Institutional/competitive funding*. Up to the 1990s, public funding for public-sector research performers in most OECD countries – including government laboratories and universities – tended to be predominantly non-competitive institutional funding. However, in a bid to raise research quality and, in some instances, to limit research spending to a few centres of excellence, governments increasingly turned to competitive modes of funding. For the most part, this has had its desired effect, providing powerful incentives for government laboratories and universities to improve their research effectiveness and efficiency. However, the use of competitive funding modes may have gone too far in some instances and jeopardised the preservation of capabilities needed for a healthy research environment.
- *Manufacturing/services focus*. Although much innovation policy claims to be "sector-neutral", programmes tend to reflect a manufacturing bias. Services have traditionally – but wrongly – been seen as less innovative than manufacturing and as playing only a supportive role in the innovation system. As a result, national innovation policies have paid scant attention to services, and service firms have not been active participants in government-sponsored innovation programmes (Tamura *et al.*, 2005).

Innovation system governance

The term "governance" is used in a variety of ways and has a variety of meanings (Pierre and Peters, 2000; Stoker, 1998). Here, STI governance is defined as the set of publicly defined institutional arrangements, incentive structures, etc., that determine how the various public and private actors involved in socio-economic development interact in allocating and managing resources devoted to innovation. STI governance involves developing capacities in three related areas: agenda setting, implementation and adaptive learning. These are represented in Figure 6.3 as three sides of a triangle divided into four areas. At the heart of the triangle are the structural factors that operate over space and time, framing and influencing the governance system's approach to agenda setting, implementation and adaptive learning. While these shaping factors both enable and constrain actions, they are not immutable. Instead, they are dynamically constructed through the interplay of endogenous and exogenous factors, including by actions on which they have an influence. The shaping factors and key tasks and functions outlined in the governance triangle can also be applied at lower "layers" of governance, including at the agency/department level and at the level of research- and innovation-performing actors (Figure 6.4).

Figure 6.3. The STI "Governance Triangle"

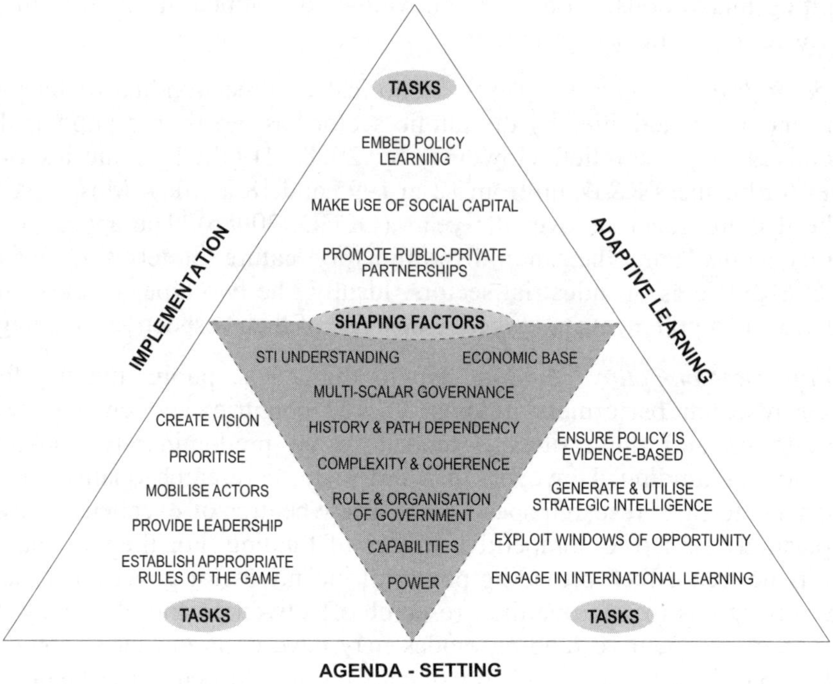

Source: The authors.

Figure 6.4. "Layers" of STI governance

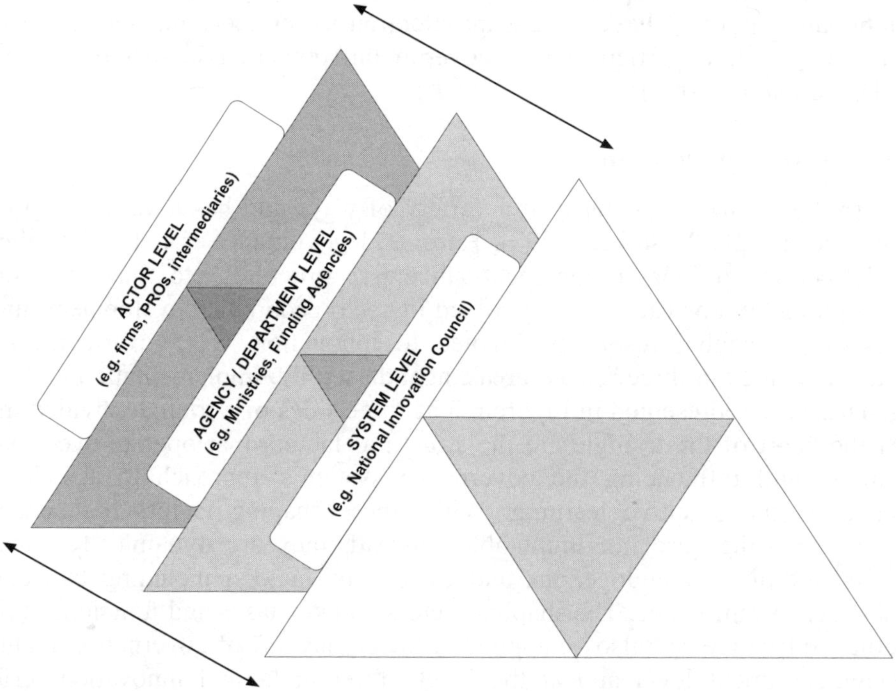

Source: The authors.

It is not possible to explore the STI governance triangle in any detail here. Instead, a few of its elements are discussed below to illustrate the changing nature of STI governance arrangements. These changes are driven by a broader concept of innovation; a recasting of institutional functions and relations; a spatial opening towards the global and sub-national/regional levels; and a strategic shift that pays greater attention in innovation policy to longer-term goals and developments. These drivers are not particularly new, but their dynamics and interplay are evolving in new ways in various settings.

The contours of a broadened concept of innovation have been outlined above and include the importance of non-R&D capabilities, the role of technological diffusion and sophisticated demand, the availability of appropriate skills, and the distributed nature of innovation. This broadening has created new rationales for innovation policy, with a wider variety of actors involved in innovation policy, and has led to the formulation of new types of policy instruments in the innovation policy mix.

Taken together, these changes have profound implications for the functions and relations of institutions in the innovation system. Innovation system actors are explicitly expected to play multiple roles, perhaps more than ever before. For example, universities have extended their traditional function of basic/strategic research to technology development and even further downstream to design, engineering and entrepreneurship. This means that similar functions may be undertaken in different types of organisations. For instance, part of the process of creating scientific and technological human capital for innovation systems is carried out by specialised education and training organisations, but a very important part is also carried out by business enterprises via large expenditures on education and training and by active management of the process of accumulating experience.

The challenge of multi-functionality may be most acute in government laboratories. Recent years have seen intensified discussion in many OECD countries about the role and mission of these laboratories in the innovation system. From a historical point of view, these research organisations were set up to compensate for market or systemic failures in countries' innovation systems. Accordingly, they have a wide range of functions: conducting "strategic", pre-competitive research; technological support to business; support of public policy; support in creating and establishing technical norms and standards; and constructing, operating and maintaining key facilities. This diversity has often contributed to "fuzziness" and a lack of clarity about the role of this sector and has placed many institutes under considerable pressure to justify not just their performance, but also their existence. In some instances, active involvement and co-financing by the business sector has become mandatory as a result of privatisation or through public-private partnerships, a development with major governance implications.

The extension of the role of universities has led to considerable convergence with the activities of government laboratories, and thus stronger competition but also greater collaboration. Competitive funding plays a much greater role in financing all public-sector research performers, but particularly government laboratories. Even core institutional funding is often based on terms and targets set out in performance agreements, a natural side effect of the turn towards new public management. Government laboratories are trying to adapt to these new environments, for example through the introduction of new business models based on open innovation. Many institutes are also taking steps to internationalise their operations by opening overseas branches and/or through cross-ownership arrangements (Hofer *et al.,* 2007).

Besides research performers, the actors involved include the research funding bodies, government ministries, and the various innovation intermediaries, and policies are formulated and implemented across a wide array of ministries and agencies. Policy coherence therefore requires horizontal co-ordination, yet co-ordination and coherence are difficult for governments, which are generally ill equipped to deal with cross-cutting policy issues such as innovation.

A further complicating factor concerns the "stretching" of innovation policy towards the global and sub-national/regional levels. National STI policy is increasingly framed in global terms, reflecting a growing sense of global identity, the global nature of many problems and issues, and the globalisation of markets and production. But many OECD countries have also been affected by a growing regionalism, with more control over policy and resources devolved to sub-national authorities. With a view to promoting local socio-economic development, regionalism has seen the emergence of innovation, and increasingly science, agendas in sub-national regions. These tend to focus upon nurturing regional clusters and capability building among knowledge producers and users, and regional policy makers are perhaps better placed than their national-level counterparts to understand local landscapes and tailor interventions accordingly. However, there is no neat division of labour between the levels, and overlaps and gaps are often evident. Moreover, multi-scalar arrangements are rarely well co-ordinated, despite their often obvious interdependence. This may constrain the effectiveness of policies at different levels and constitute a significant source of inertia.

Given this complex landscape, governments have adopted arrangements to improve coherence between different policies and programmes. This can be achieved at different points in the governance system, *e.g.* by implementation through joint programming. It can also be achieved in agenda-setting processes, for example through the establishment of high-level policy councils, the merger of institutions, and/or the formulation of strategic, long-term policies and visions that set the direction for priority setting. Such measures are part of a strategic shift which gives more attention to longer-term goals and developments in innovation policy.

STI policy governance in many OECD countries has become more strategic in the sense that it is increasingly informed by explicit expectations of outputs and outcomes, which are themselves framed in a broader framework of appropriateness. Appropriateness is determined not only by assessing current needs and opportunities, but also by anticipating what these might be in the medium term. Such anticipation is not passive; rather, it takes an active view by assuming that future innovation can be shaped for the better through deliberate, well-informed decisions today. In segmented and fragmented institutional landscapes, a strategic vision can contribute to the steering of a governance system, by mobilising and aligning system actors. It can also be a useful means for coupling S&T activities (and what they promise) to current and emerging socio-economic agendas. In some countries, such visions have become an important means of seeking to achieve system co-ordination and to mobilise actors. Broad participation in the vision-building process can help induce strong commitment to a vision and to its paths to action.

Prioritisation has been an important component of this strategic shift. Traditionally, the prioritisation of areas of science has often been implicit, with a large dose of path dependency and lock-in shaping budget portfolios. In some respects, this has been due to the autonomy of the scientific community and its reliance upon peer review for project selection. While this certainly has strong merits, its conservatism is a well-known weakness, as is its unsuitability for selecting among areas of science. More recently, many OECD countries have adopted more explicit prioritisation exercises to guide the selection of research areas for special attention. Such exercises are often justified in terms of the need to identify and exploit emerging opportunities and to consolidate research efforts so as to carry out a critical mass of activities. Yet, several tensions tend to appear: the desire to specialise *versus* the desire to diversify; a concern with harnessing high-technology sectors *versus* a concern with meeting the innovation needs of established industries; adopting a science-driven *versus* a market-/society-driven approach; and a focus upon meeting short-term needs *versus* a focus upon long-term opportunities.

A further component of the strategic shift in innovation policy concerns the increased use of evaluation. The demand for evaluation and other information tools for policy making has increased owing to organisational changes, new management practices and various countries' commitment to step up investment in science, technology and innovation, in spite of tight budgetary constraints. Evaluation is a necessary ingredient of evidence-based policy making and modern management, although countries differ considerably with respect to the extent to which their policy making is evidence-based. This has a significant impact on the demand for, and thus on the level of, development of domestic strategic intelligence.

An exemplary triad: Chile – Norway – Switzerland

This section examines further some of the issues raised above, drawing on the OECD reviews of innovation policy of three countries: Chile, Norway and Switzerland (OECD, 2007d, 2008a and 2006). This triad and some pairs (notably Norway – Switzerland and Chile – Norway) may have more in common than is apparent at first glance. All three are small open economies. Norway and Switzerland are high-income countries, with a significant record in science and technology. Chile and Norway are, to a considerable extent, resource-based economies. These similarities can be extended to other dimensions, some of which will be discussed in the following sections.

At the same time, these countries are sufficiently varied to show that new global developments in the area of science, technology and innovation – such as the internationalisation of R&D or the changing nature of market-driven R&D – lead to different challenges and opportunities, depending on countries' specific conditions. For example, the effect of the emergence of new players in global R&D is different for countries with a strong R&D base and for emerging actors in R&D. In addition, differences in initial conditions (state of development, economic and innovation specialisation and performance) and political and cultural institutions and characteristics (as reflected in the role of government, accumulated social capital, etc.) result in different sets of constraints on policy responses to the perceived challenges and opportunities.

This section therefore begins with an exploration of the initial conditions in these three countries, in terms of their economic performance and processes of structural change, their innovation performance, and the role of public policy. It then considers the nature of STI policy responses to three challenges: the need for economic diversification, the ability to maximise the benefits from the internationalisation of R&D, and the means for improving the governance of the innovation system. The discussion shows why and how similarities between countries but also country-specific conditions and constraints should be taken into account in learning from international experience.

Initial conditions

Economic performance and structural change

Norway and Switzerland are both small economies closely interlinked – in different ways – with the European Union and the global economy more generally. They are among the countries with the highest income per capita in the world. Chile, in contrast, is a high medium-income country which – following a sustained period of successful institution building and economic development – is now a candidate for accession to the OECD. In other respects, Chile and Norway seem to have more in common. As major world exporters of copper and other raw materials and oil and gas, respectively, their economies are resource-based to a significant extent. In some areas, such as salmon farming, the two countries are competing locations supplying world markets, linked through related FDI and knowledge flows (*e.g.* Norwegian engagement in Chilean aquaculture).

Medium-term economic performance has been strong in Chile and Norway, but rather weak in Switzerland. From the early 1990s, Switzerland has had more than a decade of slow growth of productivity and GDP per capita which rebounded for only a short interval before the current economic crisis. In contrast, Norway's and Chile's performance was robust over an extended period of time, as both countries have benefited from strong demand for raw materials. However, their economic success has not been limited to the exploitation of natural resources. Norway's performance in terms of per capita GDP growth has been consistently good for a long time. The growth of the offshore hydrocarbons sector has been a major factor, but even excluding this, per capita GDP in purchasing power parity (PPP) in mainland Norway is comparable to that of Finland and higher than in the major EU countries. For more than three decades from 1970, labour productivity grew at an average rate of about 3% a year, the same range as Finland and Japan. Norway has had exceptionally high productivity growth in the private services sector. Such high levels of productivity growth cannot have been maintained without significant innovation. More generally, Nordic labour market organisation (known as "flexicurity" in Denmark) has made it possible to introduce new processes and products without too much disruption and with the acceptance of related change in the workplace.[6] Chile has recorded high economic performance over the two decades preceding the current economic crisis, underpinned by economic reforms and modern and stable institutions; it has the best performance in the Latin American region.

All three economies have been undergoing significant structural change, most strikingly in Norway, with the sudden emergence of a strong hydrocarbon extraction sector following the discovery of North Sea oil and gas reserves. This massive shock has overshadowed significant other shifts in the Norwegian economy. Since the discovery of oil and gas reserves, Norway has been successful in building new competences and comparative advantage not just in but also around the oil and gas extraction industry, in equipment, engineering, other related services, etc. Within the hydrocarbon sector, Norway successfully acquired knowledge from foreign partners participating in the

6. There are also downsides to this model. Among others, the Nordic organisation of the labour market and the welfare state may tend to reduce incentives to work. For example, Norway has a very high incidence of sickness absence and disability pensioners. Also, corporatist arrangements tend to be biased towards vested interests and incumbents, as opposed to the new entrants which are of key importance for creating an innovative and dynamic environment. It is therefore important to maintain effective competition mechanisms to counter this tendency.

exploitation of Norwegian oil and gas fields and established a large national company, Statoil, as a major player. At the same time, Norway has been active in other areas, *e.g.* by further developing its strong and varied services sector, extending from marine transport to telecommunications. Norway's economic development has been facilitated by a social compromise, with the labour unions agreeing to high rates of technology deployment, rationalisation and productivity increases to allow the financing of a welfare state.

Chile has also benefited from the boom in raw material prices of the recent past and has not been strongly affected by increased competition from emerging economies. Its export structure is largely complementary to that of China and other emerging countries. In addition, Chile has diversified its economy, notably by developing export-oriented businesses in the agro-food sector, including the wine, fresh fruit and salmon industries, but also services such as air cargo transport. However, Chile's economy still remains relatively undiversified and strongly dependent on raw materials. Manufacturing has remained weak, and product variety and intra-firm trade low.

Structural change in the Swiss economy may appear less spectacular in some respects. Switzerland has not shifted, for example, towards the production of ICT-related goods and services to the same extent as some comparator countries, but there has nevertheless been significant structural change, including within industries, *e.g.* in the composition of the chemical industry. The reinvention of the Swiss watch industry is a spectacular example of how innovation can provide for revival while redefining a traditional comparative advantage.

Innovation performance

Norway and Switzerland both have mature innovation systems, although they differ significantly in their composition and relative performance, owing to different paths of economic development and patterns of specialisation. While Swiss innovation performance is high in international comparison in almost every respect, Norway performs rather poorly in a number of traditional S&T indicators in comparison with high-income countries. However, these may be less well adapted to capture certain aspects of innovation that are more important for Norway than for other countries (see below). Chile has an innovation system that in many respects is still maturing, and innovation performance still lags compared to most OECD countries.

Switzerland has strong industrial R&D and a strong record in scientific research, notably in its famous federal institutes of technology, ETH Zurich as well as EPF Lausanne. The scale and scope of Switzerland's market-led industrial innovation is extraordinary for a country of its size, spanning pharmaceuticals, food, instruments, machinery and equipment, and financial services. A mix of large Swiss-based multinational enterprises (MNEs) (such as Novartis, Roche, Nestlé, ABB, UBS, Credit Suisse and others) and a strong segment of highly innovative SMEs allow for economies of scale and scope in R&D and a high degree of flexibility. While the Swiss education system faces specific challenges, it provides well-skilled workers at all levels, including through professional training, and thus supports broad-based innovation. Switzerland is one of the most internationalised innovation systems in the world. However, global competition is becoming more of a challenge.

The three countries differ markedly with respect to the role of public research organisations (Figure 6.5). While Switzerland is characterised by very strong market-led industrial innovation and strong ties between industry and its excellent universities, relatively little public research is performed outside the higher education sector. In contrast, Norway's economic structure (at least outside the petroleum sector) is not conducive to large-scale R&D (Figure 6.6). Universities and a strong public research sector compensate for this weakness. Public research organisations have so far contributed little to the performance of the Chilean innovation system.

Figure 6.5. Archetypes of innovation systems

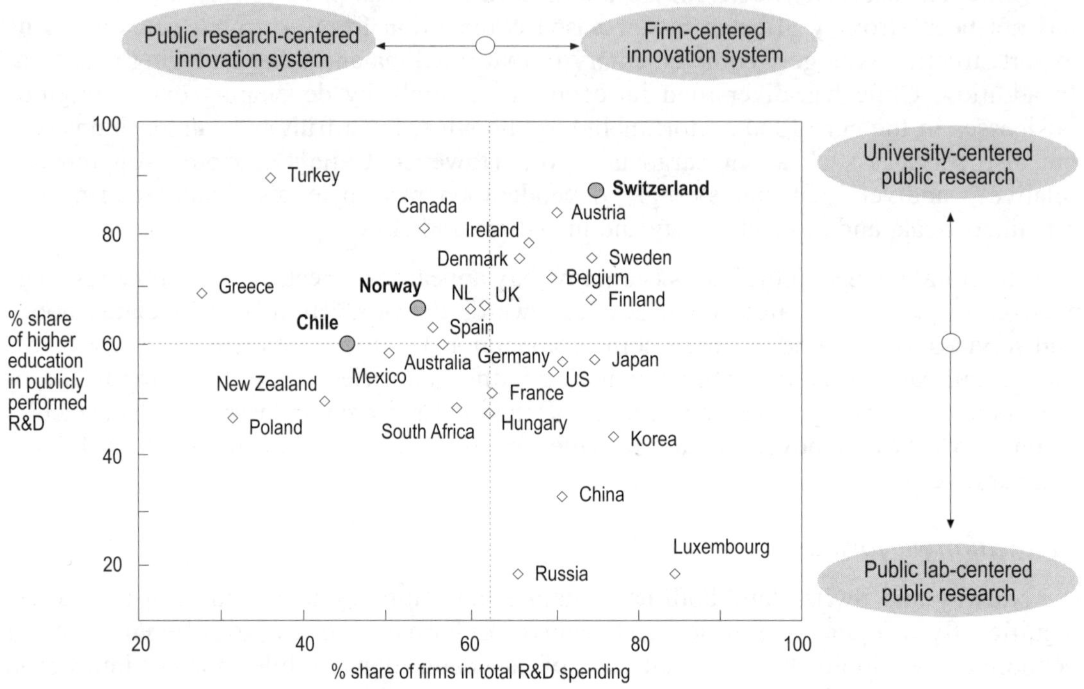

Source: The authors.

In recent times, the "Norwegian puzzle" – Norway underperforms against conventional S&T and innovation indicators despite its persistently high economic performance – has received a fair degree of attention. However, closer examination shows that the true level of innovation activities may be significantly higher than the conventional indicators suggest. Conventional innovation indicators fail to capture important contributors to innovation and economic performance, such as education and social capital. Relative underperformance is concentrated in some parts of manufacturing. Low business-sector R&D expenditure can be largely "explained" by the industrial structure's smaller share of R&D-intensive industries than the OECD average. Innovation that is not based on R&D, such as innovation in firms' organisation and business models, which is difficult to capture via available quantitative indicators, seems to underlie the exceptional productivity performance of the private services sector, which would otherwise be hard to explain. In addition, many innovation activities in or around the hydrocarbon sector may not be fully accounted for. For example, important aspects of large-scale engineering-based activities associated with major undersea oil and gas extraction projects, which are technically demanding and highly knowledge-intensive, appear only partly in R&D expenditure figures.

Figure 6.6. Trends in R&D expenditures and R&D intensity

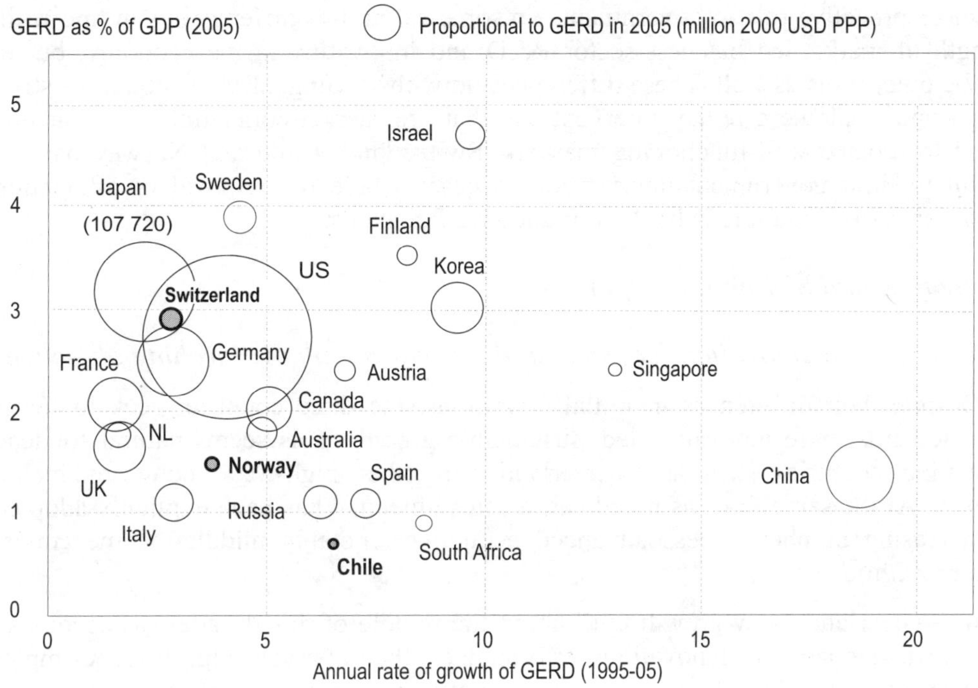

Source: The authors.

Chile's innovation performance is relatively weak. Many elements of a modern innovation system are missing or at least less well developed than in most OECD countries, and these shortfalls are acutely felt when the process of catching up with high-income countries requires more innovation. A legacy of inequality and, despite various efforts, shortfalls in the education system also hold back the shift towards more innovation-driven development. A further obstacle is the weak links within the innovation system, notably between business and academia, which is common to many low- and middle-income countries. Chile has not so far benefited significantly from the internationalisation of R&D. Notably, inward and outward FDI in industrial R&D is negligible; this is not surprising given the prevailing structure of the Chilean economy. However, Chile has become an important location for international scientific research (observatories).

The role of public policy

Despite the differences among them, all three countries understand that basic framework conditions for innovation (international openness, competition, the regulatory regime, intellectual property rights, etc.) are of key importance for a country's innovation performance and the success of policy measures targeted to foster innovation. As regards the latter, the approaches differ: Switzerland has a deeply rooted reluctance to use direct government support for business sector R&D (although subsidies are readily used in other areas of economic activity). Support is concentrated on funding knowledge and technology transfer to business firms, in addition to creating favourable framework conditions for innovation and continued adaptation of the education system to changing needs. While the prudent Swiss approach has its merits, it limits the choice of instruments to an extent that may not always be optimal (notably for countering problems faced by the SME sector). In comparison, Norway and Chile implicitly or explicitly assign a greater and more proactive role to government. To some extent this difference is in line with the strength of market-led business-sector R&D and innovation in the economy, but there may be other roots as well. These differences notwithstanding, all three countries strive to design and implement policy interventions that are market-compatible, *i.e.* that do not intend to replace well-functioning markets. Switzerland, Chile and Norway have been careful to limit government intervention to cases where the removal or alleviation of market or systemic failure is likely to induce welfare gains.

Challenges and STI policy responses

Achieving more innovation-led, sustainable growth – an overarching objective

Despite their differences in initial conditions, the three countries seek to foster or move towards more innovation-led, sustainable growth. This seems natural for leading innovating countries such as Switzerland and other high-wage countries including Norway. At the same time, as noted earlier, this objective has been explicitly adopted by an increasing number of less advanced economies, notably middle-income countries, including Chile.

In Switzerland, slow growth until about the middle of this decade was accompanied by relative stagnation in innovation performance, albeit at a very high level. Comparator countries and a number of newcomers have moved faster in terms of investment in R&D and innovation during this period. There were concerns that slow growth of investment in R&D (Figure 6.6) may have persistent negative effects, eroding Switzerland's position as a top performer in the longer term. Under these conditions it seemed appropriate to promote investment in innovation. Towards the middle of the decade, R&D expenditure indeed picked up, supported by a renewed commitment by government. For Norway, the key strategic task for the long term is to maintain high, sustainable growth even after oil and gas production has peaked. Any foreseeable restructuring of the Norwegian economy compatible with this goal will entail a shift towards other knowledge-based activities and STI policy will have to contribute to this process. For Chile the agenda is twofold: to continue the successful process of catching up and to foster diversification of the economy in the longer term. The achievement of these objectives requires boosting innovation capabilities.

All three countries share a common understanding of the role of knowledge, notably technology diffusion, and understand that the latter is not a substitute for generating knowledge. Chile was successful in its efforts to extend its comparative advantages and acquire the complementary skills required to realise this strategy. Fundación Chile, a private non-profit organisation for the promotion of innovation, has been instrumental in these attempts.

Economic diversification

Creating a positive feedback loop between a high level of adaptability and innovation allows an economy to diversify and to move up the value chain. The need for diversification is perhaps most evident for resource-based economies. Natural resource-based countries need to manage a number of tasks – which require adequate governance mechanisms – to avoid being afflicted by the "resource curse":

- Stabilisation by smoothing disposable revenues from raw materials (such as oil and gas, metals), while avoiding the risk of an adverse impact on other sectors of the economy ("Dutch disease").
- Facilitating and fostering structural change towards new economic activities, preparing for a future of declining revenue from natural resources.
- Providing adequate incentives for long-term change towards more innovation-driven growth. This can be a difficult task when rents have become a major source of income for major parts of the economy.

Norway has been an excellent example of how to manage the large revenues from its petroleum sector prudently and with a long-term perspective. It has combined good management of its oil and gas revenues and seized opportunities for knowledge-intensive activities in and around this sector, using it as a platform for developing technological capabilities which have led to the development of marketable goods and services. Chile also actively manages its revenues and allocates part to long-term investment in innovation (see below). Chile has recently embarked on developing a cluster initiative which is expected to strengthen and broaden the scope of its competitive advantage.

Opportunities also arise from concerted efforts in areas of strength in science and the economy. For example, more than many other countries, Norway has nurtured strong social support for action to solve problems of global relevance, such as sustainable development, and related issues, such as clean energy. Large-scale programmes to address such topics can be very efficient focusing devices for public support to innovation and can have widespread impact on industries and fields of science and technology.

Switzerland, as an innovation-based economy with few natural resources – aside from an impressive landscape – stresses excellent framework conditions, education and knowledge, and technology transfer, mainly from higher education institutions and industry. Cluster approaches have been adopted at the regional level, and there has been a fair degree of priority setting in science to pursue critical mass and excellence.

All countries show an interest in fostering innovation in services in order to achieve more innovation-led growth. Chile's cluster initiative includes services. Switzerland has initiatives in tourism, and its universities contribute to providing a knowledge base for financial services. Norway shows special interest in innovation in its comparatively large and high-quality public sector. Yet the promotion of innovation in services is generally less well understood and sometimes in the shadow of more spectacular initiatives in areas of "high technology". This leaves much scope to derive innovative, customised policies in countries at different levels of economic development.

Maximising the benefits from the internationalisation of R&D

Switzerland is one of the world's most internationalised innovation systems. International recruitment of highly qualified personnel is indispensable for operating Switzerland's top higher education institutions and corporate laboratories at their current level of performance. At the same time the presence of excellent education and R&D facilities, as well as comparatively high salaries for S&T personnel, are major advantages in the international competition for talent. Switzerland has indeed succeeded so far in attracting a sufficient number of students, researchers and other human resources for science and technology from abroad. Large Swiss-based MNEs are global in their orientation and invest heavily in R&D at home and even more abroad; many innovative

SMEs operate at the international level. Switzerland has been very active in maintaining these advantages. Policy has provided favourable framework conditions (including through a dense web of agreements with the European Union, concerning freedom of movement and full participation in European research programmes, among others). University recruitment policies, *e.g.* at ETH Zurich, can be considered best practice. Challenges remain, in particular for the SME sector which needs to stay abreast of international developments and to strengthen its integration in global innovation networks. In addition, since the establishment of the IBM research laboratory in Zurich (Rüschlikon) in the 1950s, competition for the location of research facilities has become much fiercer, with a number of new competitors worldwide. Swiss-based MNEs can also be expected to invest massively in R&D labs located in centres of knowledge production around the world (such as Novartis in the Boston area).

Norway's record in the internationalisation of STI is somewhat mixed. Science is well linked internationally to the European Research Area and beyond. Universities attract students and researchers from abroad. Norwegian MNEs are well positioned in global innovation networks. At the same time a rather large segment of companies is insufficiently linked to world markets and networks, and the contribution of inward FDI to the national innovation system is rather low.

Chile is in a much less favourable position and therefore has to observe tighter constraints in devising policies. One drawback is the lack of R&D-intensive Chilean enterprises, both large and small. Chile has made efforts to establish itself as a location for international research facilities but linkages to the local economy seem limited. It has quite successfully exploited certain channels in order to acquire knowledge and technology from abroad, for example through students studying at universities abroad. But a more strategic use of scholarships could help increase the contribution of returnees to economic and social objectives. The development of innovative clusters should provide new opportunities to develop international linkages in innovation and R&D.

Improving the governance of the innovation system

As discussed earlier, boosting innovation requires efficient governance mechanisms in the innovation system. Continuously improving these mechanisms is a challenge for almost all countries. The formulation of comprehensive ("whole-of-government") strategies and co-ordinated implementation of innovation policies has become a priority in many. Again, initial conditions differ widely and so do the approaches countries take in designing and operating relevant institutions. For example, the co-ordination of science and innovation policy across government is institutionalised and implemented in different ways. Science and innovation policy councils have become a preferred institutional arrangement in such co-ordination efforts. They are often modelled after the Finnish example, but their actual role and composition differ widely in practice.

In order to foster the transition towards a more innovation-based development path, Chile has put in place new governance structures and funding mechanisms for innovation, and has made considerable progress in a relatively short time. A major institutional innovation was the establishment of the Innovation for Competitiveness Fund (set up along with a levy on mining revenues). Second, Chile has established an advisory National Innovation Council for Competitiveness which has been successful in developing a national STI strategy and deploying a cluster initiative. The Council was recently complemented by an Inter-ministerial Committee for Innovation, the Council's counterpart in the executive branch.

Chile has taken a relatively formal approach to institution building. This is partly due to the legalistic traditions of the country, but may be more closely related to the fact that an appropriate institutional setup for the envisaged new phase of developing the Chilean innovation system was largely missing. In comparison, Norway has rather informal co-ordination mechanisms and lacks an institutionalised co-ordination arena. While some changes in the governance of the Norwegian innovation system might facilitate prioritisation and efficient delivery of co-ordinated policies, the perceived lack of institutionalised co-ordination seems at least partly compensated for by relatively strong informal co-operation among ministries involved in science, technology and innovation policy.

A high level of trust seems to facilitate co-operation and informal co-ordination of innovation policy in Norway. Chile, in contrast, shows a rather low level of mutual trust among business firms on the one hand and between the latter and public research organisations on the other. This is one of the obstacles to lifting the persistently low level of co-operation between industry and academia. The development of trust and networks is a long-term endeavour. However, science, technology and innovation policy is, in many countries, less controversial than other policy areas, and may therefore offer opportunities to create coalitions that could be instrumental for building trust.

The advisory Swiss Science and Technology Council is largely centred on science and higher education, although its mandate is broad. Unlike comparable councils in other countries it is comprised exclusively of representatives of academia.

The composition of STI policy councils should reflect the strategic tasks to be fulfilled within the respective national innovation system. This includes ensuring an adequate degree of openness to the outside world (*e.g.* through the nomination of members operating beyond national boundaries). Surprisingly, this is often not the case in practice.

The increasing importance of the regional dimension of STI policy, which is nurtured by regionalisation movements, has implications for governance. Among the countries discussed here, Chile is a centralised country, and Switzerland has a strongly federal system and a more even distribution of research and innovation capabilities across the country. There is evidence that both excessive centralisation and federalism come at a cost. Norway – sometimes compared to Switzerland, but in some respects topographically more similar to Chile – assigns an important role to its regions, including in STI policy. There is a trade-off between regionalisation and concentrating sufficient resources to achieve economies of scale. Internationally, there appears to be some tendency towards convergence, as countries need to counteract fragmentation but acknowledge the benefits of empowering regions to design and deliver policies that are best dealt with at that level (*e.g.* regional cluster policies).

Conclusions

A striking similarity among the three countries discussed above is an emerging consensus on the critical importance of innovation for economic and social development. Yet, innovation systems often have "failures" – such as innovation capability failures, particularly among SMEs, and network failures, especially between industry and academia – which justify similar sorts of policy interventions, *e.g.* cluster policies. There is therefore a growing convergence in policy, driven by common policy rationales and an increasingly codified policy instrument toolbox, at least among high- and middle-income countries. Whether such convergence extends to low-income economies is debatable, with the real possibility of a widening innovation divide emerging.

At the same time, there are a number of differences in policy responses and governance arrangements among the countries covered by the OECD country reviews, including among the three small open economies examined here. Some of these differences may be due to a certain degree of lock-in, with international good practices yet to be adopted in place of more established (sub-optimal) practices. But it is perhaps more likely that variations in policy and governance can be explained by differences in initial contextual conditions. One example is the difference in governance arrangements between Chile and Norway. The Chilean government has adopted a formal institutional approach as the institutional and cultural conditions for more organic, Norwegian-style arrangements are largely absent. Such conditions cannot be changed quickly or easily, and sometimes this should not be attempted. Instead, policy and governance arrangements need to be designed to work within given constraints and conditions. The policy instrument toolbox remains an important source of ideas and good practices, even under varying conditions. This gives considerable scope for international learning around good practices. But the appropriate mix of policies and instruments will vary and will shift over time as innovation systems evolve. This not only means that "one-size-fits-all" policy mixes are unsuitable, but also that notions of an optimal policy mix are inappropriate as well. The OECD country review process recognises the evolutionary nature of innovation processes and policies and promotes informed, contextualised learning from international experience rather than the "mechanical" international diffusion of national "best" practices.

References

Abramovitz, M. (1986), "Catching Up, Forging Ahead, and Falling Behind", *The Journal of Economic History*, Vol. 46, No. 2, pp. 385-406.

Arnold, E. (2004), "Evaluating Research and Innovation Policy: A Systems World Needs Systems Evaluations," *Research Evaluation*, Vol. 13, No. 1, pp. 3-17.

Coe, D., E. Helpman and A. Hoffmaister (2008), "International R&D Spillovers and Institutions", IMF Working Paper No. 08/104, Washington, DC.

Cohen, W.M. and D.A. Levinthal (1989), "Innovation and Learning: The Two Faces of R&D", *Economic Journal*, Vol. 99, September, pp. 569-596.

Eaton, J. and S. Kortum (1996), "Trade in Ideas: Patenting and Productivity in the OECD", *Journal of International Economics*, Vol. 40, No. 3-4, pp. 251-278.

Eaton, J. and S. Kortum (1999), "International Technology Diffusion: Theory and Measurement", *International Economic Review*, Vol. 40, No. 3, pp. 537-570.

Fagerberg, J. (1994), "Technology and International Differences in Growth Rates", *Journal of Economic Literature*, Vol. 32, September, pp. 1147-1175.

Gerschenkron, A. (1962), *Economic Backwardness in Historical Perspective*, Belknap Press, Cambridge, MA.

Gong, G. and W. Keller (2003), "Convergence and Polarization in Global Income Levels: A Review of Recent Results on the Role of International Technology Diffusion", *Research Policy*, Vol. 32, No. 6, pp. 1055-1079.

Hofer, R., W. Polt, B. Nones, E. Jantscher and H. Wiedenhofer (2007), *Europäischer Benchmark der Entwicklungstrends ausseruniversitärer Forschungsorganisationen*, Wien-Graz.

Hollenstein, H. and G. Hutschenreiter (2001), "Innovation, Productivity and Economic Growth – A Survey", in Aiginger, K. *et al.*, *Innovation and Productivity of European Manufacturing*, Austrian Institute of Economic Research, Vienna.

OECD (1998), *The OECD Jobs Strategy: Technology, Productivity and Job Creation. Best Policy Practices*, OECD, Paris.

OECD (2005), *The Measurement of Scientific and Technological Activities – Oslo Manual: Guidelines for Collecting and Interpreting Innovation Data*, 3rd edition, OECD, Paris.

OECD (2006), *Country Review of Innovation Policy: Switzerland*, OECD, Paris.

OECD (2007a), *Country Review of Innovation Policy: Luxembourg*, OECD, Paris.

OECD (2007b), *Country Review of Innovation Policy: New Zealand*, OECD, Paris.

OECD (2007c), *Country Review of Innovation Policy: South Africa*, OECD, Paris.

OECD (2007d), *Country Review of Innovation Policy: Chile*, OECD, Paris.

OECD (2008a), *Country Review of Innovation Policy: Norway*, OECD, Paris.

OECD (2008b), *Science, Technology and Industry Outlook*, Chapter 5, OECD, Paris.

OECD (2008c), *Country Review of Innovation Policy: China*, OECD, Paris.

OECD (2008d), *Country Review of Innovation Policy: Hungary*, OECD, Paris.

OECD (2009a), *Country Review of Innovation Policy: Korea*, OECD, Paris, *forthcoming.*

OECD (2009b), *Country Review of Innovation Policy: Mexico*, OECD, Paris, *forthcoming.*

OECD (forthcoming), *Country Review of Innovation Policy: Greece*, OECD, Paris.

Pierre, J. and B.G. Peters (2000), *Governance, Politics and the State*, Macmillan, London.

Stoker, G. (1998) "Governance as Theory: Five Propositions", *International Social Science Journal*, Vol. 50, 155, pp. 17-28.

Tamura, S., J. Sheehan, C. Martinez and S. Kergroach (2005), "Promoting Innovation in Services", *Enhancing the Performance of the Services Sector*, OECD, Paris, pp. 113-177.

World Bank (2007), "Mexico's Competitiveness: Reaching Its Potential", Report No. 35388-MX, The World Bank, Washington, DC.

Chapter 7

Different innovation strategies, different results: Brazil, Russia, India, China and Korea (the BRICKs)

Carl Dahlman[1]

A key aspect of the emergence of the BRICKs has to do with the use of smart innovation policies to tap existing global knowledge and adapt it to local conditions through investments in R&D and education, reliance on reverse engineering and the diaspora as well as ICT technologies. In each country, the broader economic and institutional regime has played an important role. This chapter discusses country-specific experiences to highlight that a "one-size-fits-all" technology policy cannot facilitate innovation in developing countries.

1. Carl Dahlman is Luce Professor of International Relations and Information Technology at the Edmund A. Walsh School of Foreign Service, Georgetown University.

Introduction

There is increasing interest in innovation and growth around the world. This stems from the realisation that we are in a period of rapid creation and dissemination of knowledge and that innovation is critical for competitiveness and growth. Furthermore, there is an increasing understanding that the most appropriate innovation strategies may differ for countries already at the world frontier and for catch-up countries.

Innovation in developing countries should not be defined just in terms of shifting global frontier technology, but in terms of what is new to the country, the sector or the firm. An innovation strategy should include policies and mechanisms that affect a country's ability to draw on global knowledge as well as domestic R&D effort. Such a strategy will be affected by policies that include trade, foreign direct investment (FDI), technology transfer, domestic research and development (R&D), human capital and education, as well as the broader economic incentive and institutional regime more generally (what the OECD calls the framework conditions).

This chapter aims to contribute to the understanding of the relation between innovation and growth in developing countries by analysing the innovation strategies of the BRICKs[2] (Brazil, the Russian Federation, India, China, and Korea). Brazil is an upper-middle-income country which industrialised rapidly in the 1960s and 1970s under protectionist policies. It stayed at that level in the 1980s and 1990s, and remains primarily a commodity exporter with some islands of technological excellence, but a poor record of growth (Rodriguez *et al.*, 2008). China is a lower-middle-income manufacturing powerhouse that has moved towards a market economy. It has leveraged its rapid growth by integrating the global economy and efficiently tapping global knowledge. It is now investing heavily in its own innovation capability (Dahlman and Aubert, 2001; OECD, 2008b). India is a low-income country which was quite autarkic for the 40 years following independence in 1947. Its recent explosion of growth is being powered by knowledge-intensive service exports facilitated by information technology (Bosworth and Maartens, 2008; Dutz, 2007; Singh, 2007). Korea is a recent entrant to high-income status which relied on a strong export orientation and large domestic conglomerates (the chaebol). It moved rapidly from a labour-intensive manufacturing exporter to an exporter of medium- and high-technology products, with little reliance on FDI (see Chapter 3 and Suh and Chen, 2007). Russia is a former technology and military superpower whose economy contracted through a wrenching transition towards a market economy. It finally began to grow on the basis of petroleum-related exports and regained middle-income country status, but its manufacturing sector has been contracting as a share of GDP (Goldberg and Desai, 2008).

These five countries therefore provide rich material in terms of different innovation and development strategies. A comparison of the varied experience of these five countries shows that innovation strategies differ across countries depending on their specific conditions. It illustrates the complexity of the relationship between innovation strategies and performance. Their experiences also show that the most appropriate focus of innovation strategy, even among developing countries, will vary over time.

2. The term BRICs was coined in a Goldman Sachs report in 2003 in an analysis of the four largest emerging market economies. This chapter has added Korea, and thus the acronym BRICKs, which represents a very specific innovation strategy. Because of its excellent performance it rapidly moved from a very poor low-income country in the 1960s to a high-income economy by 1996 when it joined the OECD.

The following section develops a brief conceptual framework for analysing innovation in developing countries. This includes the need to look at the broader economic and institutional regime as well as education and skills. The discussion then turns to the growth and performance of the BRICKs and their key economic parameters. The following sections summarise some of the countries' broader economic and institutional characteristics and their education and training strategies before analysing the extent to which they have used various modes to tap global knowledge. Next, their domestic R&D inputs and outputs are reviewed, followed by an examination of the dissemination of knowledge. A final section summarises the lessons that can be drawn from these comparisons and some implications for other developing countries.

Innovation in the context of developing countries

The rate of creation and dissemination of knowledge has accelerated. This can be seen, for example, in the increase in scientific and technical papers and patenting activity, the proliferation of new products and processes, the shortening of product life cycles, and rapid advances in information and communications technologies (ICTs). In addition, there has been a trend towards the globalisation of knowledge. The share of non-OECD economies in total world R&D increased from 11.75% in 1996 to 18.4% in 2005 (OECD, 2008a) largely as the result of increased R&D in the BRICs. In addition, multinational enterprises (MNEs) are setting up foreign R&D centres and thus undertake an increasing share of R&D abroad. There also has been an increase in joint research articles and patents involving researchers from different countries (OECD, 2007).

Because the global stock of knowledge is increasing rapidly, innovation in the context of developing countries should be considered not just in terms of the creation of knowledge that is new to the world, but also in terms of products, processes, services or forms of organisation[3] that are new to local practice, not necessarily to global practice. Furthermore innovation can be new to the country, new to the sector, or at a more micro level, new to the firm. Therefore it is useful to distinguish three sources of innovation. One is the acquisition of technology that already exists abroad. A second is the domestic creation of relevant new knowledge.[4] The third is the dissemination and effective use of this new knowledge throughout the economy, whether it has been created locally or imported from abroad.

3. An example of an innovation in organisation which significantly reduced costs is the elimination of middlemen by various computer manufacturers, whose customers directly specify the features they want and the computers are then mailed directly to them. Another is the linking of sales information directly to orders from suppliers and the integration of production and distribution chains by mass retailers such as Wal-Mart. Both result in considerable savings in inventory and middlamen costs which are passed on to the consumer. An example of a non-technological process innovation is the use of containerisation in shipping which has drastically reduced transport costs by cutting the time that ships, airplanes or trucks need to wait to be loaded and unloaded.

4. Technology, whether domestically created or imported from abroad, often has to be adapted to local conditions. This is particularly true in agriculture, where new technologies such as hybrid seeds are very sensitive to specific local conditions. Thus, further research and experimentation is often required to adapt them to specific temperature, soil and water conditions as well as local pests. To a lesser extent, industrial technologies also have to be adapted to local conditions: availability of raw materials, special characteristics or other local idiosyncrasies such as sources of power or local standards, and climate or health conditions.

Figure 7.1. Scheme of the innovation system in a developing country

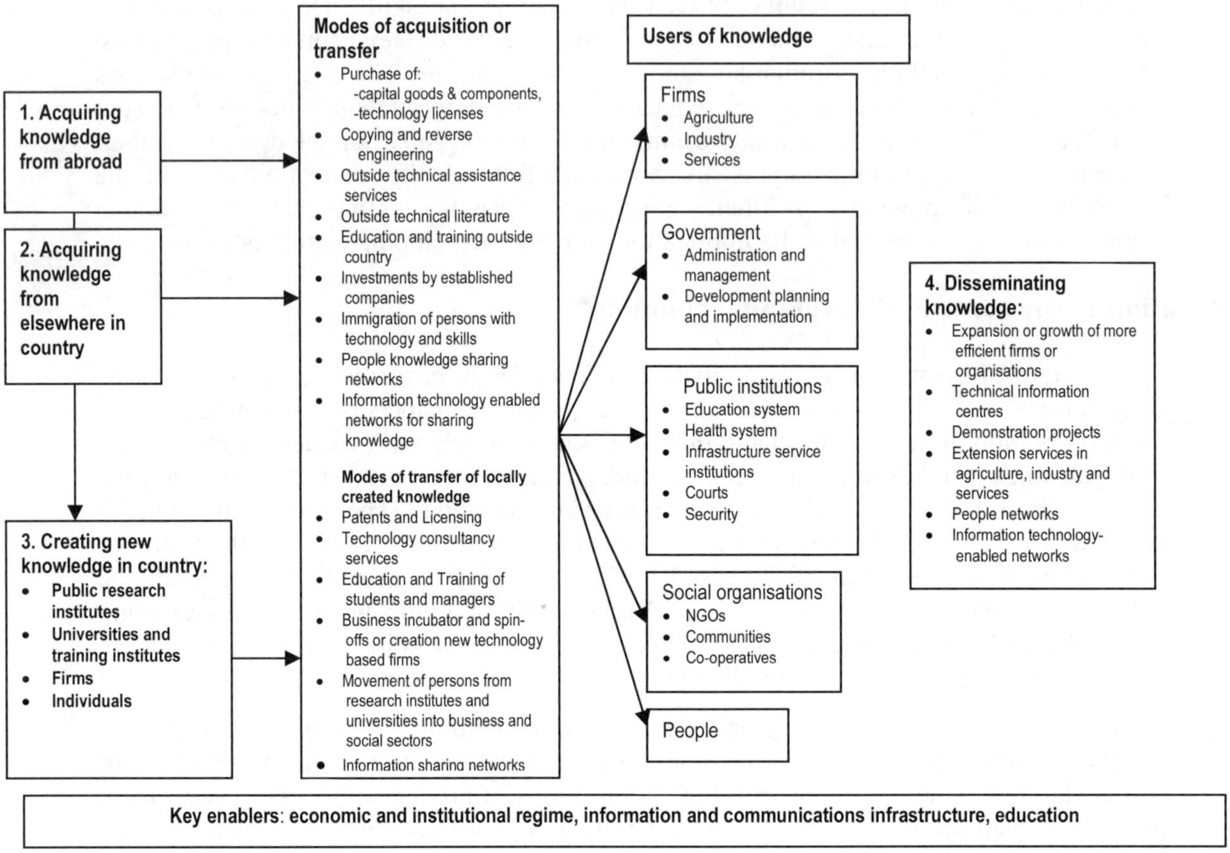

Figure 7.1 presents a schematic diagram of an innovation system, as broadly defined in this chapter, in developing countries. Innovations may come from abroad, from other users in the same country, or be created by public and private R&D labs or firms in the same country (first column). They may be transferred to users (third column) which are not only firms, but government, social organisations and people through many modes (second column). These modes range from investment or formal purchases of technology, capital goods, components or products, to movement of people and informal sharing of information via people or information-enabled networks (second column). Dissemination occurs through the growth of more efficient firms, as well as through special institutions or programmes such as technological information centres, productivity and extension agencies, people, and information-enabled networks (fourth column).

Innovation is affected by the broader economic and institutional regime. This includes macroeconomic conditions (inflation, interest rates, exchange rates); the business environment (rule of law, quality and effectiveness of government, including appropriate or excessive regulation and competition policy); and the quality and efficiency of the physical and information and communications infrastructure. The education and skills of the population and the workforce are also important. The technological capability of a country also depends on the strengths of its technology-related institutions. These include research laboratories in firms, universities and government laboratories, but also the capability of production, engineering, management and service firms; the strength and quality of universities and government institutions relevant to innovation; and technology and knowledge networks. All of these should be considered when comparing countries' innovation strategies.

Characteristics of the BRICKs

Characterisation of growth strategies and performance

China and India are the world's two most populous countries, accounting for 20% and 17% of the world population (Table 7.1). Brazil is the fifth largest, accounting for almost 2.9%, and Russia is sixth with 2.2%. Korea is much smaller with a population of less than 50 million. China's economy has grown faster and for a longer period of time than that of any of the other countries. As a result, although its gross national income (GNI) was smaller than that of all the others except Korea in 1980, it is now three times that of any of the other four countries (Table 7.1).

In nominal per capita terms, based on the World Bank classification, India is a low-income country. China is a lower-middle-income country. Brazil and Russia are upper-middle-income developing countries. Korea is classified as a high-income country. In purchasing power parity (PPP) terms, per capita income more than doubles for China, India and Russia. It increases by 60% for Brazil and only by 25% for Korea, so the variance among the five countries decreases significantly.

Korea leads the other four countries in having already made the transition from developing to developed country. It followed an export-oriented development strategy based on the Japanese experience, including heavy reliance on large conglomerates (the chaebol). Its growth rates in the 1970s and 1980s were similar to the very high Chinese growth rates. It was able to move rapidly up the technology ladder owing to the deep pockets of its large chaebol which could cross-subsidise risky new ventures from strong profits in other parts of their operations. However, Korea was hit hard by the Asian financial crisis of 1997 but recovered quickly, thanks to strong government action (for details, see Suh and Chen, 2007). Its growth since 2000 has been more than 50% above the global average. Like Japan and unlike China, however, it tightly restricted FDI inflows, and it was not until the 1997 financial crisis that it opened up more to FDI.

Brazil had high growth rates from the second half of the 1960s to until about 1981, when it was severely affected by the Latin American debt crisis and the first oil shock. It basically lost two decades of growth as a result of the ensuing macroeconomic imbalances and only managed to stabilise its economy in the current decade (Rodriguez *et al.*, 2008, analyse Brazil's performance over time from the knowledge perspective). It began to grow by over 5% from 2000 owing to the boom in commodity exports driven by China's massive commodity imports. However, even now, its growth rate significantly trails that of the other BRICKs. Furthermore it is heavily dependent on the high commodity prices fuelled by China's voracious appetite for raw materials and commodities.

India grew at a yearly average of 2-3% from its independence from the United Kingdom in 1947 until the 1980s. In the 1980s and 1990s it had average rates of growth of 5-6% except for a severe financial crisis in 1991 which forced it to liberalise its economy. This put it on a higher growth path. India had been the most technologically autarkic until it began to grow at near Chinese rates owing in large part to its integration in international trade through exports, enabled by information technology, of highly skilled services. Such growth, led by a skilled services sector in a low-income country, was completely unanticipated by traditional development theory.

China has had the highest growth rates and for the longest time. Between the Communist takeover in 1949 and the mid-1980s, China went through boom and bust cycles, which started with a heavy industrialisation drive and included the disaster of the Great Leap Forward as well a major famine in which an estimated 30 million people died. China's more impressive growth period started with the opening up to the world in the late 1970s under Deng Xiao Ping. Since then China has increased its integration in the world economy. It started slowly with the establishment of four export processing zones. The pace then speeded up and China eventually joined the World Trade Organization (WTO) in 2001.

Table 7.1. The BRICKs: basic economic indicators

	Brazil	China	India	Korea	Russia
GDP (2007)					
GNI (2007 nominal billion)	1 122	3 126	1.071	956	1.070
GNI as share of global GNI (%)	2.12	5.91	2.02	1.80	2.02
GNI/capita (2007 nominal)	5 860	2 370	950	19 730	7 530
GNI (trillion 2007 PPP)	1 776	7 151	3 083	1 204	2 037
GNI as share of global GNI(2007 PPP)	2.70	10.87	4.69	1.83	3.10
GNI/capita (2007 PPP)	9 270	5 420	2 740	24 840	14 330
Growth of GDP (1980-2006)					
1980-1990 av. annual growth	2.7	10.2	5.8	9.4	..
1990-2000 av. annual growth	2.7	10.6	5.9	5.84	-4.7
2000-2007 av. annual growth	3.3	10.3	7.8	4.7	6.6
Exports (2007)					
Merchandise exports (millions)	161 649	1 21 7 776	145 325	371 489	355 175
Merchandise exports(% of world total)	1.16	8 73	1.04	2.66	2.54
Commercial service exports (millions)	22 555	121 654	89 746	61 536	39 119
Service exports (% of world total)	0.67	3.62	2.67	1.83	1.17
People (2007)					
Population (millions, 2007	192	1 318	1 125	48	142
Population as share of global population	2.90	19.93	17.02	0.73	2.15
Life expectancy at birth(2007)	72	72	64	78	66
Human Development Index					
1995	.737	.681	.545	.852	.779
2005	.800	.777	.619	.921	.802
Poverty and inequality					
% below USD 1.25/day poverty line (2005)	7.8	15.9	41.6	2.0[1]	<2
% below USD 2/day poverty line (2005)	18.3	36.3	75.6	2.0[1]	<2
Gini coefficient (2004)	55.0	41.5	36.8	31.6	37.5

Note: The GDP in PPP figures are based on new PPP series published in the December 2007, which decreased the estimates for both China and India by 40%

1. 1998 for Korea.

Source: World Bank *World Development Indicators* 1999, 2009; and *World Bank KAM* 2008.

What distinguishes China the most is the very rapid growth of its exports. Starting with less than a 0.5% share in world merchandise exports in 1980, by 2007 they accounted for nearly 9% of global exports, roughly eight times the exports of Brazil or India, and more than three times those of Korea or Russia. In 2007 its merchandise exports surpassed those of the United States and were second only to Germany's. In a sense, China has become the world's manufacturing workshop. India's strength has been in the export of services. Its share of global service exports is almost three times that of its merchandise exports, and its strength is in exports of ICT-enabled services. Therefore, in terms of the two unbundlings referred to by Baldwin (2006), China benefited the most from the unbundling of production, while India benefited most from the unbundling of tasks facilitated by rapid advances in ICT. Brazil has not really taken advantage of either. It remains primarily an exporter of commodities, although it has made impressive improvements in agricultural productivity and has developed some islands of excellence, including exports of airplanes, and deep water oil exploration.

Russia was a military and technological super power. However its innovation strategy was very isolationist, and in spite of high levels of R&D and of technical human capital it had poor economic performance until the last decade when it started to grow rapidly with the boom in petroleum prices. In addition, the transition was a very difficult period of economic contraction. Moreover, its growth over the last decade was unrelated to previous R&D strength, as R&D actually shrank considerably during the transition to a market economy. Life expectancy fell and poverty increased significantly during the transition.

However, Korea and Russia are more developed than the other BRICKs, all of which have lower per capita incomes and significant pockets of poverty. In Brazil 18% of the population still lives on less than USD 2 a day,[5] a share that jumps to 36% for China and to 76% for India.

Innovation, as broadly defined in this chapter, is a critical factor for the growth of any economy. A quick appreciation of the role of innovation, broadly defined, in the growth of these economies can be inferred from differences in the contribution of total factor productivity (TFP).[6] In the countries which have grown fastest for a longer period of time – Korea, China and India – TFP accounted for as much as half of the total growth of output. Brazil had very low TFP growth over much of the last 25 years. Russia's TFP growth was negative for much of the 1990s and only turned positive recently. This is not to say that higher investment rates are not part of the explanation for more rapid growth. They are. But the simple distinction between factor growth and the TFP residual misses the key point that new investments usually incorporate innovations in product, processes, or new machinery and equipment to produce new products and services. Also, high investment that does not incorporate new technologies or inefficient investment will not contribute as much to growth. This was the case during the period when the Soviet Union had very high, but inefficient investment.

5. This is because Brazil has one of the world's most unequal income distributions as revealed by a Gini coefficient of .57 compared to .47 for China and just .37 for India.

6. Total factor productivity is the residual growth which cannot be explained by growth in factor inputs. It tracks improvements in productivity which may be attributed to better use of resources to produce goods and services and can be taken as a rough proxy for innovation. See Bosworth and Collins (2007) for China and India; Suh and Chen (2008) for Korea; and Goldberg and Desai (2008) for Russia.

Economic and political characteristics

The five countries differ in their economic structure, degree of global integration through trade, structure of trade, consumption and investment, political systems, and the role of government (Table 7.2).

Economic structure. China's structure is very heavily skewed towards industry. At 49%, the share of industry in GDP is one of the highest in the world (it is 39% in Korea, 38% in Russia, 29% in Brazil and 30% in India). On the other hand, at 40%, the share of its services sector is relatively small for an economy with its GDP. India is still primarily an agricultural economy. More than half of its labour force is in agriculture, and the share of agriculture in GDP is still 18%, compared to 11% in China, 6% in Brazil, 5% in Russia and 3% in Korea. However, at 52% the share of India's services sector is above average for a county of its level of development, owing in part to the rapid growth of its ICT and business sectors as will be detailed below. (See Table 7.A1.1 for details on the economic structure of the five countries.)

Degree of global integration. Korea is the most globally integrated of the BRICKs. Imports and exports account for 90% of GDP. China is the second most globally integrated. India has more than doubled its global integration between 1990 and 2007, largely because it is much more open for trade in services. Service exports account for more than one-third of its total trade. Brazil continues to be the least globally integrated, with a share of trade in GDP of only 27%. However, its average levels of import tariffs are a little lower than India's. Russia's share of trade in GDP has not changed much in the last 15 years, although the structure of trade has changed.

Trade structure. China and Korea are mainly exporters of manufactured products, India is mainly an exporter of services and manufactured products. Korea has been mostly an exporter of manufactured products since the 1980s. The share of manufactures in merchandise exports in China has increased to 93%, compared to 64% in India. In Brazil, 53% of merchandise exports are primary commodities, consisting of food (26%), ores and minerals (12%), fuels (8%), and agricultural raw materials (4%). Russia's share of manufacturing exports has shrunk from 26% in 1990 to 17%. The share of fuels has increased from 43% to 61%, as it has become mainly an exporter of petroleum products.

Consumption and investment. China has been a heavily investment-driven economy. Over 2002-06, investment as a share of GDP averaged 43% compared to 29% for India, 23% for Korea, 21% for Russia and only 16% for Brazil. India had been a heavily consumption-driven economy. However, in India the share of investment in GDP has increased over the past four years to 39% in 2007. Brazil has the lowest investment rate at just 18% in 2007. At 21% it also has the highest share of government consumption in GDP, compared to 18% in Russia, 15% in Korea, 14% in China and just 11% in India. In Korea and Russia, the share of investment in GDP has fallen over the last decade and a half. China has grown faster than the other countries in part because it has had more or less twice the investment rate for the last 20 years. While it is true that a lot of this investment has been inefficient, it is also true that high rates of investment allow for the embodiment of new technology.

Political system. China is an authoritarian one-party state since the Communist takeover in 1949. India has been a democracy since independence from the British. Brazil has been a democracy since its independence although it has gone through various military dictatorship regimes, the last of which (1964-85) coincided with part of Brazil's miracle growth years (1968-80) until the economy ran into trouble and the military regime was ousted. Korea had a series of military governments until 1992. Russia was an authoritarian state until 1991 and has been democratic since then, although authoritarian trends are resurfacing.

Table 7.2. The BRICKs: basic economic and political characterisation

	Brazil	China	India	Korea	Russia
Economic structure	Industry and services	Heavily industry-dominated	Agriculture and services	Industry and services	Industry and services with large mining sector (petroleum)
Global integration through trade (2007)	Relatively high tariff and non-tariff barriers Not very globally integrated: trade only 27% of GDP	Lowest tariff and non-tariff barriers Very globally integrated: trade 76% of GDP	Highest tariff and non-tariff barriers Moderately integrated: trade 46% of GDP	Low tariff and non - tariff barriers, but not as low as China Most globally integrated trade 90% of GDP	High tariff and non-tariff barriers, approaching India's Relatively globally integrated: trade 52% of GDP, because of oil exports
Trade structure (2007)	Still heavily primary-based Merchandise exports: 52% primary/ 47% manufacturing Service exports: very low, only 1.72% of GDP, with 53% of this total ICT-related	Very heavily manufactures-based Merchandise exports: 93% manufacturing, 7% primary Service exports: low at 3.8% of GDP, with 43% of this total ICT-related	Heavily service-based, more than 30% of its total exports .Merchandise exports: 64% manufacturing, 36% primary Service exports very large: 7.63% of GDP, with 74% of this total ICT-related	Very heavily manufacturing-based Merchandise exports: 89% manufacturing, 11% primary (with recent increase in fuels) Service exports: 6.3% of GDP; with 29% of this total ICT-related	Very heavily petroleum-based Merchandise exports: 61% fuels in 2007 up from 43% in 1995, decline in manufactures to only 17%. Services exports, only 3.03% of GDP; with 41% of this total ICT-related
Consumption and investment orientation (2007)	Heavily consumption-oriented: 81% of GDP, with 20% government consumption. Investment only 18% of GDP	Heavily investment-oriented: 43% of GDP. Total consumption only 47% with government consumption only 14%	Previously heavily consumption-oriented: 75% of GDP in 1990, but down to 64% with 10% government consumption. Investment increased to 39% of GDP in 2007 from 27% in 1990	Increasingly consumption-oriented from 63% in 1990 to 70% in 2007 Investment has fallen from 38% in 1990 to 29% in 2007	Consumption decreasing from 71% in 1990 to 67% in 2007 Investment decreasing from 25% in 1990 to 20% in 2007 Exports increasing from 29% in 1990 to 35 % in 2007
Average gross investment to GDP 2002-06	16.4%	43%	29.4%	23.2%	20.7%
Government	Multi-party democratic (except for military regime (1964-80)	Single-party authoritarian since Mao Zedong took power in 1949	Multi-party democratic since independence from United Kingdom in 1947	Multi-party democratic since 1987	Multi-party democratic since 1991

Note: See Annex Tables 7.A1.1- 7.A1.3 for more details on economic structure, merchandise and service exports from the BRICKs.

Role of government. In China's centrally planned economy, the government has had and continues to have the largest role. However, starting in the 1980s the Chinese government began to move slowly towards a market economy. This process accelerated with its entry into the WTO. Many state enterprises have been shut down or restructured, private enterprises have been allowed to develop (private property was officially recognised in 2007), and the private sector is estimated to account for over 60% of economic activity. Although a democracy since its independence, the early years of the Indian economy were marked by a strong role of the state in the economy, with the institutionalisation of five-year industrialisation plans originally structured on the Russian model. In addition government was very opposed to big business. It restricted the growth of large private businesses and reserved several hundred products to very small-scale businesses. Controls on private business began to be lifted in the 1980s. Further liberalisation in the 1990s has continued today. While there has been some privatisation, the state still has a strong presence in the industry and services.

Brazil was the most market-oriented economy. However, during the military period the state played a strong role in the economy and created many large state-owned enterprises (SOEs) in critical areas. The 1990s saw considerable privatisation and Brazil is the most market-oriented of the five economies. However the state still has an importance presence in the services sector.

In Korea the state's role has diminished rapidly since the 1980s. In Russia it also diminished in the 1990s and the beginning of this decade, but it is now strengthening.

Economic and institutional regime

This section examines some of the broader macroeconomic and institutional regime issues: inflation, interest rates, the exchange rate, physical infrastructure, telecommunications infrastructure, and bureaucratic transactions cost (Table 7.3).

Inflation

Brazil has had the most unstable macroeconomic conditions of the five countries. It suffered several bouts of very high inflation between 1980 and 2000. High inflation made it very difficult for business or government to plan for the long term. In addition, financial engineering, managing supplier credits and receivables, and pricing were much more important than reducing production costs, improving quality, or developing new products. Russia also had a great deal of macroeconomic instability and a sharp recession from the first phase of its transition in 1991 until stabilisation in 1998. Between 1991 and 1995 GDP fell 50%. A large part of Russia's technical human capital left the country and expenditures for R&D were drastically reduced. China has had the most stable macroeconomic regime of these three countries, with generally low inflation since the 1980s. In addition government plans have been quite clear and stable and this has facilitated long-range planning by domestic and foreign investors. India has also had a relatively stable macroeconomic regime over the last two and a half decades, except for the financial crisis at the beginning of the 1990s. However, the macroeconomic environment has not been as stable and predictable as in China. Korea has been relatively stable except for the financial crisis of 1997 which was followed by a quick recovery.

Table 7.3. The BRICKs: broader economic regime variables

	Brazil	China	India	Korea	Russia
***Inflation*: CPI:**					
1990-2000	199.5	8.6	9.1	5.1	108.0
2000-07	7.7	1.8	4.4	3.1	12.9
Cost of capital	Extremely high crowding out by government, very inefficient banking	Very low real interest rates, thanks to high saving rate	Low interest rates, relatively efficient banking sector	Low interest rates; relatively efficient banking sector	Very variable
Real interest rate:					
1995	65.5	-1.5	5.9	1.5	72.3
2007	38.1%	0.0%	7.8%	5.3%	-3.1%
Banking spreads					
2007[1]	33.1%	3.3%	n.a.	1.4%	4.9%
Undervalued exchange rate	No, after a sharp devaluation in 1999 the real exchange floated upwards, gradually at first and then appreciated rapidly in 2003-06 to pre-devaluation levels.	Yes, Chinese yuan pegged to USD has become undervalued as dollar has depreciated. Chinese current account surplus soared from 1% to 9% of GDP (2001-06)	No. Real rupee rates have been relatively constant since a major devaluation in 1991, but have begun to appreciate slightly in real terms in 2007.	No. Purchasing power parity conversion factors have been relatively constant between 1995 and 2007.	No. Exchange rate has become overvalued as result of strong petroleum exports. Real effective exchange rate in 2007 was 173% of 2000 level
PPP conversion factor: local currency to international USD,					
1995	0.7	3.4	11.1	690.0	1.5
2007	1.4	3.6	15.3	747.9	15.8
Ratio of PPP conversion factor to market exchange rate					
2007	0.7	0.5	0.4	0.8	0.6
Cost of doing business (global ranking based on IFC surveys)**					
2008	126	90	120	22	112
2009	125	83	122	23	120
Cost and availability of physical infrastructure	High cost, moderate availability. high-cost international transport	Very low cost, good availability, good logistics with low-cost international transport	High cost, very poor availability except for cell phones. High-cost international transport	Low cost	Very high cost
Cost to export 20' container**	USD 1 240	USD 460	USD 945	USD 767	USD 2 150
*% of sales lost owing to electricity interruption***					
Survey 2002-05	1.6	1.3	6.6	..	2.0
Fixed and mobile phones/1 000					
1995/2007	90/840	37/700	13/250	449/1,360	170/1 460
Computers/1 000					
1995/2007	17/161	2/57	1/33	108/556	18/133
Internet users/1 000					
1995/2007	1/352	0/161	0/72	8/759	2/211

1. Lending rate minus deposit rate in percentage points.

Source: All from world Bank, *World Development Indicators 2009* except those marked with ** from IFC *Cost of Doing Business* 2009 website, *www.doingbusiness.org/CustomQuery/.*

Cost of capital

Low cost of capital is a critical variable for inducing greater investment. Thanks to its very high savings rate, the cost of capital in China has been extremely low for a very long period. The Chinese investment rate has been one of the highest in the world and increased to over 44% in the last three years. However, the bulk of the investment is directed to state enterprises and the capital output ratio has been rising. In Korea the cost of capital was also very low for much of the 1980s and 1990s. Easy credit and high leverage was one way for the government to encourage the rapid growth of the chaebol. This allowed them to buy new plant and equipment embodying more advanced (usually foreign) technology. The real cost of capital in India has been more in line with the real interest rate in developed economies. India's banking system is also more developed and more efficient than China's. The cost of capital in Brazil has been one of the highest in the world and is a major disincentive to investment and growth. The high cost of capital, plus strong restrictions on the import of more efficient foreign capital goods helped to make Brazilian industry uncompetitive except in sectors which used foreign capital goods for export industries such as pulp and paper. The high real interest rate is due to large government borrowing to service the domestic debt. In addition the banking sector has one of the highest interest rate spreads between deposit rates and lending rates in the world. In Russia, real interest rates were very high during the 1990s and there was little investment. Since stabilisation and growth the cost of capital has been much lower, even negative in 2006, and there has been investment in the expansion of petroleum-related industries.

Exchange rate

China has pegged the yuan to the US dollar for many years. As the dollar has been depreciating relative to most currencies since 2001 the value of the yuan also has been depreciating. There was a one-time 2.5% appreciation of the yuan against the dollar in mid-2005. Since then the yuan has appreciated by almost 20% against the dollar. Some economists argue that the yuan may be still 20% to 30% undervalued relative to the dollar. It is also argued that China has an explicit policy of undervaluing the yuan in order to support its dramatic export expansion. The real value of the Indian rupee has been maintained more or less constant since the major devaluation of 1991. However, with the recent boom in foreign and domestic investment in India the rupee started to appreciate in 2007. The Brazilian real has been appreciating significantly since 2003 and this is having a negative effect on exports of manufactures. In the context of the current international financial and economic crisis, the exchange rates of all the BRICKs except China have depreciated against the dollar as investors have fled to the perceived greater security of US Treasury Bonds. However, the exchange rates of all the BRICKs except China devalued sharply against the dollar as part of the 2008-09 global financial and economic crisis.

Cost and availability of infrastructure

China has more modern and efficient trade infrastructure than either Brazil or India. Brazilian infrastructure and transport services are more expensive than those of either country, but better than India's. The availability and cost of infrastructure services (except cellular telephone calls) are worse in India and are a major source of competitive disadvantage. The ICT infrastructure has been built up very rapidly in China and India. Not surprisingly, as the most developed country, Korea has the best physical and telecommunication infrastructure. Russia's physical and trade infrastructure are not well developed, and its trade costs are the most expensive. However, Russia has been rapidly upgrading its telecommunications and information infrastructure. By 2007 it had more telephones per 1 000 persons than the other four countries thanks to very rapid expansion of cell phones. However, it still lagged Korea and Brazil in computer and Internet use.

Bureaucracy and transaction costs

All four emerging countries have a great deal of government red tape. Brazil, India and Russia have the most onerous regimes as shown by their low ranking in the IFC's cost of doing business report for 2008 and 2009 (*www.ifc.org*). China is 40 positions higher, but Korea ranks 23rd in the world.

In sum, the BRICKs differ markedly in terms of their broader economic regimes. Particularly notable are the greater long-term stability, very low interest rates, very high investment rate and high degree of trade integration of China. This has facilitated its rapid assimilation and use of global knowledge. At the other extreme is the much greater macroeconomic instability, very high real interest rates, very low investment rates, and much less global trade integration of Brazil, all of which help to explain its historically much lower growth rates. Korea is similar to China except for the 1997 crisis, and it also used to have much higher investment rates. India has also been more stable than Brazil, and until the last four years its investment rates were in the low 20% rather than over 30%, so a large part of it recent rapid growth is due to high investment rates and drawing more on global knowledge through greater openness to imports. Russia is a special case, because of a wrenching big bang transition which had vey high economic and social costs. These are all the more striking in comparison to China's much more successful gradual move towards a market economy.

Education[7]

People need education and new skills to use new technologies. To produce new knowledge, they need specialised tertiary education. People with higher levels of education tend to adopt new technologies faster. Despite rapid increases in the supply of higher education in most of the world's countries, the gap has not narrowed between the earnings of college graduates and of those who only finish secondary school because higher-order skills are necessary to make effective use of new technologies. Empirical studies show that more education leads to higher productivity and higher growth. They also show that it leads to higher earnings for individuals.[8]

7. This section draws on work by the author for Dahlman *et al.* (2007).

8. There are other benefits from more education. Many analyses show a positive causal relationship between higher educational attainment, better mental and physical health, lower rates of unemployment, and jobs with higher income. People with higher incomes can spend more on health. Education can increase a person's self-esteem, problem-solving and social skills, and capacity to respond to adversity. In addition, research has shown a positive relation between literacy and participation in voluntary community activities, and between greater civic knowledge and higher levels of civic participation (OECD, 2005).

Table 7.4. The BRICKs: education

	Brazil	China	India	Korea	Russia
Literacy rate, population 15 and above					
1995	83.2	80.8	53.3	97.0	99.4
2007	90.5	93.3	66.0	97.9	99.5
Av. ed. attainment of adult pop. (2000)	4.88	6.35	5.06	10.84	10.03
Basic education	Basically universal but poor quality	Universal but spotty quality	Incomplete and poor quality	Universal and good quality	Universal and good quality
Secondary education enrolment ratio (%)					
1980	34	46	30	78	96
2007	105	76	55	98	84
Higher education enrolment ratio (%)	Was leader among developing countries in 1980s, but has not increased as fast as others and is falling behind	Initially very low compounded by damage of cultural revolution period, but has had rapid ramp-up since 1998	Initially high through Indian Institutes of Technology, relatively little expansion and high variance and much low quality	Initially low, but had very rapid ramp-up and is currently one of highest in world, significantly above OECD averages	Initially high but enrolment rates fell in first part of transition period before recovering
1980	11	2	5	15	46
2007	24	22	11	93	70
Skilled labour	Relatively underdeveloped	Well-developed training market inside and outside firms	Very under - developed. Very poorly developed training market	Well developed. Large active training market including firm training	Not well developed for the needs of a market economy

Source *World Development Indicators* various years, and author's assessments.

Higher productivity

Better educated workers are generally more productive and may also raise the productivity of co-workers. Higher stocks of human capital facilitate investments in physical capital, enhance the development and diffusion of new technologies, and raise output per worker (OECD, 2005). Studies based on neoclassical growth theory which trace the relation between the stock of education and the long-run level of GDP find that a one-year increase in average educational attainment raises the level of output per capita between 3 and 6 percentage points (OECD, 2005). Studies based on the new growth theory which examine the relation between the stock of education and the rate of growth of GDP find that an increase of one year of education raises the growth rate of GDP by around 1 percentage point (Wosmann, 2003), although research suggests diminishing effects above an average of 7.5 years of education (Krueger and Lindale, 2001). The cumulative impact of this increase in the growth rate soon exceeds the one-time increase in output. Rising labour productivity accounted for at least half of GDP per capita growth in OECD countries from 1990 to 2000 (OECD, 2005).

This section compares education across the BRICKs and looks specifically at basic educational attainment, basic and secondary enrolment rates, and higher education enrolment rates. It ends with a general comment on the role of education and lifelong learning.

Average educational attainment

Korea and Russia have the highest average educational attainment rates. Russia invested in education earlier and more massively than any of the others. Korea started to invest heavily in education in the 1970s and 1980s as part of its drive to move up the technology ladder. Brazil had the next highest average educational attainment. However, China, which was further behind in the 1980s invested heavily in education and has now surpassed Brazil with an average of more than eight years of educational attainment among the adult population. India has the lowest level of average educational attainment, and the highest levels of illiteracy. In 2006 illiteracy was 51% among women, and 26% among men.

Basic and secondary education enrolment rates

India's system of basic education is still very poor with tens of millions of primary school children out of school. Its secondary enrolment rates are very low at just over 54%. In 1990, Brazil had low secondary enrolment rates for its level of per capita income, but in the 1990s it invested heavily in secondary education and was able to significantly increase its secondary enrolment rates. It still has a problem of poor quality and many repeaters, which explains why its secondary enrolment rate is over 100% (Rodriguez *et al.*, 2008). China has near universal primary and lower secondary enrolment rates and it is now working on increasing enrolments at the upper secondary level. Korea and Russia already had high secondary enrolment rates in 1980. Korea continued to increase these rates but in Russia they dropped to 84%.

Tertiary education

India set up seven Indian Institutes of Technology (IIT) starting in the 1950s and later several Indian Institutes of Management (IIM). These have produced a critical mass of well-educated English-speaking professionals who have been instrumental in India's emergence in software and ICT. Although it expanded tertiary enrolments in the 1980s it made little progress after 1995. Also, the quality of its tertiary educational system, except in these institutes and a few other prestigious institutions, is low. Starting in the late 1990s China undertook a massive expansion of its tertiary education system to make up for the damage to its higher educational system during the Cultural Revolution (1965-75). By 2006 its enrolment rate reached 22% (twice that of India's). Because of its large population, China had more students at the tertiary level in 2006 than the United States, 40% of whom were in engineering and sciences. However, like India, aside from some key prestigious institutions, China's higher educational system has problems of quality. Based on a survey, McKinsey reports for example that only 10% of Chinese and only 25% of Indian tertiary graduates are ready to be hired by MNEs (Farrell *et al.*, 2005). Brazil had been considerably ahead of both China and India in tertiary education, but China has caught up. The Brazilian higher education system also has problems of quality and relevance (Rodriguez *et al.*, 2008). Russia had very high tertiary enrolment rates before its transition. With the strong recession during the transition, tertiary enrolment rates fell. Since the recovery they have climbed back up and now stand at 70%. Korea invested very heavily in higher education in the last decade as part of its strategy to become a knowledge-based economy. Its tertiary enrolment rate, now at 93%, is among the highest in the world, considerably above the United States (82%) and the average for high-income countries (67%) (World Bank, *World Development Indicators*, 2009).

The low levels of basic education in India have constrained it from tapping into and assimilating foreign technology. However, in spite of low overall tertiary enrolment rates, India's early investments in quality higher education were critical in positioning it to take advantage of the ICT revolution and the potential for exporting information-enabled services. This shows that it is important to distinguish between overall educational attainment and critical mass when looking at large countries.

Korea's and China's rapid expansion of basic education was part of their strategy to take advantage of global knowledge. Korea also expanded its higher education system because it needed to compete at higher levels of technology. This required more qualified technical human capital, particularly in order to develop its technological capability without much participation from FDI. China has been expanding its higher education system as part of its strategy to become a knowledge economy (Dahlman *et al.*, 2007).

Russia had a very strong higher education base as part of its technological superpower strategy. However with the transition to a market economy and a sharp decline in the need for military technology, much of its highly skilled base emigrated to Israel, the United States, Europe and elsewhere. Moreover, the Russian higher education system's specialisation in narrow scientific disciplines was ill-matched to the skill needs of the emerging market economy and has been restructuring to meet the new demands of the economy. Part of the increase in higher education enrolments has been due to persons who see higher education as a way to emigrate.

Thus the levels of education have been very different across the five BRICKs and have changed at different speeds as part of their development strategy and other objectives. Most notable have been first Korea's and later China's rapid expansion of education in order to make more effective use of knowledge and, more recently, to be able to do more innovation (Table 7.4). Moreover, both of these countries have developed systems for continuously upgrading the skills of their labour forces in order to keep persons who have left their formal educational systems up to date (for further detail, see Dahlman *et al.*, 2007). This move towards lifelong learning is a global tendency that is more developed in more advanced countries.

Acquiring foreign knowledge

Countries that are behind the technological frontier in specific industries get higher increases in productivity and improvements in welfare from acquiring existing knowledge than by doing R&D to push back the technological frontier. Creating new knowledge is generally more difficult and risky and requires much more technological capability. The main means of tapping into global knowledge are trade, FDI, technology licensing, foreign education and training, copying and reverse engineering, and accessing foreign technical information in print and through the Internet. On all these counts, China has been more aggressive and systematic than the other countries.

Trade

Trade is an important way of acquiring global knowledge. Imports of capital goods and components are a very effective way of acquiring embodied technology. Equipment suppliers also provide relevant product and process information. Importing foreign products or services also gives ideas for copying or reverse engineering similar products or services. Exporting forces firms to be aware of what is happening at the global frontier in terms of products, designs and processes since exporters have to compete with the best the world has to offer. In addition, foreign buyers often provide product designs as well as production process assistance. Therefore participation in world trade is a good proxy for accessing global knowledge.

Korea began opening up much earlier than the other countries as part of its outward-oriented strategy in the late 1960s. In 1980 the share of imports and exports was already 71% of GDP. In 1990 it was 59% of GDP, and by 2007 it was 90%. In its early years Korean companies produced many products to the specifications of foreign buyers who provided production and design specifications as part of the process (Rhee *et al.*, 1981). The Korean government used increasing exports as a yardstick to measure the performance of the chaebol, which therefore were forced to keep up with the rapidly changing global technological frontier. As the chaebol grew and invested in their own R&D, they developed strong technical capability and launched their own brand names.

Learning from the experience of Asian neighbours such as Korea, Hong Kong (China), Singapore and Chinese Taipei, China began opening up much earlier than India and has become much more integrated into the global economy. The share of imports and exports in China was 44% of GDP by 1990 against 16% in Brazil and 23% in India (Table 7.5). By initially protecting its industries from imports, China developed basic technological capability. Then by opening up to foreign investment in special economic zones with close to free trade status it gained access to world-class technology and inputs. This worked very well. It not only began to modernise China, but also provided needed foreign exchange and employment. The special economic zones were expanded from the initial 4 to 19 and then to many more. The programme was so successful in generating employment and foreign exchange that by the late 1990s China decided to significantly generalise this free trade status by joining the WTO. This involved committing to a major programme of reduction of tariff and non-tariff barriers and opening up to foreign investment not only in the manufacturing, but also in financial and other service areas. For a good analysis of China's progressive entry into the global system, see Naughton (2007, pp. 375-424).

Unlike Korea and China, which have largely removed tariff and non-tariff barriers to trade, India is still one of the world's most closed economies in terms of tariff and non-tariff barriers. In the 1950s India followed a very autarkic policy of self-reliance, mostly relying initially, like China, on massive capital goods imports from the Soviet Union. However, unlike China, India maintained a strongly inward-oriented nationalist policy through the 1980s. During this period technology policy focused very much on self-reliance. It was only after trade liberalisation in the early 1990s that India began to open up more to foreign technology imports. There were also very strong restrictions on FDI and on the licensing of foreign technology (Dutz, 2007).

Brazil's autarkic policy was not as extreme as India's. It has a period of rapid economic growth from 1965 to 1980 when it continued a period of rapid import-substituting industrialisation. Even in the mid-1990s after India's liberalisation, Brazil had lower tariff and non-tariff barriers than India or China (Table 7.5). However since China joined the WTO it has become more open. Furthermore, as Brazil is the least integrated of the BRICKs in global trade in terms of imports and exports as a share of GDP, its industry has been and continues to be relatively protected from global competition and thus has felt less pressure to create new products and processes and to lower costs.

Table 7.5. The BRICKs: acquiring global knowledge

	Brazil	China	India	Korea	Russia
Trade as a share of GDP (%)					
1990	16	44	23	59	55
2007	27	76	46	90	52
Merchandise exports (% of GDP)					
1995	6.1	20.4	8.6	24.2	20.5
2007	12.2	38.0	12.4	38.3	27.5
- Manufactured exports (% of merchandise exports)					
1995	54	84	74	93	26
2007	47	93	64	89	17
- High-technology exports (% of manufactured exports)					
1998	9	15	5	27	12
2007	12	30	5	33	7
Commercial service exports (% of GDP)					
1995	0.8	2.5	1.9	4.3	2.7
2007	1.7	3.8	7.6	6.3	3.0
- Computer, information, communication & other business services (as % of service exports)					
1995	23.6	24.4	31.4	34.5	22.8
2007	52.8	42.7	73.7	28.5	41.2
Tariff & non-tariff barriers[1] (2006)					
1995	56	20	0	69	52
2008	71	70	51	66	44
Average tariffs 2006 (in %)					
Average simple tariffs	12.3	8.9	17.0	8.5	9.9
Average weighted tariff	6.8	5.1	13.8	8.0	7.2
Foreign direct investment	Much FDI is for producing for protected domestic markets rather than for export	One of the main means of rapidly modernising China	Much less FDI overall, as India has only recently liberalised FDI regime	Very little FDI until the aftermath of 1997 Asian financial crisis	Very little FDI except for oil and gas sector
Average gross FDI/GDP					
2000-05	3.4	3.2	0.9	0.9	1.6
Royalty and licence fee payments, 2007 (USD millions)	2 259	8 192	949	5 075	2 806
Royalty and licence fee payments, 2007 (per billion GDP)	1.72	2.56	0.80	5.23	2.18

1. Normalised such that 0 is very high protection and 100 is very low protection.

Source: World Bank, *World Development Indicators 2007*, and *World Bank KAM* 2008.

In Russia the share of trade in GDP was higher in 1990 than in China, India or Brazil. However that share has declined slightly in Russia but has increased in the other four countries. Its tariffs have fallen, but it still has relatively more non-tariff barriers than the others as can be seen from its lower rating in Table 7.5 for 2006.

Foreign direct investment

The inflows of FDI into China were several multiples of those into India and higher than those to Brazil over the last 15 years. This is the result of several factors. First, China opened up its regulatory regime as regards FDI more than ten years earlier and more broadly than India. Second, China's larger and richer market has been an important pull factor and it has therefore surpassed even Brazil. Third, China has many cost advantages over India and Brazil, even though its labour costs are now generally higher than India's. Transport is more efficient, service infrastructure is more developed, and the red tape for trade in physical products is less burdensome (Table 7.5). As a result, China has been very attractive not just as a production platform for global markets, but also for producing for the Chinese market as it is the world's fastest growing market. This strong pull has allowed the government to encourage strong competition among foreign MNEs to bring their best technology when they locate in China, even though they are well aware of China's poor intellectual property protection and the risk that their technology will be pirated.

In China, the government also negotiated with the large MNEs that wanted access to the Chinese market. Initially they forced companies to enter into joint ventures with domestic firms. They also negotiated local content and training requirements. Motorola, for example was obliged to develop an extensive training programme for the management of the 1 000 largest Chinese SOEs (Dahlman *et al.*, 2007).This greatly helped them develop technological and management capability. In the auto industry, the government was able to oblige both Honda and Toyota to undertake joint ventures with the same Chinese manufacturer. This allowed the Chinese company to use the best of both systems to develop its own brand and production. The government was able to achieve this because of the attractiveness of the domestic Chinese market to the foreign manufacturers. Once the cost advantage of producing in China became apparent to both the government and the MNEs, the government relaxed the joint venture requirement in order to encourage the foreign firm to bring its best technology.

In China, the most important contribution of FDI has not been capital since China has a high savings and investment rate. More important has been access to advanced technology and management. Equally important has been entry into global markets as foreign investors integrate their Chinese operations into their global supply chains (Gil and Kharas, 2007). This means that they do not have to own production plants in China but can simply source from China. For example, Wal-Mart sources over USD 25 billion directly from China into its retail stores without using middlemen.

India only began to open to FDI in the 1990s and then only slowly and selectively. As a result inflows were very small. In the last five years India has liberalised FDI inflows and trade inflows but these remain very small compared to China. Indian industrial policy has protected domestic industry for too long and did not take advantage of the technology it could get from abroad or from economies of scale and scope by pushing its firms to operate globally. Relatively little FDI has been attracted because of high transactions costs and poor infrastructure. The exception has been in software and ICT-related services which have not been constrained by the regulatory regime or the physical infrastructure.

Brazil attracted FDI earlier than China. Until China began to aggressively court FDI in the late 1980s and 1990s, Brazil was the largest developing country host for FDI. In Brazil, a large percentage of total FDI came either for the protected domestic market or to exploit domestic natural resources. In the last decade a lot has come as part of the privatisation of state-owned enterprises. In addition, although the government set various export and local content requirements on FDI in manufacturing, it had less leverage than China, mainly because it was not as attractive a location for exports.

As noted, Korea followed the Japanese model with respect to FDI. It restricted inflows of FDI to induce foreigners to license their technology rather than to invest in Korea to compete with domestic countries on their home soil. It was only as a result of the 1997 financial crisis, when Korea was very short of foreign capital and its large chaebol were in trouble, that it opened up more to FDI. Also as it has become very clear that the large MNEs are the main generators and disseminators of applied knowledge, the large Korean firms have entered into strategic alliances with large foreign firms.

Russia opened up to FDI later than the other four countries. Moreover foreign companies have found it difficult to operate in the Russian environment. Therefore even when the foreign investment rules have been relaxed, relatively little foreign investment has flowed into Russia except to the very attractive petroleum-related areas. However even there, there have been concerns since the renationalisation of Yukos in 2004. These are being fanned by British Petroleum's difficulties with the Russian government in 2008. Overall Russia has not acquired as much technology through the FDI route as China or Brazil, as can be seen from the lower ratio of FDI inflows to GDP (Table 7.5).

Technology licensing

Korea, rather than allowing FDI, relied very heavily on licensing foreign technology. Relative to the size of its economy, Korea is by far the most licensing-intensive of the five. In terms of GDP, technology licensing payments are USD 5.23 per billion of GDP compared to USD 2.56 in China, USD 2.18 in Russia, USD 1.72 in Brazil, and USD 0.80 in India.

Foreign education and training

Sending students abroad for education and training is a very effective way to get access to foreign knowledge. It is hoped that these students will then return to their home countries. In 2006, there were 2.75 million tertiary students studying outside their home country. The two countries with the most students abroad were China (15.1% of the total number of tertiary students in world studying abroad) and India (5.1%). This is not surprising given that these two countries have the largest populations, but it is noteworthy that China's share is much larger than India's although their populations are not very different. A more telling indicator is tertiary students studying abroad as a percentage of total tertiary students studying in the home country. The share is highest for Korea (3.7%)[9] followed by China (2.0%), India (1.1%), Russia (0.4%) and Brazil (0.4%) (UNESCO, 2008). It is clear that Korea and China have been the most active in tapping global knowledge by sending students abroad. Many of these students stay abroad after finishing their studies to get practical work experience. Some never return. However Korea and China early developed programmes specifically designed to attract back students and nationals living abroad.

---

9. This ratio is one of the highest for other than small countries with limited domestic tertiary educational facilities.

Diasporas

Korea, China and India have benefited enormously by drawing on their respective diasporas. China had done this more systematically and for longer than India. The first export processing zones were set up opposite Hong Kong, China, and Chinese Taipei, as most of the foreign investment came from those two economies. More than half of the FDI in China has come from Chinese Taipei, Hong Kong, China, and Singapore. These are market economies highly experienced in operating in global markets. They were already well plugged into global supply chains. They initially moved their more labour-intensive operations to China. As China has moved up the technology ladder they have moved more technology-intensive operations. Companies from Chinese Taipei, in particular, are now putting some of their most advanced production facilities in China. In addition, China has set up special high-technology parks specifically targeted at attracting back experienced Chinese abroad to set up high-technology companies. Several of China's more than 100 high-technology parks cater specifically to this diaspora. China has also made a sustained effort to attract back Chinese professors and former foreign students to staff its expanding tertiary education sector.

India was much slower than China to use take advantage of its diaspora. It was joked that NRI (non-resident Indians) were not required Indians. However India eventually learned the importance of harnessing its diaspora and since then has make special efforts to attract people back with special tax breaks and other fringe benefits. However, even today its efforts and policies in this area are not as well developed as China's. India's success in exporting ICT-enabled services has nevertheless been largely due to linkages to its diaspora in high-technology sectors in the United States and Europe. India has had less success than China in bringing back professors to help with faculty shortages in certain fields because it is more constrained by regulations that do not allow universities to pay professors competitive salaries.

The diaspora was also very important for Korea. Starting as early as the 1960s when it was setting up the Korea Advanced Institute of Science and Technology (KAIST), the government created special incentives to attract back to Korea researchers, professors and managers to help it develop its scientific and technological capital base.

As Brazil has a very small diaspora this has not been an important way to gain access to global knowledge. It has not made any significant efforts to attract back its diaspora.

Russia lost a large stock of its most qualified high-level human capital, including hundreds of thousands of scientists and engineers, during its transition in the early 1990s. Many went to Israel, the United States and Europe. Russia has not been very successful in attracting these highly skilled people back home except very recently as its economy began to grow strongly.

Copying and reverse engineering

Both of these modes have been very important for rapid catch-up countries. While there are no hard data, it is quite certain that Korea and China have been much abler than Brazil, India or Russia in this area. In China, greater access to foreign knowledge through all the formal channels listed above, higher levels of human and technological capital, and a policy (now changing) of ignoring intellectual property rights laws have given China an advantage in copying and reverse engineering foreign technology.

Undertaking R&D

The creation of knowledge is the process of inventive activity. It is usually the result of explicit research and development effort normally carried out by scientists and engineers. The key institutions involved in the creation of knowledge are public R&D laboratories, universities and private R&D centres. However, not all creation of knowledge is the result of formal R&D. Sometimes the invention comes from experience in production or from informal trial and error. Sometimes it comes from serendipitous insight. This raises a measurement problem because not all R&D activity results in an invention, and not all inventions come from formal R&D activity.

Moreover, invention is just the process of discovery. If the discovery is basic knowledge, it is often published in a scientific and technical journal. If it is applicable, it may be kept as a trade secret. If it is novel enough, and its inventor is so inclined, it may patented. But even when it is patented it is not a contribution to productive activity. Normally further development and engineering work is required to use the discovery in some concrete application which involves more costs, often larger than the original costs of the invention.

The main actor in the creation of basic knowledge is the government or university research lab. However, the main actor in the creation of applied commercialised knowledge or innovation is usually the productive enterprise. Moreover, the main actor in the creation and dissemination of applied knowledge is the MNE. More than half of all global R&D is done by transnational companies (UNCTAD, *World Investment Report 2005*). The R&D budgets of just one of these large MNEs are larger than the total R&D expenditures of all but the very largest R&D spending developing countries. For example the expenditure of Toyota (USD 8.4 billion in 2007) or General Motors (USD 8.1 billion was larger than the total R&D spending of India.[10]

This section compares the innovative efforts of the BRICKs in terms of two inputs (R&D and scientists and engineers) and two outputs (scientific and technical publications). It also assesses commercialisation of technology and protection of intellectual property rights.

Figure 7.2 puts the R&D expenditures and the relative intensity of scientists and engineers in R&D in perspective for the G5 and the BRICKs. Korea leads the BRICKs in terms of the two key normalised inputs of expenditures of R&D as share of GDP (x-axis) and scientists and engineers in R&D per million population (y-axis). However, absolute scale matters in the area of knowledge since, unlike other goods, knowledge is not consumed when it is used. Once knowledge is produced, it is potentially available to all.[11] China leads the BRICKs in R&D spending. By the end of 2007, in PPP terms, China was the world's third largest spender on R&D.[12] This is the result of an explicit strategy by the Chinese government to go beyond acquiring global knowledge through copying, reverse engineering, FDI and technology licensing to invest in innovation on its own account. In 2006 the government announced a 15-year plan to increase expenditures on R&D to 2%

10. Jaruzelski and Dehoff (2008) list the global top 20 firms and analyse the spending of the top 1 000.

11. Provided that it is made public in a scientific or technical journal or in a patent. However, insufficient education to understand the article or IPR restrictions may limit is actual use.

12. In the old PPP series, China's R&D expenditures surpassed Japan's by the end of 2006, as was reported by the OECD. However, the new PPP series released in December 2007 reduced the PPP estimates for GDP in China and India by 40% each.

of GDP by 2010 and to 2.5% (the average level of more advanced developed countries) by 2025. By 2006 it had already increased it to 1.42% of GDP. In addition, as part of the global outsourcing trend, many MNEs are increasing their R&D work in developing countries, particularly China and India. By 2006 there were more than 750 MNE R&D labs in China and over 250 in India. It is estimated that 25% of business R&D spending in China is foreign, owing to the rapid increase of R&D done in China by foreign firms (OECD, 2008a, p. 166).

Figure 7.2. Relative R&D expenditure and scientists and engineers, G5 and BRICKs

Expenditures in 2006 in USD billions in PPP

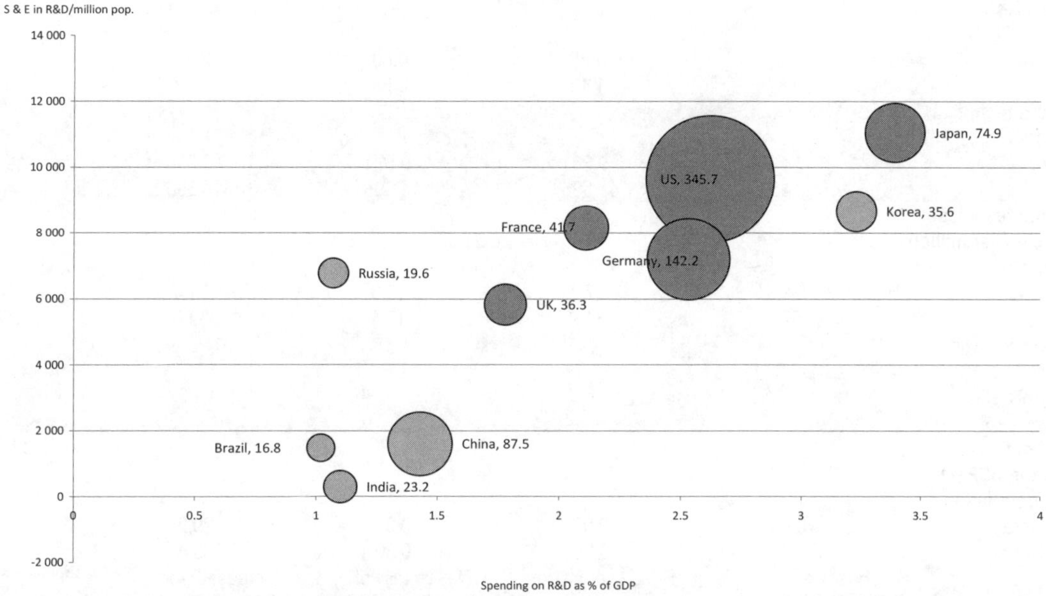

Source: Author's calculations based on data in OECD (2008a) on R&D spending as a percentage of GDP and scientists and engineers in R&D combined with data on 2006 GNI in PPP of respective countries (except for input data on India which is based on author's estimates).

In India, additional R&D investment by MNEs and increased investments by the domestic private sector have raised total R&D spending. This has been concentrated in pharmaceuticals, ICT, electronics and auto parts and has raised Indian R&D expenditure from a 20-year average of 0.88% of GDP to 1.1% in 2005 (for further details, see Dutz, 2007).

In 1990 Russia was a superpower in R&D and spent 2% of its GDP on R&D. However this share fell to 0.74 by 1992. It has increased in the last ten years to 1.08% of GDP in 2006. The stock of scientists and engineers doing R&D fell from over a million in the pre-transition area to 622 000 in 1994 and further to 465 000 in 2006.[13] The bulk of R&D is still done by the public sector. However, foreign firms have been increasing their R&D in Russia to take advantage of the relatively lower cost of Russia's scientists and engineers. The share of foreign funding in Russian R&D increased from 2% in 1994 to nearly 10% in 2006 (OECD, 2008a, pp. 172-173).

13. Watkins (2003) analyses the Russian innovation system and discusses reforms needed to improve its performance.

Table 7.6. The BRICKs: R&D inputs and outputs

Indicator	Brazil	China	India	Korea	Russia
Researchers in R&D					
1995	26 578	531.997	145 115	118 640	650 000
2006	84 971	926 252	117 528	179 812	464 577
R&D researchers per million population					
1995	168	445	157	2 636	4 439
2006	461	926	111	4 162	3 255
Spending on R&D (USD billions)					
USD billions nominal 2007	9.1	37.5	7.7	27.4	8.9
USD billions in PPP 2007	14.6	101.5	21.3	38.9	22.0
Spending on R&D (percentage of GDP)					
1995		0.55	0.80	2.50	
2006	0.82	1.42	0.69	3.23	1.08
Scientific and technical journal articles					
1995	3 471	9 261	9 591	3 806	19 974
2005	9 889	41 596	14 608	16 396	14 412
Scientific and technical journal articles per million population, 2005					
1995	21.5	7.7	10.3	84.4	134.8
2005	52.9	31.9	13.4	339.5	100.7
Patents granted by the USPTO					
Average 1991-95	65	56	36	1 322	74
Average 2002-06	135	448	316	4 233	194
Patent applications granted by the USPTO per million population					
1991-95 average	0.40	0.05	0.04	29.2	0.50
2002-06 average	0.75	0.35	0.30	88.4	1.34
R&D	Has been near 1% of GDP, slightly higher at 1.1% recently; about 50/50 split between government and private sector	Rose over 5 years from 0.55% to 1.4% of GDP in 2006; more than 69% by productive sector and 25% of business R&D by foreign firms	Traditionally at 0.85% of GDP but increased to 1.1% in last two years with significant increase in private and now 50/50 government/private	Increased from 0.5% of GDP in 1965 to 2.5% by the mid-1990s when companies had trouble getting technology; 75% by private sector	Fell sharply from more than 2% of GDP before 1991 to just above 1% now; roughly two-thirds still done by government
Science parks and business incubators	Few science parks, few business incubators, few spin-offs as commercial firms	Rapidly expanding science parks and business incubators with many spin-offs to productive high-technology firms	Few science parks and few business incubators, few spin-offs as commercial firms	Several science parks and business incubators, but less successful than expected.	Several science parks and business incubators; few successful technology-based companies
Intellectual property rights	Historically stronger than China or India; now lags updated regimes in those countries; enforcement weak but less serious than in other countries	Very weak until China was required to update as part of WTO accession; enforcement is weak	Weak until completed compliance with WTO requirements in 2005; enforcement weak, but considered less of a problem than in China	Weak until relatively recently but now the strongest of BRICKs; enforcement is strongest	Still weak on regulatory and enforcement
Protection of IPR WEF Survey 1=least; 7=most					
2008/09 GCR	3.27	3.88	3.70	*4.98*	*2.93*

Sources: *World Development Indicators* 2009; *World Bank KAM* 2008, World Economic Forum 2008/2009 report.

The history of Korea's spending on R&D shows how its relative importance changes as it moves closer to the world frontier. In the mid-1960s Korea was a very poor developing country and was exporting labour-intensive products, including wigs and clothing and transistor radios assembled from imported components. Total R&D spending was just 0.5% of GDP and 80% was from the government. Even in those early days the government was intent on developing greater domestic R&D capability. It set up the KAIST to develop technical human capital for R&D and various public research institutes. It also offered the private sector many incentives to carry out R&D, including tax exemptions, grants and exemptions from military service for scientists and engineers. However it was not very successful in getting firms to increase R&D because they found it easier to get technology by importing capital goods and licensing foreign technology. By the late 1970s though, Korea was moving very quickly up the technology ladder and becoming a competitive threat to companies that had been licensing technology to its firms. The foreign licensors then became more reluctant to license their technology, or would only do it on the condition that Korean firms would not export to their key markets.

In light of these restrictions, the chaebol began to invest heavily in R&D in order to have more technological capability and obtain better terms or to invent around foreign patents. The government also created new research institutes in areas such as telecommunications and information technology. By the mid-1990s Korean spending on R&D had increased to 2.5% of GDP and 80% was being done by the private sector (Kim, 1997). In the last ten years Korea has become an important world-class innovator. Some of its companies (such as Samsung, LG electronics, Daewoo, SK telecom and Hyundai) have become global household names, and some, like Samsung, have become innovation leaders. Since 2006, Samsung has been the second largest patentor in the United States (behind IBM) (USPTO, 2008).

The evolution of the focus of innovation policy in Korea shows that countries or firms that are near the global frontier need to put a lot of effort into developing new knowledge to remain competitive as other countries or companies catch up. Countries therefore have to place a premium on strengthening the institutional infrastructure for developing new knowledge. This means emphasising public and private research centres, university research, and training of scientists and engineers. For companies, it means scanning the world for relevant technical knowledge, developing strategic alliances with other firms that can contribute knowledge, and interacting closely with government and university laboratories to obtain access to basic knowledge, and undertaking work to create new knowledge.

Disseminating knowledge

For the dissemination of knowledge it is useful to distinguish the commercialisation of new knowledge from that of existing knowledge, as they require different mechanisms. Once created, knowledge has to be commercialised, particularly if it has been created in government research labs or universities as opposed to firms that can directly apply the technology. This requires an appropriate technological commercialisation infrastructure which includes: adequate intellectual property rights (IPR) protection, technology transfer offices at universities and research institutes, science-industrial parks, business incubators, early-stage technology finance, and venture capital. Table 7.6 provides some comparative information on these elements across the BRICKs.

The dissemination of the knowledge that exists in an economy takes place by the expansion of the enterprises that developed the knowledge, the sale or transfer of that knowledge, as well as by various forms of imitation or replication by other firms or organisations (Figure 7.1). For the dissemination of knowledge it is important to have appropriate mechanisms for educating potential users about the benefits of the technology. This often involves more than providing technical information. In agriculture for example it involves showing potential users the actual performance of the new technology in their domestic conditions. In manufacturing, much dissemination occurs not simply by the expansion of the innovating firm but by the sale of new machinery or other inputs that embody the new technology. There is also explicit training, demonstration projects, or technical assistance on how to use the technology. In services, the technology is disseminated through direct interaction with the user of the service. However the new service may require new equipment, procedures, or inputs, and training is also usually required to disseminate it.

Efforts to disseminate knowledge have been quite systematic in China. They have included not only extension work in agriculture, but also special programmes such as Spark for rural innovation and Torch for high-technology innovation. In India and Brazil they have mostly concerned agriculture – the green revolution in India and the extensive research and dissemination efforts through Embrapa in Brazil. Although both Brazil and India have made dissemination efforts in manufacturing and services, these have not been as systematic or successful as those in China. It is also likely that the more competitive environment in China has led to more rapid diffusion through imitation as well as through the rapid expansion of the more innovative firms.

In Korea the government has had many programmes aimed at the dissemination of technology to small and medium-sized enterprises. It also made a special effort to promote the dissemination of ICT, including massive training programmes in stadiums to teach people how to use the Internet.

In Russia the dissemination of new knowledge has been a challenge. Some of the mechanisms and institutions that are standard in market economies did not exist in the former Soviet system. A good example is engineering firms which, by their very nature, are an excellent way to apply and disseminate knowledge across different sectors. In the former Soviet system plants were built by the engineering departments of their respective ministries. However dissemination improved as new market-oriented capital goods and equipment suppliers and consultants developed and began to spread new technologies throughout the economy. Nevertheless, many barriers to the spread of more efficient firms remain. In addition, much of the plant and equipment in traditional industries is outmoded if not obsolete (Goldberg and Desai, 2008).

To appreciate the gains that can be obtained from dissemination and effective use of existing knowledge it is useful to examine some empirical data. A recent study on Brazil shows that the value added per worker can be as much as 300 000 times greater in the most efficient than in the least efficient firms in the machinery and equipment sector. The average for the nine sectors studied was 57 000 times greater. To obtain a measure less influenced by outliers, the maximum was adjusted by "eyeballing" the distribution of dispersion and taking as a maximum the value when the distribution began to have some density. That adjusted maximum averaged 53% of the distance to the recorded maximum. Even with these conservative adjustments, it appears that if average productivity could be raised to the adjusted maximum level, average productivity would increase by a factor of 10.

Using a similar methodology, the average level of productivity was estimated to rise by a factor of five in India. It is surprising that productivity dispersions are, on average, twice as large in Brazil as in India, considering that dispersions in the latter already exceed those in most of the countries to which it has been compared (Dutz, 2007).

This analysis suggests just how much national output could be raised – at least in principle – if all Brazilian or Indian firms were to adopt technology that other firms already are using. Obviously, moving to higher-productivity technologies is not costless. The firms that use them are likely to be much larger, use other modern equipment, generally employ more up-to-date management practices, use better inputs, and have better educated and more highly skilled workers. Yet the larger point is that these production technologies are used by at least by some firms in the country, while typical firms that do not are operating far behind their more efficient counterparts. Far more must and can be done to disseminate and effectively employ domestically existing knowledge across the board.

Lessons and implications

Lessons

Table 7.7 presents a summary matrix comparing basic characteristics, innovation strategies and performance of the BRICKs. Some useful lessons can be drawn from their comparative experience.

A first lesson is that technology and innovation are an important component of growth. Higher growth in China, India and Korea compared to Brazil and Russia is due to higher total factor productivity in those economies over long periods of time. This is not to deny the importance of investment for growth. Investment is certainly necessary, but higher rates of investment also permit the incorporation of new equipment embodying better production processes or allowing the production of new or better goods and services, hence the importance of innovation for growth.

The second lesson is the importance of the economic incentive and institutional regime. The key elements of the economic incentive and institutional regime are macroeconomic and political stability, high investment rates, and outward orientation.

This was the case for China and Korea but not for Brazil, India and Russia. India's growth did not accelerate until after the trade liberalisation reforms in 1991, and then not again until the last five years as it opened up more to foreign investment and to trade in goods and services. Although Russia has strong R&D and human capital it performed poorly because of the lack of competition and/or outward orientation. Russia has been growing rapidly since the devaluation and reforms in 1998. In fact, growth has speeded up because of the rapid increase in petroleum prices until mid-2008. However the lesson is not as simple as liberalisation and greater global integration. Korea, China and India are not likely to have performed as well as they have if they had not developed strong domestic capability. Much of this capability was developed when they were more protected from imports. And it is likely that if they had completely open economies, they would not have had a chance to develop that capability. Thus industrial policy was important for their development (Dahlman, forthcoming; Pack and Saggi, 2006). The critical issue is to balance policies to help domestic firms develop this capability and avoid the risk of creating permanent infants. The sequencing of such policies is important, as is the efficiency and effectiveness of government in implementing them.

Table 7.7. The BRICKs: summary comparative matrix

	Brazil	China	India	Korea	Russia
	Commodity exporter with islands of excellence	Manufacturing workshop for the world	Knowledge-intensive service exporter	Conglomerates lead export growth	From industrial superpower to growth led by natural resources and exports
Tap global knowledge through trade and technology transfer	Low, relatively closed economy, but some use of technology licensing	Massive imports of capital goods and components, technology licences	Very poor except for ICT sector	Excellent with much knowledge from foreign buyers	Poor until last decade when started strong imports, but exports still largely petroleum-based
Tap global knowledge through FDI inflows	High, but fewer externalities than China	Very high and strong positive externalities for local firms	Very low until quite recently because of restrictions	Very low until after 1997 financial crisis	Low but increasing especially more recently into petroleum sector
Human capital	Weak especially in quality, limits on absorption capability	Increased quickly, has surpassed Brazil in average years of education and United States in tertiary education (numbers)	Low tertiary enrolment rates, poor quality except for IITs and ITMs	Very strong but too rigid and too academic, not very good at working in groups	Very strong but lost much high-level human capital. Also gap in more market-oriented disciplines
ICT hardware	Very low	High	Very low	Very high	Low
ICT software	Low except for financial services	High	Very high	Moderate because of language	High, but not using it well
R&D	Moderate: 1.1% in 2007	High: 1.42% of GDP in 2006	Moderate: 1.1% of GDP in 2007	Very high: 3.3% of GDP in 2006	Moderate: 1.08% of GDP in 2006 from 2% of GDP in 1990
Overall innovation system	Active S&T, but not so successful in growth or welfare. Problems with low human capital and poor economic incentive regime	FDI-dependent, active S&T, now moving from imitation to innovation; very successful	Autarkic until early 1990s active S&T policy, now tapping global knowledge and innovating; less successful than China	Autonomous; little FDI, but tapping global knowledge, investing in own R&D and education	Isolated, had strong R&D sector and high human capital, but poor economic and institutional regime. Lost much high-level technical human capital and reduced R&D effort in transition to market economy
Broader policy regime	Inward-oriented	Outward-oriented	Very inward until recently	Outward-oriented since late 1960s	Inward-oriented except for petroleum exports
EIR macro	Weak	Very strong	Strong	Moderate	Moderate/variable?
EIR rule of law	Weak	Weak	Weak	Moderately weak	Very weak
EIR stability	Moderate	Very strong	Moderate	Moderate	Very strong
Performance	Very weak until after 2000 commodity boom	Excellent throughout	Weak until acceleration in late 1980s; major jump since 2003	Strong until 1997 financial crisis, but quick recovery	Very weak until post-1998 recovery and rapid growth based on natural resource

The third lesson is the importance of effectively tapping into global knowledge. The global stock of knowledge is growing very rapidly and is much larger than even the most technologically advanced countries (even the United States) can hope to produce. Therefore it is essential to tap into global knowledge. It is a large part of the reason for the extremely rapid growth of China, Korea and the ICT sector in India. China has made extremely good use of all the means of doing so. India has not made much use of any of the channels except education to some extent, its diaspora, and more recently FDI and trade. Brazil, like India, has not made as much use of trade. Unlike India, however, it attracted high levels of FDI but without as strong results as China. Until it restarted growth in the late 1990s, Russia relied on its strong domestic R&D and knowledge base and did not make much use of global knowledge. In large part, it is why it was not competitive. In 2000 it began to import capital goods and components and to license foreign technology. However, most of it has gone to the oil-related sectors. Russian manufacturing is not very competitive as can be seen from the rapid shrinking shares of manufactures in its exports (Table 7.A1.2). Korea tapped into global knowledge through all the methods mentioned except FDI. Because it opted to develop without relying much on FDI, it had to invest earlier in R&D and human capital in order to develop it own domestic capability.

The fourth lesson is that it is not just a matter of large amounts of FDI. It is important to use it effectively. Brazil had a great deal of FDI, but it was less globally competitive because much of the FDI focused on the protected domestic market which did not require the best technology. China got more advanced technology because its domestic market was very competitive. Initially its attractiveness was its low-cost labour for export-oriented FDI. Subsequently it was the lure of the very large and rapidly growing domestic market. Both of these gave the government strong negotiating power with respect to FDI. FDI has also been important for the Indian ICT sector. Korea initially made up for not obtaining technology through FDI through heavy investment in human capital and domestic research. An important issue is that most countries cannot get access to FDI or are not able to manage it effectively. Also, most countries do not have Korea's ability to compensate for the lack of FDI by heavy investment in high-level human capital and R&D to generate knowledge in order to compete on global markets

The fifth lesson is the importance of education. Human capital is obviously crucial for developing knowledge and even for tapping into global knowledge. In India the low average educational level combined with little access to global knowledge (because of inward-oriented government policies) has constrained its assimilation of foreign knowledge. However, in India the creation of a critical mass of highly educated engineers and MBAs was essential for the successful development of India's software and ICT-enabled services. In Brazil workers' low average educational level has also limited making effective use of global knowledge. However, investments in high-level human capital have been vital for Brazil's islands of excellent in airplanes, deep oil exploration and agricultural research. High literacy levels were part of the attractiveness of China's low labour costs.

However, Russia offers an important cautionary note. It shows that high levels of human capital, even when associated with high R&D capacity and spending, as was the case before the transition, are not enough to guarantee growth or competitiveness. The economic regime must provide incentives to improve performance and reallocate assets to more efficient use. In addition it is important to take advantage of the tremendous and rapidly growing stock of knowledge outside the country.

The sixth lesson is the importance of the diaspora. It was important for Korea to strengthen its knowledge base in the early years of creating local technological capability. It was very important for China's very rapid development. It has also been very important for the creation of the Indian information-enabled software and knowledge services export industry. Both Brazil and Russia have benefited less from their diaspora. Brazil has not sent many students abroad for study, and most of those who go tend to return. Russia suffered a rapid drop in its stock of high-level technical human capital during 1991-95, and many students used their higher education degrees to emigrate until recently, when the country's high growth rates made it more attractive for them to stay in Russia.

The seventh lesson is the importance of copying and reverse engineering. This has meant low enforcement of IPR. This was been a key element of Korea's and China's rapid catch-up. However, as Korean firms invested more in their own R&D, the IPR regime has improved (Table 7.6). China now places more focus on domestic innovation. In India the explicit lack of protection of process patents was a key to the development of a strong indigenous pharmaceutical industry. Domestic firms were able copy foreign products and to produce them with slightly different processes. This had the double benefit of producing pharmaceuticals at low prices to improve the welfare of Indian citizens and developing a strong domestic pharmaceutical industry. When India extended process patents to pharmaceuticals in 2005, it had developed a strong domestic industry that could compete with foreign firms. In Brazil and Russia, there is still a problem with respect and enforcement of IPR.

The eighth lesson is that the focus of innovation strategy changes over time. Countries at early stages of development have little technological capability, and the easier trajectory is to import knowledge in as many ways as possible and to use it to improve their productivity. As they acquire more capability, they become able to search for global knowledge effectively, to acquire it on better terms, and to adapt it to local conditions. As countries acquire more capabilities, they can go from imitation to innovation. This is the pattern in Korea, China, Brazil and India.

There also seems to be some convergence towards outward-oriented and domestic R&D except in Russia. This is linked to a strong focus on producing high-level technical human capital as was the case in Korea. It is also the case in China and to a lesser extent in India. Brazil is also focusing on increasing the number of scientists and engineers and plans to increase its investments in R&D.

Russia has been an exception to many of these patterns. It actually went from strong domestic R&D with little formal tapping into global knowledge to smaller domestic R&D capability and greater use of global knowledge. Now it has opened up more to acquiring global knowledge through trade and formal technology licensing and even to FDI. However FDI has come primarily to the oil and gas sectors. Furthermore, Russia has not developed strong linkages with the global R&D system aside from the emigration of many of its scientists and engineers. In addition the domestic R&D system continues to have few strong linkages with the domestic productive sector.

A ninth lesson is the importance of strong diffusion efforts. As argued above, acquiring or creating technology and having it used by just one firm is not enough. There is a tremendous potential to increase productivity if the technology diffuses throughout the economy. This can be accelerated by explicit government or other programmes to demonstrate the gains to be had by adopting new technology.

Implications for strategies of developing countries

First, a priority for all developing countries is to acquire and use knowledge that already exists. This is less costly and less risky than creating new knowledge. While some of this knowledge is protected by intellectual property rights and would have to be purchased, a tremendous amount is in the public domain. Policies of opening to global knowledge are therefore critical. This includes trade, foreign direct investment, technology licensing as well as copying and reverse engineering. In addition there is great scope for acquiring foreign knowledge through foreign education, travel, participation in foreign trade shows, Internet access to information and databases, attracting back the diaspora, setting up R&D antennae abroad, buying foreign companies to get access to their technology, brand name or market.

How well countries use this form of innovation will depend not only on policies but also on the support institutions of the country as well as the capability of the institutions and the people in them. Therefore, much more attention needs to be given to how to scan for relevant global knowledge. This means an explicit effort to develop capabilities to assess it, to acquire it, and to adapt it to local conditions. Most developing countries have not put a great deal of effort into this so there is great scope for improvement.

Multinational companies are the principal creators and disseminators of knowledge. In addition they control most of the global supply and distribution chains. Therefore developing countries need to seek not only to attract foreign investment but, very importantly, to leverage it. This involves not just production, employment and exports, but access to technology, development of backward linkages, standards, worker and management training, and access to global markets. However, foreign investment does not even go to many countries, so that careful thinking is needed about ways to improve conditions that may be hindering foreign investment (poor investment climate, corruption, poor infrastructure, poor skill base, etc.). It also requires proactive strategies to develop the externalities that foreign investment can bring.

A second key priority for all developing countries is to disseminate and make effective use of knowledge that exists in the country, whether it has been brought from abroad or developed locally. As noted, all countries have a tremendous dispersion of productivity within sectors, and raising average productivity to local best practice (or even better to global best practice by acquiring more knowledge from abroad) can result in high returns. This requires strong public policies on dissemination and use, including standards and quality control legislation. It also requires an institutional infrastructure with technical information services; extension services for agriculture, industry and services; productivity organisations; metrology standards and quality control institutions; and industrial clusters. In terms of capabilities, people need to have not just basic literacy and numeracy but job- and technology-specific skills in order to use the new technologies and production organisational techniques. Some of these require improving the basic education curriculum. Others have to be provided through specialised training at vocational centres or technology suppliers or as part of firm-specific training. Again, this is an area with tremendous potential, but most developing countries do not have the policies, institutions or capability to exploit it. This merits much more effort.

Third, developing countries have a big challenge in the area of R&D given the importance of new knowledge for competitiveness. This is most relevant for larger countries, even low-income countries such as India, because of the critical mass created by their large size and potential use of scientific and technical resources. Even smaller poor countries have to have some R&D capability. At the very minimum, they need enough to assess global knowledge, to negotiate to acquire it, and to adapt it to local conditions.

A critical policy issue here is how limited public resources are allocated and how effectively they are used. Unfortunately, in most developing countries R&D resources are limited and are typically not allocated or used very well. A policy priority is therefore to improve the allocation of public resources. This includes better definition of the areas the government should support when budgets are small. A second priority is to manage these resources effectively to make a contribution to the economy. It is difficult to justify pure academic research in countries with pressing social and economic needs where more applied R&D can make an important contribution. However, some basic research capacity may be the only way to get a country plugged into global knowledge networks that may be important for the country. Wagner (2008) explains what it takes to enter global research networks.

Beyond R&D by the public sector, another key issue is to get the private sector to do more R&D. The private sector needs to be encouraged to undertake more R&D, not only to be able to keep up to date with new developments and incorporate them, but to also carry out cutting-edge research in areas critical for their competitiveness. Because of the problems of market failure in the appropriability of the returns to private R&D, this calls for public support programmes such as matching grants and tax subsidies to stimulate R&D by the private sector.

While it makes sense for developing countries to invest in areas in which they have a comparative advantage in order not only to maintain but to enhance that advantage, they also need to invest in new technological areas such as genetic engineering, biotechnology and nanotechnology. The public sector will have to play a greater role in carrying out this type of riskier and more uncertain research. It should be seen as part of an investment portfolio strategy of exploring new areas with potentially high returns. These investments are necessary to have the capability to move rapidly into areas that begin to show promising results.

It is necessary to put in place not only appropriate policies but also the supporting institutions, public and private, to create new knowledge and to support the acquisition and dissemination of knowledge. A key problem in most developing countries is that even when relevant knowledge is created in public labs or universities, it is not commercialised. This calls for the supportive infrastructure of technology parks, business incubators, technology transfer centres and venture capital to commercialise knowledge as is happening in East Asia, particularly in Korea, Chinese Taipei and China. The country also needs to develop the human resources to conduct and manage R&D and to commercialise the results (techno-entrepreneurs).

Obviously how much a country should invest in an R&D and commercialisation infrastructure will depend on its resources and its size. The richer and more developed in terms of institutions and human capital, the more it can do. Smaller and poorer countries are going to be more limited. They will nonetheless have to develop innovation strategies in order to improve their opportunities to use knowledge to leverage their development. They should take into account some of the lessons that can be learned from the BRICKs, tailoring them to their specific human, institutional and natural resource endowments. There is no universal approach. Each country will have to see how it can increase its ability to take advantage of existing knowledge and to increase its capability to develop its own appropriate knowledge. Improving education and technological capability will have to be an important element of this strategy. However, innovation is not just about education or technology. It also depends very much on the broader economic and institutional context and on how effectively a country inserts itself into a very demanding global context and adjusts to changing challenges and opportunities.

Annex 7.A1.

Structure of output and exports of merchandise and commercial services in the BRICKs

Table 7.A1.1. The BRICKs: structure of output, 1980, 1995, 2005

	Brazil	China	India	Korea	Russia	World
Agriculture						
1980	11	30	38	14	..	7
1995	6	20	26	6	7	4
2007	6	11	18	3	5	3
Industry						
1980	44	49	24	40	..	38
1995	28	47	28	42	37	30
2007	29	49	30	39	38	28
Manufacturing						
1980	33	41	16	28	..	25
1995	19	34	18	28	..	20
2007	18	32	16	28	19	18
Services						
1980	45	21	38	46	..	55
1995	67	33	46	52	56	65
2007	66	40	52	58	57	69

Source: World Bank, *World Development Indicators 2008*, I2000.

Table 7.A1.2. The BRICKs: structure of merchandise exports, 1980, 1990, 2006

	Brazil	China	India	Korea	Russia	World
Food						
1980	46	..	28	7	..	12
1995	29	8	19	2	2	9
2007	26	3	9	1	2	7
Agri. raw materials						
1980	4	..	5	1	..	4
1995	5	2	1	1	3	3
2007	4	0	2	1	3	2
Fuels						
1980	2	..	0	0	..	
1995	1	4	2	2	43	6
2007	8	2	16	7	61	10
Ores and minerals						
1980	9	..	7	1	..	11
1995	10	2	4	1	10	3
2007	12	2	8	2	8	4
Manufacturing						
1980	37	..	59	90	..	66
1995	52	84	73	92	26	75
2007	47	93	64	89	17	72

Source: World Bank Institute (2008), *World Development Indicators 2000*.

Table 7.A1.3. The BRICKs: structure of commercial service exports, 1990, 1995, 2005

	Brazil	China	India	Korea	Russia	World
Transport						
1990	36.4	47.1	20.8	34.7	..	26.6
1995	43.3	18.2	28.1	41.9	35.8	26.9
2007	18.0	25.7	10.2	54.9	30.2	23.7
Travel						
1990	37.3	30.2	33.8	34.5	..	35.2
1995	16.2	47.4	38.2	23.3	40.8	32.5
2007	22.0	30.6	11.9	9.4	24.6	26.6
Insurance and financial services						
1990	3.1	3.9	2.7	0.1	..	6.6
1995	16.9	10.1	2.5	0.4	0.6	5.9
2007	7.2	0.9	4.2	7.3	4.0	8.4
Computer, information and other commercial services						
1990	23.2	18.7	42.7	30.7	..	32.1
1995	23.6	24.4	31.4	34.5	22.8	36.2
2007	52.8	42.7	73.7	28.5	41.2	42.1

Source: World Bank, *World Development Indicators*, 2005 and 2008.

References

Baldwin, R. (2006), "Globalization: the Great Unbundling(s)", paper contributed to "Globalization Challenges to Europe and Finland", organised by the Secretariat of the Economic Council, Prime Minister's Office, Canada, June.

Bergstein, C.F., G. Bates, N.R. Lardy and D. Mitchell (2006). "China. The Balance Sheet: What the World Needs to Know Now about the Emerging Superpower", Center for Strategic and International Studies and Institute for International Economics, Washington, DC.

Bosworth, B. and S.M. Collins (2007), "Accounting for Growth: Comparing China and India", National Bureau of Economic Research Working Paper No. 12943, Washington, DC.

Bosworth, B. and A. Maertens (2008), "The Role of the Service Sector in Economic Growth and Employment Generation in South Asia", World Bank draft paper for a volume on Trade in Services in South Asia.

Dahlman, C. (forthcoming), "Growth and Development in China and India: The Role of Industrial and Innovation Policy in Rapid Catch-Up", in M. Cimoli, G. Dosi and J.E. Stiglitz (eds.), *Industrial Policy and Development: The Political Economy of Capabilities Accumulation*, Oxford University Press, New York.

Dahlman, C., D. Zhihua Zeng and S. Wang (2007), *Enhancing China's Competitiveness through Life Long Learning*, The World Bank, Washington, DC.

Dahlman, C. and A. Utz (2005), *India and the Knowledge Economy: Leveraging Strengths and Opportunities*, The World Bank, Washington, DC.

Dahlman, C. and J-E. Aubert (2001), *China and the Knowledge Economy: Seizing the 21st Century*, The World Bank, Washington, DC.

Desai, M. (2005), "India and China: An Essay in Comparative Political Economy", in W. Tseng and D. Cowen (eds.), *India's and China's Recent Experience with Reform and Growth*, Palgrave MacMillan, New York.

Dutz, M. (ed.) (2007), *Unleashing India's Innovation Potential*, The World Bank, Washington, DC.

Farrell, D., M.A. Laiboissiere and J. Rosenfeld (2005), "Seizing the Emerging Global Labour Market", *The McKinsey Quarterly*, No. 3, pp. 93-103.

Gill, I. and H. Kharas (2007), *An East Asian Renaissance: Ideas for Economic Growth*, The World Bank, Washington, DC.

Goldberg, I. and R.M. Desai (eds.) (2008), *Can Russia Compete? Enhancing Innovation in a Globalizing World*, The Brookings Institution, Washington, DC.

Gordon, R. and P. Gupta (2006), "Understanding India's Services Revolution,", in W. Tseng and D. Cowen (eds.), *India's and China's Recent Experience with Reform and Growth*, Palgrave MacMillan, New York, pp. 229-263.

Krueger, A.B. and M. Lindale (2001), "Education and Growth: Why and for Whom?" *Journal of Economic Literature* Vol. XXXIX. American Economic Association, Nashville, TN.

Jaruzelski, B. and K. Dehoff (2008), "Beyond Borders: the Global Innovation 1000", Booz Allen Hamilton, *Strategy and Business*, Issue No. 53.

Kim, L. (1997), *From Imitation to Innovation: The Dynamics of Korea's Technological Learning*, Harvard Business School Press, Boston.

Lall, S. (2003), "Reinventing Industrial Strategy: The Role of Government Policy in Building Industrial Competitiveness", Working Paper No. 111 (October), Queen Elizabeth House, Oxford.

Lall, S. (2004), "Industrial Success and Failure in a Globalizing World", *International Journal of Technology Management and Sustainable Development*, Vol. 3, No. 3, pp. 189-213.

Naughton, B. (2007). *The Economy of China: Transitions and Growth*, MIT Press, Cambridge, MA.

OECD (2005), *Education at a Glance*, OECD, Paris

OECD (2006), *Economic Survey of Brazil: Boosting Innovation Performance*, OECD, Paris.

OECD (2006), *Economic Survey of Korea: Sustaining High Growth through Innovation – Reforming the R&D and Education System*, OECD, Paris.

OECD (2007), *Science Technology and Industry Scoreboard*, OECD, Paris.

OECD (2008a), *Science, Technology and Industry Outlook*, OECD, Paris.

OECD (2008b), *Reviews of Innovation Policy: China*, OECD, Paris.

OECD (2008c), *Education at a Glance 2008*, OECD, Paris.

Pack, H. and K. Saggi (2006), "The Case for Industrial Policy: A Critical Survey," World Bank Research Policy Working Paper No. 3839, Washington, DC.

Panagariya, A. (2005), "India in the 1980s and 1990s: A Triumph of Reforms", in W. Tseng and David Cohen (eds.), *India's and China's Recent Experience with Reform and Growth*, Palgrave MacMillan, New York.

Rhee, Y.W., G. Purcell and L. Westphal (1984), *Korea's Competitive Edge*, John Hopkins University Press, Baltimore Maryland.

Rodriguez, A., C.J. Dahlman and J. Salmi (2008), *Brazil: Knowledge and Innovation for Competitiveness*, The World Bank, Washington, DC.

Singh, N. (2007), "Services-led Industrialization in India: Assessment and Lessons", in D. O'Connor (ed.), *Industrial Development for the 21st Century: Sustainable Development Perspectives*, United Nations, New York, pp. 235-291.

Soubotina, T. (2006), "Generic Modes of Technological Learning by Developing Countries", Draft Report, World Bank Science and Technology Program, Washington, DC.

Suh, J. and D. Chen (2007), *Korea as a Knowledge Economy: Evolutionary Processes and Lessons Learned*, The World Bank, Washington, DC.

UNCTAD (2005), *World Investment Report*, UNCTAD, Geneva.

UNESCO (2008), *Global Education Digest 2008*, UNESCO Institute of Statistics, Montreal.

United States Patent and Trademark Office (USPTO) (2008), *Patent Statistics Reports*, *www.uspto.gov/web/offices/ac/ido/oeip/taf/asgstc/regions.htm*.

Wagner, C.S. (2008), *The New Invisible College: Science for Development*, The Brookings Institution, Washington, DC.

Watkins, A.J. (2003), "From Knowledge to Wealth: Transforming Russian Science and Technology for a Modern Knowledge Economy", World Bank Policy Research Working Paper No. 2974, Washington, DC.

Wosmann, L. (2003), "Specifying Human Capital", *Journal of Economic Surveys*, Vol. 17, No. 3, Blackwell Publishing, London.

World Bank (2008), *World Development Report 2008*, The World Bank, Washington, DC.

World Bank, Knowledge Assessment Methodology (KAM) 2008, *www.worldbank.org/kam*.

World Bank (2008), *Cost of Doing Business 2009*, The World Bank, Washington, DC.

World Bank (2009), *World Development Indicators 2009,* The World Bank, Washington, DC.

Chapter 8

Technology diffusion in the developing world

Andrew Burns[1]

This chapter reports on results of a large-scale effort to estimate directly the extent to which different technologies have penetrated the economies of developing countries and the pace at which penetration has been changing. It finds that on average middle-income countries use technologies at about one-half the rate of intensity of high-income countries and that the pace of technological progress in these countries has been much faster over the past 15 years. However, the level of technology achieved and the pace of progress vary widely across countries with the most advanced middle-income countries about as advanced as the less advanced high-income countries. Increased access to foreign technologies, through foreign direct investment, imports of high-technology products and intermediate inputs have played a central role in the dissemination of technologies from high-income countries to developing countries. However, such flows are not in themselves sufficient. A country's technological absorptive capacity (the level of basic and advanced technological literacy, the quality of the regulatory environment, access to finance, and the effectiveness of proactive government policies to promote technology creation and diffusion) determines the extent to which these technologies are absorbed by domestic firms and incorporated into daily economic life. Weak absorptive capacity in Latin America suggests that it is converging towards a lower level of technological achievement than countries in Europe and Central Asia.

1. Andrew Burns is Lead Economist with the Development Prospects Group and Economic Advisor to the Director, World Bank. The findings, interpretations and conclusions are those of the author and do not necessarily represent the views of The World Bank or its Executive Directors. This paper draws heavily on work done for the 2008 edition of the World Bank's *Global Economic Prospects: Technology Diffusion in the Developing World,* which was produced by a team led by the author. It included Dilek Aykut, Antonio David, Yvan Decreux, Teng Jiang, Annette de Kleine, Mariem Malouche, Sanket Mohapatra and Bill Shaw.

Introduction

Technological progress – improvements in the techniques (including firm organisation) by which goods and services are produced, marketed and delivered to end users – is at the heart of human progress and development. At the national level, technological progress can occur through invention and innovation, the adoption and adaptation of pre-existing but new-to-the-market technologies, and the spread of technologies across firms, individuals and the public sector within the country.

Traditionally in economics, technological progress is measured by the change in total factor productivity (TFP) – the speed with which output increases after controlling for increases in inputs (capital and labour). While useful for understanding the impact of technology on growth and well-being, TFP measures offer little insight into the nature of the technology employed or the process by which it is acquired.[2] This chapter takes a different approach, reporting on efforts to establish a direct measure of the level of technology achieved and of technological progress.

Earlier efforts to measure these levels directly have tended to be based on a linear combination of a selection of factors thought to be correlated with technology.[3] A common shortcoming of those approaches is that the relative importance of each variable included in the indexes constructed is essentially determined by the subjective judgment of the individuals building them.

The research reported here builds on this earlier work, deriving an overall measure of the level of technology achieved from many of the same base indicators, but using a statistical technique (principal components analysis) to identify the weights attached to each indicator.[4] It also departs from many earlier studies by including a set of variables that capture the extent to which countries are exposed to technology generated elsewhere in the world, *e.g.* indicators capturing some of the channels of technology diffusion. Estimates are made for two points in time, the early 1990s and the early 2000s. This allows for a direct measure of the pace of technological achievement in the 77 countries for which sufficient data are available. A similar analysis is made of the principal factors underpinning technological absorption.

---

2. Typically, TFP is measured as the influence on growth of all factors other than increases in capital and labour. Thus changes in TFP could reflect changes in the composition of output (for example, a shift from agriculture to manufacturing), changes in the quality of labour or capital not reflected in the data (for example, educational levels), or any other variable that is an important determinant of growth but is not included in growth equations.

3. These include: *the index of innovation capability* (UNCTAD, 2005), which consists of an unweighted average of an index of human capital indicators and scientific activity scaled by population; *the index of competitive industrial performance* (UNIDO, 2002), which focuses on outputs and is calculated as a simple average of four manufacturing indicators; and *the technology achievement index* (UNDP, 2001), which combines as simple averages indicators of human skills, of the diffusion of old innovations, of recent innovations and of the creation of technology. See Archibugi and Coco (2005) for a summary and comparison of some of the most prominent of these.

4. A similar approach has been used in health economics (Gwatkin *et al.*, 2000a, 2000b, 2000c; McKenzie, 2003, Montgomery *et al.*, 2000; Yvas and Kumaranayake, 2006), in poverty analysis (Sricharoen and Buchenrieder, 2005); in regulatory policy analysis (Nicoletti *et al.*, 1999) and in the analysis of e-readiness in India (Ministry of Communications and Information Technology, Government of India, 2006).

The chapter is structured as follows. In the next section, a framework for understanding the process of technological progress in developing countries and the recent evolution of domestic features that facilitated technological absorption are discussed. The following section presents the various indicators that make up the index of technological achievement, along with the main features of the indicator. A presentation of the estimates of the rate of technological progress across countries based on a second estimate using data from the early 1990s. Factors that might explain the extent of technological absorption (and creation) across countries and how these results compare with actual observations are next discussed. A penultimate section looks at the determinants of technological diffusion in developing countries, and places equal emphasis on the international connections and networks that expose firms and individuals in developing countries to cutting-edge technologies and on the domestic factors that determine how well countries are able to absorb and apply those technologies.

Technological progress and economic development

Central to understanding the role of technology is the recognition that technology and technological progress are relevant to, and exist in, a wide range of economic activities, not just manufacturing or computers. For example, some estimates suggest that technological progress has boosted productivity in agriculture four times as quickly as in manufacturing (Martin and Mitra, 2001). Indeed, seemingly low-technology products such as corn or flowers can be the result of relatively high-technology production processes, while in some countries the production of ostensibly high-technology products such as computers is the result of relatively low-technology assembly activities. Finally, in many cases technology is embodied in production and management systems rather than in physical goods or software algorithms. A computer loaded with the latest software that sits unused on a desk for most of the day is a very different manifestation of technology from the same computer running a production process or managing an accounts payable system.

As mentioned, economists normally measure technological progress by the rate of change of TFP. International comparisons suggest that enormous gaps in TFP exist between high-income and low- and middle-income countries (Table 8.1). The average level of TFP in low-income countries was estimated to be only 5% of the level in the United States in 2005 and less than 25% of that level in upper-middle-income countries. While TFP gaps narrowed or remained stable over the past 15 years in low- and middle-income countries considered as a whole, gaps have widened in three of six developing regions.

The relationships between income growth, technological progress, capital accumulation and welfare are, of course, much more complex than can be summarised in a simple measure of TFP, partly because each factor of production and the technology with which factors are combined are dependent on one another. Capital goods, business processes that accompany foreign direct investment (FDI) and academic knowledge often embody significant technological progress, and there is no simple way to distinguish between the contribution that each makes to growth. Moreover, the contribution of technology to welfare is only imperfectly measured by its impact on GDP (quality improvements, health effects and social benefits are among the impacts that are imperfectly measured).

Table 8.1. Disparity among TFP levels remains wide

	TFP relative to the United States, 2005 (Index, United States=100)	Annual TFP growth, 1990-2005 (Annual % change)
Regions		
East Asia and Pacific	8.4	5.1
Eastern Europe and Central Asia	21.7	2.2
Latin America and the Caribbean	19.3	0.2
Middle East and North Africa	13.3	0.5
South Asia	5.8	2.3
Sub-Saharan Africa	5.6	0.2
Income level		
High-income OECD countries	77.1	1.3
High-income non-OECD countries	53.1	0.7
Upper-middle-income countries	23.7	1.2
Lower-middle-income countries	9.6	3.2
Low-income countries	5.2	1.7

Source: Poncet ,Sandra (2006), “The Long Term Growth Prospects of the World Economy: Horizon 2050”, Working Paper 2006–16, Centre d’Etudes Prospectives et d’Informations Internationales, p. 33.

Even relatively simple technologies can have far-reaching development impacts. In many low-income countries, fairly commonplace technologies are often in short supply because of weak capacities. Moreover, relatively simple innovations such as those of the green revolution, information about simple storage procedures, rainwater catchment systems, and the use of insecticide-treated mosquito nets can have profound effects on rural health, poverty and incomes.

A framework for understanding technological progress in developing countries

The pace at which technology progresses in a country, be it through technology creation (exploiting pre-existing knowledge) or the adoption and adaptation of technologies created elsewhere, depends on the extent to which the country is exposed to foreign technologies and the ability of the domestic economy to absorb and adapt the technologies to which it is exposed. Many developing countries do not have the skills and competences to participate meaningfully in the kind of scientific innovation and invention that occurs at the frontier. As a result, the bulk of technological progress in developing countries is achieved through the absorption and adaptation of pre-existing and new-to-the-market or new-to-the-firm technologies.

Figure 8.1 presents a stylised description of the process by which technology is created and absorbed in developing countries. As a first step, an economy and its scientific community are exposed to the technological frontier. For the domestic knowledge creation sector, this comes principally through contact with researchers in the rest of the world. For the economy, it occurs through contact with higher-technology business processes, products and services through foreign trade, foreign direct investment, and contact with the national diaspora and other communications channels including academic and international organisations (the large arrows at the top of the figure). The larger these flows, the greater the exposure of the economy to the global technological frontier. Goldberg *et al.* (2008) for example find that in Europe and Central Asia, the technological sophistication of countries varies directly with the extent to which scientists in those countries exploit the ideas and results of scientists in high-income countries.

Figure 8.1. Technological progress in developing countries: an analytical framework

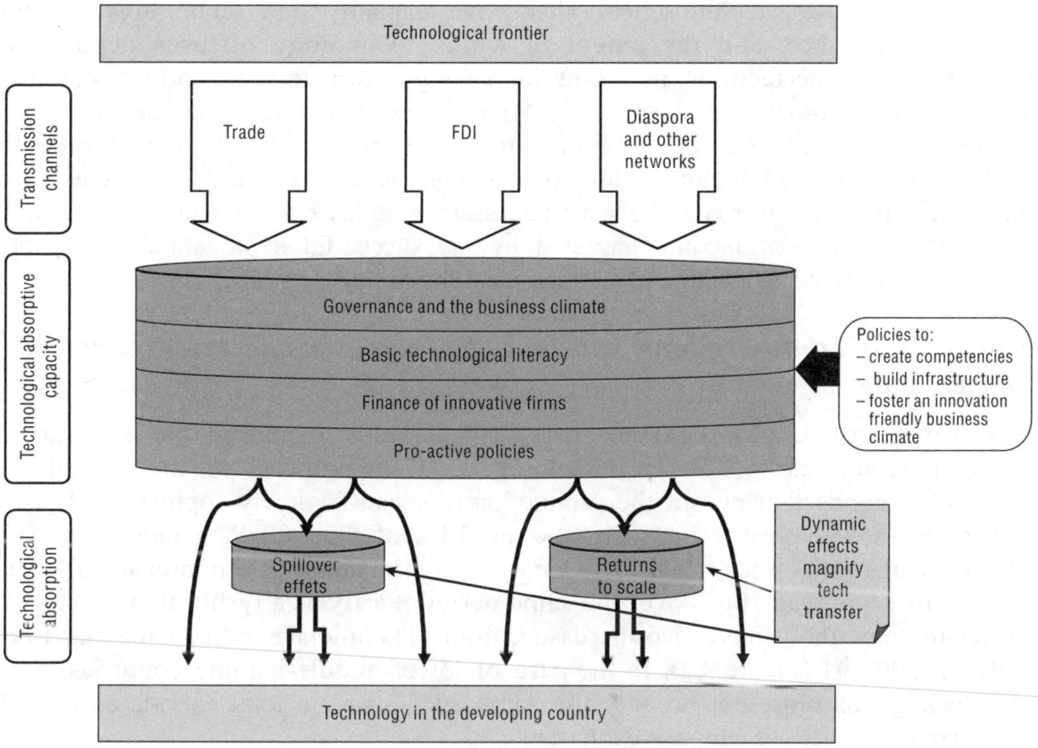

Source: The World Bank.

But exposure to new ideas and techniques is not sufficient to ensure technological progress on the ground. The extent to which this occurs depends on the technological absorptive capacity of the economy and the quality of and incentives facing the scientific community (represented by the multiple-ringed drum). These in turn, depend on the overall macroeconomic and governance environment, which influences the willingness of entrepreneurs to take risks on new and new-to-the-market technologies. In addition, the level of basic technological literacy and advanced skills determines a country's capacity to implement technologies and to do the research necessary to understand, implement and adapt imported technologies, as well as to create and develop them. Because firms are the basic mechanism by which domestic or foreign technology spreads within the private sector of an economy, the extent to which financing for innovative firms is available – through the banking system, remittances or government support schemes – will also influence the pace of progress.

Government policy also has a crucial role to play. First, governments are often the primary channel through which certain technologies (electricity, fixed-line telephones, transport infrastructure, medical and educational services) are delivered.

Moreover, government policy is largely responsible for creating a business environment that facilitates easy firm entry and exit and is not hostile to the profits to be made from exploiting new technologies. Too often, rules and/or specific features of the domestic market prevent firms from profiting from the exploitation of a new technology, which as a result does not spread within the country. Policy should also ensure that R&D and dissemination efforts give priority to creating and introducing products for which there is a market (domestic or foreign) and assisting firms to exploit them.

The overall process is, of course, much more complicated and much less mechanistic. Technological flows, technological absorptive capacity and technology creation influence each other, and the extent to which technology diffuses depends on various market imperfections, including increasing returns to scale and technological spillovers (the small rings in Figure 8.1). Here the existence of a financial sector that intermediates between savers and innovators may be necessary to overcome the initial cost of some new technologies. In particular, access to finance may be essential if innovative firms are to reach the scale necessary to unleash a potential virtuous circle, such that the additional income garnered by the successful exploitation of one new technology permits the acquisition of another, resulting in further gains.

Trade, FDI and diaspora have substantially increased countries' contacts with technology

Over the past 15 years globalisation has markedly increased the exposure of developing countries to foreign technologies. Their imports of the capital and intermediate goods that permit the production of technologically sophisticated goods and services now represent between 6% and 14% of their GDP, a more than 80% increase in this ratio since 1994. The ratio of high-technology imports to GDP has increased by more than 100% over the same period. Partly as a result, their exports of high-technology goods have also increased, from 11% of total exports in the mid-1990s to 19% in 2002-04 (Figure 8.2). In the case of lower-middle-income countries, high-technology goods represent broadly the same 23% share in total exports as in high-income countries (15% if China is excluded).

Figure 8.2. Developing countries' trade in technology goods has risen

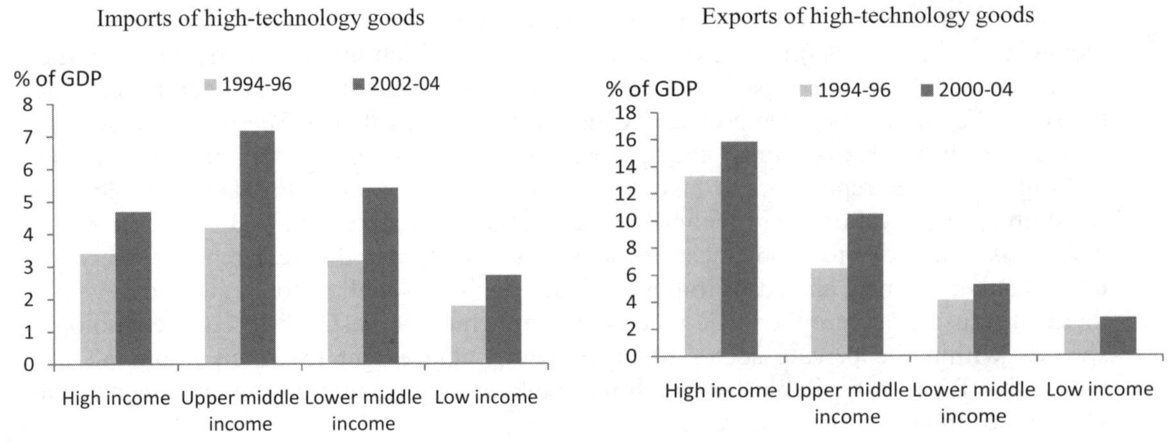

Source: The World Bank.

Foreign direct investment is a major source of process technology and learning-by-doing opportunities for individuals in developing countries. Over the past 15 years FDI inflows to developing countries has almost doubled as a percentage of GDP. In addition, foreign firms are contributing importantly to the technological capacity of host countries. They perform more than 40% of the total R&D in some countries, and the competition and standards they bring to the domestic market can have important spillover effects. Firms in developing countries have also increased their access to technology by purchasing technologically sophisticated firms domiciled in high-income countries.

Substantial technology transfers are also associated with international migration and the diasporas of developing countries. Although 93% of university-educated individuals from developing countries return or remain in their country of origin (Docquier and Marfouk, 2004), brain drain is a serious problem for some countries. And, because in most developing countries more highly skilled individuals than less skilled migrate, the resulting diasporas are much more technologically sophisticated than the home population. Given limited opportunities in many developing countries the true cost of this emigration is uncertain. Moreover, these diasporas form an important technological resource for the home country.

They contribute to technological transfers by strengthening trade and investment linkages with more advanced economies, through networks that provide access to technology and capital and through remittances. Remittances not only contribute to domestic entrepreneurship and investment, but, along with the introduction of mobile telephony services, have also greatly expanded the provision of banking and other arm's-length financial services within developing countries. Returning migrants can provide critical resources such as entrepreneurship, technology, marketing knowledge and investment capital. A single returning emigrant armed with skills acquired abroad can have (and has had) large economic and technological impacts on the country of origin.

Domestic absorptive capacity has improved more slowly

While there have been marked increases in the flows of the principal international transmitters of technology, improvement in the factors that determine the capacity of developing countries to absorb and effectively employ that technology has been much more gradual. On the positive side, most developing countries have improved their investment climates. Their macroeconomic and political environments have become more stable over the past 15 years: the number of international conflicts has fallen by more than 50% since the 1990s; median inflation has dropped from about 20% 15 years ago to less than 5%; and exchange rate volatility has fallen by more than 50% in every developing region (Figure 8.3). All of these factors reduce risk and increase the likelihood that entrepreneurs will take the chance to introduce a new technology within a country. These same factors have contributed to higher per capita GDP and a significant decline in the number of people living in absolute poverty. This has eased the constraints on these economies' ability to generate resources for investment, and enabled firms and individuals to take greater risks.

Figure 8.3. Macroeconomic stability has improved since the early 1990s

Median inflation rate

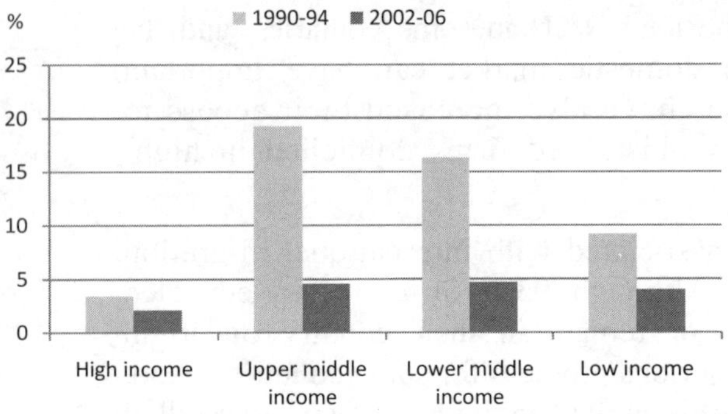

Real effective exchange rate volatility

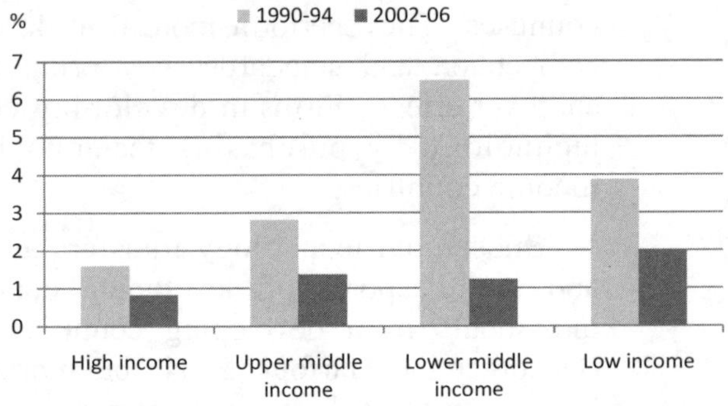

Source: The World Bank.

Figure 8.4. Literacy rates have increased in all regions

Percentage of adult population

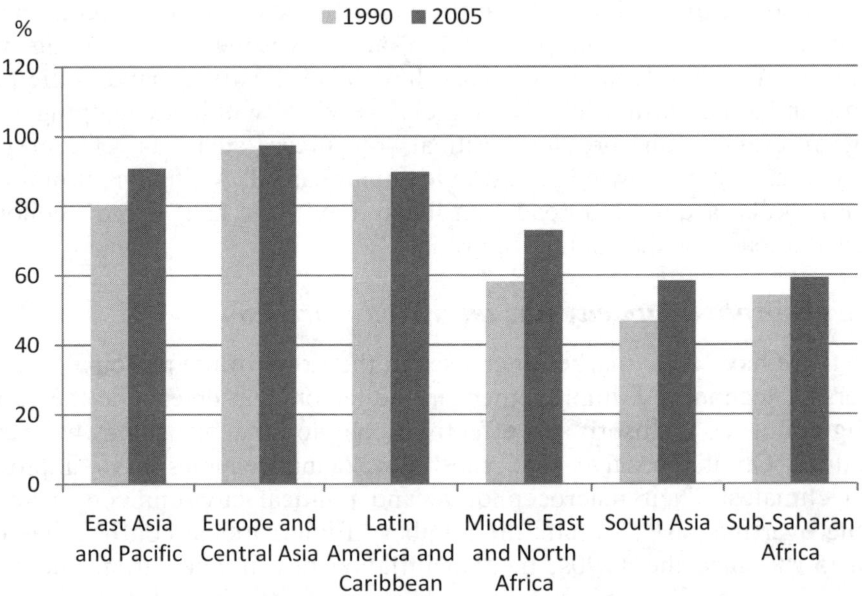

Source: The World Bank.

Improvements in the quality of human capital in most developing countries have increased their capacity to adopt and adapt technologies. Poor health is also receding as a factor impeding technological progress. Life expectancy in middle-income countries has reached 70 years and continues to rise, while life expectancy in low-income countries outside of Sub-Saharan Africa is up from 59 years in 1990 to 64 years in 2005 (in Sub-Saharan Africa very low incomes and the HIV/AIDS epidemic have led to a drop in life expectancy since 1990). The labour force in most developing countries is also becoming better educated. Adult literacy rates have increased in every developing region over the past 15 years (Figure 8.4). The share of children graduating from primary school has also increased in all regions except East Asia and the Pacific (where it already stood at 98% in 2005). Meanwhile secondary school and college enrolment rates are up across the board. Increased school enrolment has raised youth literacy rates to close to 100% in all of the predominantly middle-income regions. In Sub-Saharan Africa almost three-quarters of 15-to-24-year-olds can read and write. This compares favourably with an adult literacy rate of 60% and suggests that over time the technological literacy of the population will rise. Nonetheless, the quality of the education received by many students in developing countries remains low.

In contrast to these areas of improvement, business climate and governance indicators have shown little improvement over the past decade, on average. Governance in several countries has improved (notably in Eastern Europe and the Baltic region), proving that a motivated political leadership can make a difference. However, in many other countries the quality of governance has declined or remained stable. Overall, the time and cost involved in starting a business are higher, contract enforcement is less efficient, and the degree of corruption greater than in industrialised countries (Figure 8.5).

Figure 8.5. Developing regions have much poorer governance than OECD countries

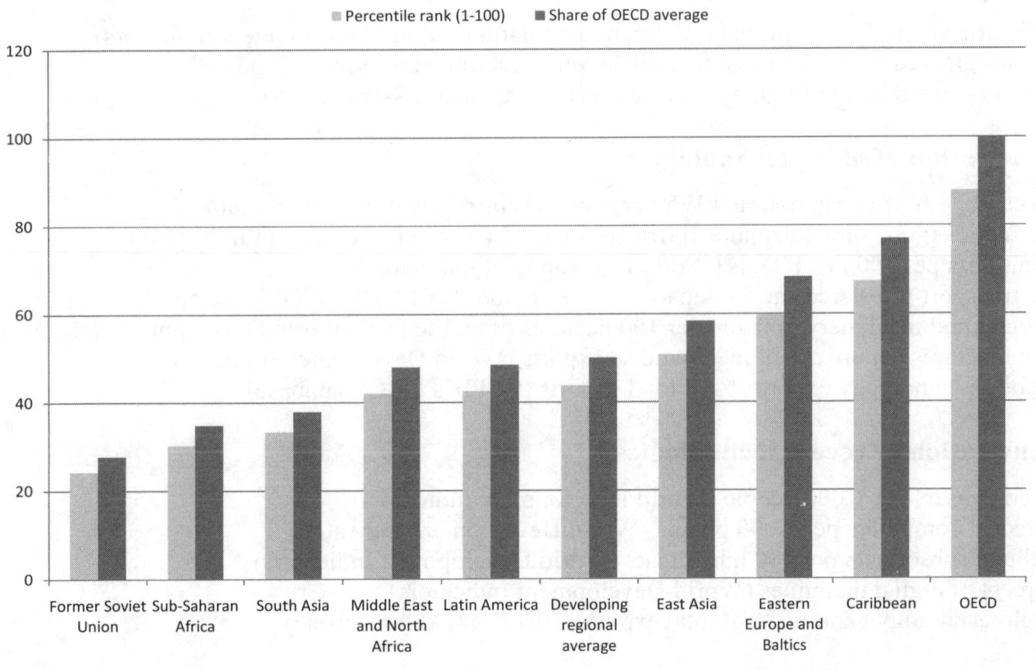

Source: The World Bank.

Measuring technological achievement

It is difficult to measure the level of technology achieved directly, mainly because, unlike pencils or automobiles, technology has no easily counted physical presence. Nor does it have a well-defined price that would allow it to be measured and aggregated as services are. Because technology is embodied in products, intermediate inputs and processes, most efforts to measure it have been forced to employ indirect techniques. Some indexes emphasise inputs into technological advances, such as educational levels, the numbers of scientists and engineers, R&D expenditures, or R&D personnel; an example is the *Index of Innovation Capability* of the United Nations Conference on Trade and Development (UNCTAD, 2005). Others also incorporate information on the diffusion of technologies and indicators of innovation such as the number of patents granted. The *Technology Achievement Index*, published by the United Nations Development Programme (UNDP) is an example. Still others focus on outputs, such as the share of high-technology activities in manufacturing value added and exports; an example is the *Index of Competitive Industrial Performance*, published by the United Nations Industrial Development Organization (UNIDO, 2002). Some focus more on the mechanisms by which technological progress is achieved (Sagasti, 2003) or by which technological learning occurs (Soubattina, 2006). For example, the National Innovative Capacity Index reflects the government and firm-level policies associated with successful innovation (Porter and Stern, 2004).

Box 8.1. Indicators used to calculate the summary technological achievement index and its sub-indices

Scientific innovation and invention

Scientific and technical journal articles, by population (World Development Indicators)
Patents granted by the USPTO, by population (Lederman and Saenz, 2005)
Patents granted by the EPO, by population (Lederman and Saenz, 2005)

Penetration of older technologies

Electrical power consumption, kWh per capita (World Development Indicators)
International outgoing telephone traffic, % of GDP (World Development Indicators)
Main lines per 100 inhabitants (World Development Indicators)
Air transport-registered carrier departures worldwide, % of GDP (WDI Development Indicators)
Agricultural machinery, tractors per 100 hectares of arable land (World Development Indicators)
Manufactures exports, % of merchandise exports (World Development Indicators)
Medium-technology exports, % of total exports (CEPII's BACI database)

Penetration of recent technologies

Internet users per 1 000 people (World Development Indicators)
Personal computers per 1 000 people (World Development Indicators)
Cellular subscribers per 100 inhabitants (World Development Indicators)
Percent of digital mainlines (World Development Indicators)
High-technology exports, % of total exports (CEPII's BACI database)

Exposure to external technology

FDI net inflows, % of GDP (World Development Indicators)
Royalties and licence fees payments, % of GDP (World Development Indicators)
Imports of high-technology goods, % of GDP (CEPII's BACI database)
Imports of capital goods, % of GDP (CEPII's BACI database)
Imports of intermediary goods, % of GDP (CEPII's BACI database)

Source: The World Bank.

Each of these approaches has its strengths, but none of them is entirely satisfactory, both because the indicators used fail to do justice to the broad definition of technology adopted here and because the method by which the index is constructed is sometimes arbitrary. To overcome these deficiencies, the indices reported in this chapter were derived from a wider than normal set of indicators, including some that have not been previously included in technology indices. Box 8.1 presents the individual indicators for which there was adequate country coverage and for which data existed both in the early 1990s and the early 2000s which make it possible to measure the pace at which the penetration of these technologies changed. These indicators are divided along three dimensions of technological achievement: the extent of scientific innovation and invention, the diffusion of older technologies, the diffusion of newer technologies and a dimension measuring the exposure of countries to external technologies. The remainder of this section discusses the degree of penetration of these and related indicators for which the data coverage was insufficient for inclusion in the aggregate indices as well as the levels of the summary indices along each of these dimensions.

Scientific innovation and invention

Both the creation of new technologies and the diffusion, adoption and adaptation of existing technologies depend on the quality, quantity and economic relevance of the national scientific community. Even though many of the world's most successful scientists and innovators have their roots in developing countries, developing countries have relatively few scientists compared with the size of their populations and are much less active at the global technological frontier than the scientific community in high-income countries. Authors from high-income countries report 7 times as many published articles as those from upper-middle-income countries and 88 times as many as authors from low-income countries. Variations are even larger for measures of patents granted and licence fees earned.[5]

Notwithstanding a still low level, patent activity has increased among middle-income countries over the past 20 years (Figure 8.6), partly because of a sharp jump in patenting among countries of the former Soviet bloc in the early 1990s. The continuous increase in patent activity among lower-middle-income countries mainly reflects activity in China, whose share in world patent applications has risen from about 1.5% in the late 1980s to a peak of nearly 10% in 2004. Excluding China, additional patenting activity in lower-middle-income countries has been relatively modest.

5. The focus on patents and scientific publications reflects what is available in academic research on technology. Patents have the advantage of being more clearly associated with processes than with products (by definition a patent is not granted on a product but on the method by which it is produced). The disadvantage is that patents exclude a number of important forms of innovation, notably software (until recently) and processes for managing multinational production and distribution networks.

Figure 8.6. Patent activity is rising in middle-income countries

Total patent applications per 1 000 persons

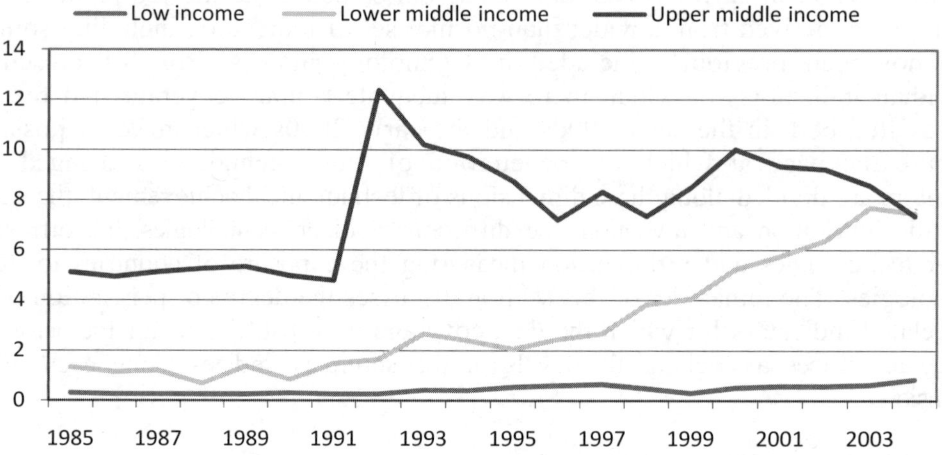

Source: The World Bank.

Table 8.2. Scientific and innovative output

	Science articles[1]	USPTO patents[2]	EPO patents[3]	Total patents	Patents to non-residents	Royalty and licence fees receipts
	Per million people				% of total	% of GDP
	2003	2006	2005	2003	2003	2004
East Asia and Pacific	17	0.7	0.01	37	77	0.02
Europe and Central Asia	90	0.9	0.40	95	28	0.06
Latin America and Caribbean	35	0.7	0.21	46	98	0.03
Middle East and North Africa	18	0.1	0.03	n.a.	n.a.	0.02
South Asia	9	0.5	0.07	1.4	60	0.00
Sub-Saharan Africa	5	1.4	1.16	157	100	0.06
World	**111**	**38.6**	**11.40**	**127**	**41**	**0.27**
High income	584	135.1	42.60	331	38	0.33
Upper middle income	85	1.4	0.40	91	42	0.04
Lower middle income	21	0.6	0.01	46	64	0.03
Low income	7	0.4	0.07	3.5	56	0.00

Note: To reduce home bias, total patents granted by the USPTO to high-income countries exclude those granted to the United States and total patents granted by the EPO exclude patents granted to high-income EU countries.

1. Scientific and technical journal articles.

2. Patents granted by United States Patent and Trademark Office.

3. Patents granted by The European Patent Office.

Source: The World Bank, United States Patent and Trademark Office, European Patent Office, and World Intellectual Property Office.

Reflecting a history of advanced scientific and engineering work in a number of countries in the former Soviet bloc, Europe and Central Asia have relatively high levels of scientific and technological innovation (Table 8.2). Publication rates there are equal to those in many high-income countries, and patent activity is more than twice the level in any other developing region. The region is also the most self-reliant of developing regions in terms of patent activity, with only 28% of patents filed by non-residents, a ratio even lower than the high-income country average of 38%. The East Asia and Pacific region also scores high in terms of patents, although its publication record is more in keeping with other developing regions. In China and India, conscious efforts to raise R&D spending have led to higher levels of scientific innovation than might be expected based on income (Lederman and Saenz, 2005), while low levels of innovation in Latin America and the Caribbean reflect an academic research tradition with few links to industry (Maloney, 2006).

While there is a strong correlation between the number of patents and scientific journal articles and GDP per capita for high-income countries, developing countries produce very few (Figure 8.7). As a result, for virtually all developing countries, technological progress mainly derives from the adoption and adaptation of pre-existing technologies. This is mainly because of the lack of advanced technological competences in these countries, but also because many citizens of developing countries are performing their cutting-edge research in high-income countries. For example, 2.5 million of the 21.6 million scientists and engineers working in the United States were born in developing countries (Kannankutty and Burelli, 2007).

Figure 8.7. Scientific innovation and invention is almost exclusively a high-income activity

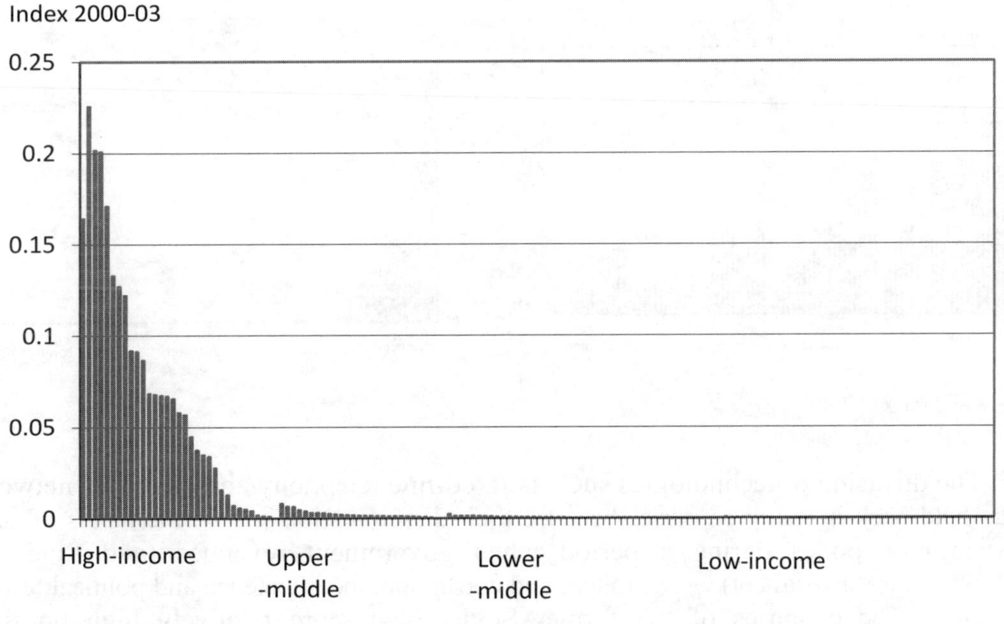

Source: The World Bank.

Penetration of existing technologies

The clear dominance of high-income countries in scientific and technical journal articles, patents granted, and licensing and royalty fees points to the relatively minor role that frontier innovation plays in determining technological progress in developing countries and the relative importance of adoption and adaptation of existing technologies. Older technologies, such as electricity, radio, television, improved roadways, vaccines and fixed-line telephony exist to some degree or other in virtually every developing country. However, the extent to which they have penetrated economic activity varies widely, even at similar income levels. Low-income countries with the highest utilisation rates of these older technologies tend to have rates as high as those of the average lower-middle-income countries (Figure 8.8). The situation is similar for lower-middle-income countries *vis-à-vis* upper-middle-income countries and these with respect to high-income countries.

Affordability, weak institutions, tax policies and a lack of domestic capacity to maintain systems explain the modest diffusion of many older technologies critical to development. In the case of electricity, roadways, fixed-line telephony and vaccinations, the way a country organised the sectors (the process technology employed) and the historical period in which the technology was introduced have had a strong bearing on the diffusion of the specific technology in the economy.

Figure 8.8. The penetration of old and new technologies depends on more than income

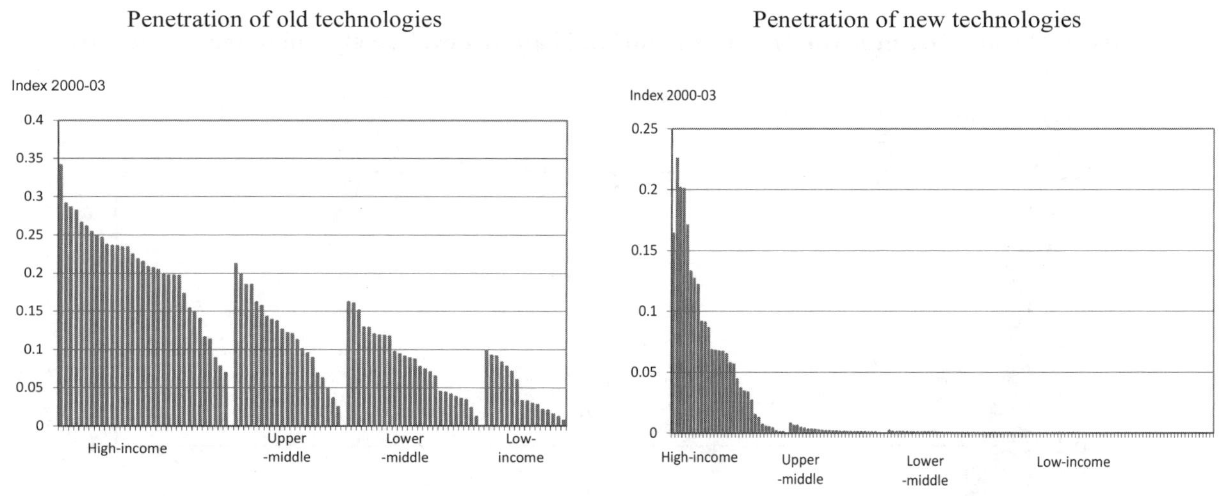

Source: The World Bank.

The diffusion of technologies such as fixed-line telephony and electrical networks was held back in many countries because their dissemination was in the hands of state-owned monopolies during a period when government revenues (and funds for infrastructure investment) were limited, and corruption, incompetence and political turmoil common. The countries of the former Soviet bloc score relatively high on these measures – even compared to other countries at similar income levels because of the strong emphasis past governments placed on providing such services to a wide range of the population. Moreover, these technologies also require large numbers of individuals with technical skills that can be quite scarce in poor countries. Similarly, past governance problems and civil strife help explain the relatively weak penetration of these technologies in many African countries, while macroeconomic turmoil and a relatively unequal distribution of incomes and skills in Latin America may have contributed to weak outcomes in that region.

The diffusion of older agricultural technologies (such as tractors, irrigation and intensive agricultural techniques) was boosted by the green revolution technologies promoted by the international community during the 1970s, 1980s and 1990s. Medical technologies, including the discovery and widespread distribution of antibiotics and the eradication or effective treatment of a wide range of previously deadly or debilitating diseases such as HIV/AIDS, have been among the most important technological achievements of the past 100 years. The diffusion of knowledge about these treatments within the medical community is generally relatively speedy and efficient, and their diffusion and application among developing countries has generally been good, with most developing regions and countries enjoy immunisation rates of 80% or more (Table 8.3). However, failures in the delivery systems in a few large countries, notably India (less than 60%) and Nigeria (less than 35%), suppress overall vaccination rates in South Asia and Sub-Saharan Africa despite rates of 90% or more in smaller countries.

Table 8.3. Older medical technologies have spread widely

Immunisation: percentage of children aged 12-23 months

	DPT	Measles
East Asia and Pacific	83	83
Europe and Central Asia	89	91
Latin America and Caribbean	90	93
Middle East and North Africa	91	92
South Asia	63	61
Sub-Saharan Africa	59	61
High income	95	92
World	**76**	**75**

DPT: Immunisation to protect against diphtheria, pertussis (whooping cough) and tetanus.

Source: The World Bank, *World Development Indicators Database*.

The health benefits of clean drinking water and sanitation facilities have been understood for centuries. Nevertheless, one in five people in developing countries lack access to improved water sources and only half have access to improved sanitation facilities (Table 8.4). In Sub-Saharan Africa and South Asia only 37% of the population has access to improved sanitation services, and only slightly more than half of the Sub-Saharan African population has access to improved drinking water (the share rises to 65% if Nigeria, where only 35% of the population has access to improved water, is excluded). The rest of the developing world does much better on these measures. For example, close to 90% of the population in Europe and Central Asia has access to improved water and sanitation sources. Nevertheless, the diffusion of these basics technologies is weak in rural areas in all developing countries, reflecting more intense affordability issues, the relative scarcity of basic technological literacy, and the competences necessary to install and maintain such systems (see Box 8.2). For example, in China and India only 44% and 33% of the population, respectively, have access to improved sanitation.

Table 8.4. Rural access to improved water and sanitation still lags

Percentage of population with access

	Improved water source		Improved sanitation	
	Rural	Urban	Rural	Urban
East Asia and Pacific	69.8	91.9	36.1	72.4
Europe and Central Asia	79.8	98.7	70.3	93.0
Latin America and Caribbean	73.0	96.0	48.7	85.7
Middle East and North Africa	80.8	96.3	57.9	92.3
South Asia	81.3	93.6	26.6	62.7
Sub-Saharan Africa	42.4	80.1	28.2	53.3
World	72.2	94.5	37.7	79.4
High income	98.5	99.8	100.0	100.0
Low and middle income	70.5	92.8	34.4	74.4
Low income	69.4	88.1	28.5	60.5

Source: The World Bank, *World Development Indicators Database.*

These old technologies have penetrated developing countries less completely than developed countries, but the gap is much less pronounced than the gap for indicators of scientific innovation and invention. Moreover, the relation between income levels and the diffusion of old technologies in the developing world is relatively weak, suggesting that the efficiency of the regulatory environment and the diffusion of basic skills within countries are more important than incomes in determining the actual level of diffusion of these technologies. Countries with the highest achievement in each income group find themselves at about the median level of achievement of the next higher income group. Once again, the level of diffusion of these older technologies tends to be higher for countries of the former Soviet bloc compared with other countries at the same income level, while both the upper-middle- and lower-middle-income countries in Latin America and the Caribbean tend to report lower levels of diffusion than other countries at similar income levels.

The striking differences between Europe and Central Asia on the one hand and Latin America and the Caribbean on the other in the diffusion of older technologies may reflect differences in income distribution and in the nature of research and development (R&D) activities. Europe and Central Asia had more equal access to education combined with greater government investment in infrastructure; this facilitated more rapid diffusion of technologies than in Latin America and the Caribbean. In addition, whereas R&D activity was clearly linked to the industrial strategy of Soviet-era firms in Europe and Central Asia, R&D in Latin America was concentrated in the universities, was oriented toward research at the global frontier (but generally not of cutting-edge quality), and had few link to firms (Maloney, 2006).

Box 8.2. Promoting appropriate technologies in Rwanda

A recent study of Rwanda identified simple technologies whose greater use could have a substantial impact on development. For example, a lack of qualified plumbers and water-sanitation technicians has been identified as a major factor holding back the implementation of simple rain water collection strategies that have helped improve the quality of drinking water supplies in neighbouring countries. Similarly, a lack of basic skills, including those necessary to manufacture stainless steel products, prevents the implementation of simple food-processing techniques, such as passion fruit pasteurisation and pulping that could reduce the share of crops lost to spoilage (sometimes as much as 30% of a crop is lost). Public-sector dissemination of best practices is hampered by poor skills and inappropriate incentives, which result in research centres producing local products that take insufficient account of users' needs and requirements. The following indicates the status of efforts to promote simple technologies in Rwanda.

Diffusion of selected "appropriate" technologies in Rwanda

Rural energy

- biogas for institutions: installations ongoing and spreading
- biogas for households: pilot programme of 163 units to start 2007
- micro hydropower: six in preparation, more in future?
- bio-fuel: no national programme/policy as yet
- wind: no programme/policy as yet
- peat: large stocks but limited exploitation
- efficient stoves for urban areas: national programme ongoing
- efficient stoves for rural areas: some programmes ongoing
- rice/ coffee husks for briquette production: limited programmes
- PV systems: technology available but slow market
- solar water heating: technology available but slow market

Water and sanitation

- roof water harvesting: only on limited scale for households
- boreholes: few and expensive
- hand pumps: imported from region or India
- VIP and Ecosan latrines technology: available, limited uptake

Agricultural technologies and transport

- irrigation through treadle and motorised pumps: limited uptake
- drip irrigation: starting
- animal traction for tillage and transport: promoted in certain areas
- small tractors for rice puddling and transport: few units imported
- rice threshing/winnowing: few machines available and locally produced
- rice hulling: opportunities for small-scale processing
- maize milling: machines imported and locally made
- oil presses for sunflower, soya, essential oils: starting
- livestock spraying: locally made machines now available

Low-cost building

- some use of rice/coffee husks and peat for brick burning
- hand brick press machines; locally made and imported
- engine brick press machines; imported

Source: Watkins and Verma (2007), p. 33.

Penetration of recent technologies

The slow diffusion of many old technologies in developing countries strikes a sharp contrast with the relatively rapid penetration that has been observed for newer technologies (Table 8.5). Macroeconomic turmoil, civil strife, and fiscal constraints limited the within-country diffusion of many older technologies during the 1970s and 1980s, but more hospitable circumstances – including low inflation, low government deficits, and a technical and regulatory environment that has better harnessed private-sector financing of new technologies – have contributed to the spread of more recent technologies. In a few cases, newer technologies have leapfrogged over older ones; an example is mobile phones, which now have higher penetration rates in some countries than fixed-line telephones.

Table 8.5. Diffusion of "recent" technologies

	Internet users per 1 000 people		**Internet bandwidth** Mbps		**Broadband subscribers** per 1 000 people	
	% change 1999-2005	2005	% change 1999-2004	2004	% change 2001-2005	2005
East Asia and Pacific	48	89	149	8 735	236	26
Europe and Central Asia	48	190	132	6 670	208	21
Latin America and Caribbean	41	156	121	4 513	89	16
Middle East and North Africa	64	89	91	899	n.a.	n.a.
South Asia	66	49	114	2 249	131	1
Sub-Saharan Africa	42	29	62	114	n.a.	n.a.
World	20	137	108	43 856	59	42
High income	14	527	107	121 433	45	163
Upper middle income	36	196	126	5 611	147	21
Lower middle income	50	95	134	5 533	187	23
Low income	72	44	120	708	143	1

	Personal computers per 1 000 people		**Cellular subscribers** per 100 people		**Digital cellular subscribers** per 1 000 people	
	% change 1997-2004	2004	% change 1995-2004	2004	% change 999-2004	2004
East Asia and Pacific	26	38	58	24	54	257
Europe and Central Asia	20	98	79	44	43	512
Latin America and Caribbean	17	88	51	32	42	337
Middle East and North Africa	17	48	73	13	70	142
South Asia	29	12	87	4	88	40
Sub-Saharan Africa	11	15	61	8	47	83
World	14	130	37	28	29	284
High income	12	579	28	77	19	768
Upper middle income	18	113	58	48	37	521
Lower middle income	23	45	61	24	54	255
Low income	25	11	92	4	88	43

Note: Period growth rates are compound annual growth rates.

Source: The World Bank.

Although arbitrary, the distinction between "old" and "recent" technologies is useful because the factors that impeded the diffusion of old technologies within developing countries are often qualitatively different from those that impede the distribution of more recent technologies. For instance, the diffusion of many of the older technologies was dependent on the creation and maintenance of expensive infrastructure at a time when many governments were grappling with severe budget constraints and weak technical and governance capacity. Penetration rates of newer technologies have risen rapidly, partly because the infrastructure for newer technologies such as mobile phones, computers and the Internet is generally less expensive to create and requires fewer (though more skilled) workers to maintain. Moreover, in most countries regulatory reform has meant that these services have been offered by private-sector investors with access to ample funds and outside expertise.[6] As a result, supply has often been more responsive to market demand and less restrained by government or budget constraints on state-owned enterprises. Finally, end-user costs tend to be lower, in part because the technologies lend themselves to sharing.

The rapid diffusion of these technologies also reflects demand-side factors. These include the use of microfinance techniques (a form of process technology) to help deploy cell phones in very poor areas (Sullivan, 2007). Another is the innate divisibility of the costs of these technologies both because of per-call billing and the possibility of sharing the fixed costs of a cell phone or Internet connection and its monthly subscription. Partly as a result, there are more than twice as many Internet users as computers in developing countries (four times as many in South Asia and the Middle East and North Africa), whereas the ratio is 1:1 in high-income countries. Taken together, these factors have contributed to the very rapid diffusion of the Internet and cell phone technology in developing countries. Cell phone ownership rates in low-income countries have almost doubled in the space of four years. Indeed, new subscribers are signing up at such a fast pace that the data in Table 8.5 are already largely out of date. In contrast personal computer ownership rates are much lower and three-quarters of low-income countries have 15 or fewer personal computers per thousand population, compared to more than 500 per thousand in developed countries.

Although delivery of these services has been provided and financed by the private sector, governments have played an important role ay. Government infrastructure is often a necessary pre-condition for the successful spread of technology – either by privately owned or state-owned firms. For example, despite rapid growth, Internet penetration in Sub-Saharan Africa remains very low, in part because there is no high-speed, low-cost backbone to connect eastern and central Africa to the rest of the world. As a result, Internet transactions must be made via satellite, which is both slower and more costly than fibre optic cable,[7] and African firms have difficulty competing internationally for call-centre contracts.

Technology is also providing solutions for overcoming infrastructure costs. In a number of countries, wireless broadband connections are outpacing DSL and cable as a means of distributing Internet access to customers. So-called 3G mobile phones already provide reasonable bandwidth in many countries, while more advanced standards

6. While initially lack of competition and difficulties encountered by innovative entrepreneurs in getting licences slowed the diffusion of mobile phone technology, much has changed in recent years (Sullivan, 2007).

7. Kenyan call centre operators pay USD 7 000 per Megabyte of bandwidth compared to around USD 500 for operators connected by fibre optic cable in India.

offer hope for even faster implementation and diffusion. Some 23 developing countries are planning to, or already have begun to, deploy WiMax systems, a wireless broadband Internet standard touted as the successor to today's WiFi and 3G systems. The Dominican Republic, Pakistan, South Africa and Uganda already have WiMax implementations.

Because the market is evolving so rapidly, with new applications of cell phone technology found on a regular basis, it is difficult to evaluate its overall impact. Penetration rates in Latin America and the Caribbean and in Europe and Central Asia are already high, rivalling those observed in high-income countries less than ten years ago. Penetration rates in East Asia and the Pacific average somewhat lower. However, for middle-income countries in the region, excluding small island economies, the average penetration rate there is higher than in Latin America and the Caribbean. Penetration rates in low-income countries are much lower on average, although some have reached levels comparable to middle-income countries. As of 2005 six Sub-Saharan African countries (Botswana, Gabon, Mauritius, the Seychelles, Sierra Leone and South Africa) had mobile phone penetration rates above 30%. While penetration rates in South Asia are also low, the large populations of these countries and the pace at which firms are adding customers means that, globally, a substantial proportion of new mobile phone subscribers are in developing countries.[8]

The technological and economic implications of the rising penetration of mobile phones are only now being assessed. In poor rural areas, where the transport of goods and people is heavily constrained by poor infrastructure, the introduction of cheap personal communications may be of great value both as a substitute for moving people and to ensure that the movement of people or goods is worthwhile. In particular, the availability of relatively cheap and efficient communications has reduced information asymmetries in a number of sectors, increasing the revenues of producers and lowering costs for consumers (albeit at the expense of middlemen). In addition, this technology is increasingly used to enable a degree of arm's-length financial intermediation which many argue is critical to development but which has largely been unavailable in the past because of a lack of infrastructure (Box 8.3).

The spread of the Internet, other modern communications technologies and the diffusion of computers, coupled with quality improvements in transport services, have combined to greatly improve the rapid and efficient delivery of goods and services, enabling just-in-time inventory processing and more efficient supply chain management. Although causality is uncertain, countries with good communications and transport sectors (and a relatively efficient customs administration) have lower logistics costs, larger manufacturing sectors and more diversified exports (World Bank, 2007a). Initially offshoring services were concentrated on lower-end software services and business processes as well as call centres. More recently offshoring has moved into areas such as investment and financial services, human resources, health services, retail functions, logistics and customer support functions (World Bank, 2005). In addition to increasing demand for labour and boosting export revenues, offshoring of services to developing countries can improve their incentives for education and training, help improve the quality of services provided domestically, encourage technology

8. According to the Cellular Operators Association of India, there were over 233 million subscribers in the country in September 2008.

and knowledge transfers, and minimise, compared with manufacturing, the environmental consequences of economic growth.

Box 8.3. Innovative use of communications technology is improving financial access for the poor

The poor face considerable challenges for gaining access to well-functioning savings and payments services. Financial institutions do not exist in many rural areas, and those that do often impose high minimum balance requirements (reflecting high unit transaction costs for small accounts) which are well beyond the reach of poorer households. However, the adaptation of technology has allowed some innovative financial institutions in Africa to extend financial outreach to the poor.

For example, the Equity Bank in Kenya has outfitted a series of vans with laptops and telecommunications facilities to act as mobile banking units. It has also designed flexible savings mechanisms (*jijenge*) with emergency loan facilities. Teba Bank of South Africa has developed a smart card (Bank A-Card) which uses existing cell phone technology to provide low-cost, electronic banking services (savings and payments) for low-income customers. The programme was originally developed to handle wage payments of migrant workers. The value of the cards can be topped up (or used to make purchases) at any of the simple wireless terminals that have been placed in shops frequented by low-income clients. A similar, but more sophisticated scheme is being introduced by Remote Transaction Systems in Uganda. A system developed by Celpay allows clients in Zambia and the Democratic Republic of Congo to use their cell phones to pay bills. The client establishes an account with Celpay and then can make purchases by texting a request to Celpay, which will transfer money to the merchant's account; security is provided by use of a personal identification number (PIN), which is needed to complete the transaction.

In a series of surveys of banking services in three middle-income and four low-income countries, Bankable Frontiers Associates (2007), found that although only 1.5% of the adult population in South Africa was using mobile-phone banking, the potential for the service was large. Between 7% and 41% of the unbanked populations of the countries surveyed (Botswana, Kenya, Namibia, South Africa, Tanzania, Uganda and Zambia) had access (including shared access) to a mobile phone and these penetration rates are rising.

Source: The World Bank (2007), *Making Finance Work for Africa*, Bankable Frontiers Associates (2007).

Technology diffusion

The observation that new technology has spread among and within developing countries more rapidly than older ones is borne out by longitudinal data collected by Comin and Hobijn (2004). The data indicate that the pace at which technologies have diffused across the globe has accelerated over time. For example, the number of years it has taken for a given technology to reach 80% of countries currently reporting data for that technology[9] has decreased from more than 100 years in the 18th century to less than 20 years today (Figure 8.9).

When technology does succeed in penetrating it tends to do so slowly initially but then much more rapidly. For example, in the first half of the 20th century, it took, on average, 52 years for a technology to reach the 5% threshold, but only an additional 13 years to reach 25%. For technologies introduced since 1975 (a group dominated by

9. The data set includes an estimate of the date of discovery for each technology.

electronics and information technologies), on average it took 16 years from its invention for a technology to reach the 5% threshold in a given country but only another 3 years to reach the 25% threshold. While the pace of diffusion is somewhat slower in developing countries than in high-income countries, it follows the same pattern. This pattern is consistent with the existence of significant economies of scale and barriers to entry among these technologies, such that once the barriers fall and the technology is in place, scaling up occurs relatively quickly.

Figure 8.9. The speed with which technologies reach countries has accelerated

Years until 80% of countries report having the technology

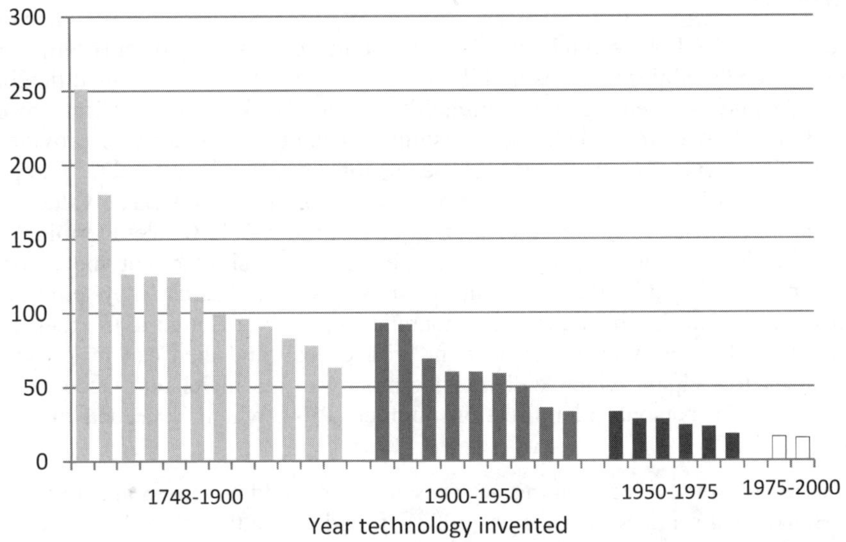

Source: World Bank calculations using *CHAT Database* from Comin and Hobijn (2004).

While diffusion has occurred relatively rapidly among technologies that diffuse successfully, successful diffusion is the exception rather than the rule. For example, of 102 country-technology pairings first recorded in the 1975-2000 period, only 55% have reached the 25% threshold and only about 34% have reached the 50% threshold (Figure 8.10). For developing countries the pace (and extent) of diffusion is significantly slower (lower) than in high-income countries. None of the technologies tracked by Comin and Hobijn discovered since 1950 have reached the 25% penetration rate, while for lower-middle-income countries very few have. Among upper-middle-income countries the result is somewhat more encouraging, as more than 40% of the country-technology pairs have reached a level of penetration that is 15% of that of the ten countries with the highest penetration. The fact that penetration rates remain low for even very old technologies suggests that affordability and absorptive capacity are holding back technological achievement.

The failure of technologies to diffuse within countries is at the root of the very low penetration rates observed. Indeed, while new technology arrives and diffuses within the leading firms and cities of most countries relatively rapidly, they spread much less quickly to the broader countryside. As a result, the overall level of technological diffusion in countries such as India and China is not much higher than in other countries at similar income levels.

Figure 8.10. The penetration of most technologies never exceeds 25% in developing countries

Share of technologies to reach 25% threshold

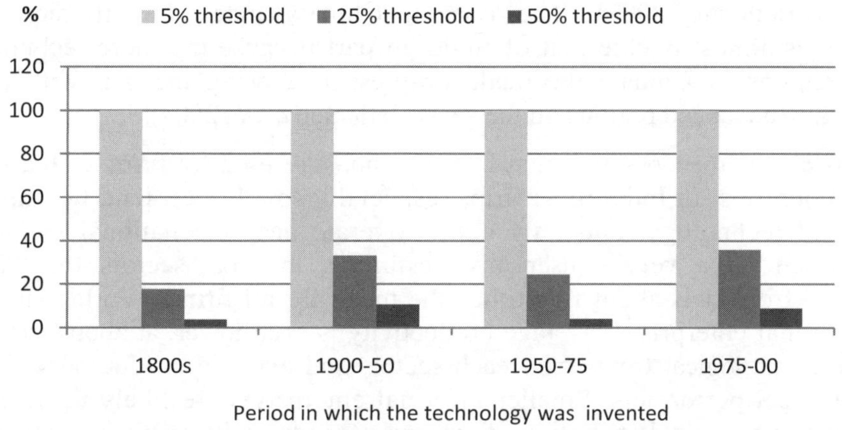

Source: World Bank calculations using CHAT Database from Comin and Hobijn (2004).

Although India is one of the world's largest markets for telecommunications technology and a leader among developing countries in exports of software and information-technology-enabled services, on a per capita basis it continues to lag middle-income countries in the rate of technological diffusion, R&D expenditure, basic and higher education attainment levels, availability and quality of logistics services, and size of revenues and employment in software and other high-technology industries. India does not score substantially better than many Sub-Saharan African countries in terms of the overall diffusion of technologies. This contrasts with popular perceptions, which are based on the relative technological sophistication of some of India's major cities and trading centres and arises because technologies have not penetrated very deeply in many parts of thc Indian countryside (Mitra, 2007). Rural telephone densities are less than 10% versus more than 50% in urban centres (Figure 8.11).

Figure 8.11. Urban and rural teledensity (fixed and mobile) in India, 1998-2007

Percentage of household with phone service

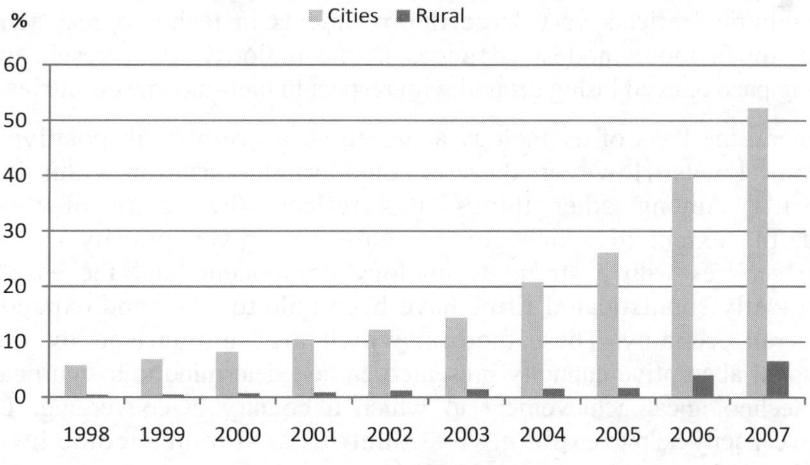

Source: Telecommunications and Regulatory Authority of India.

Here, the challenge is to put in place a basic infrastructure in the countryside that can support the kind of sophisticated technologies available to the country's elites. As one observer put it, an ET (energy technology) revolution must precede any IT revolution (Friedman, 2007). The rise in China's index of diffusion of new technologies is almost double that of India, in part because the more technologically backward regions in China have made progress in closing the gap with the more technologically advanced regions on the coast (Jefferson *et al.*, 2007).

There is also tremendous variation in the technology used by firms within sectors in individual countries. In India most firms, especially small ones, tend to employ very low levels of technology, and only a few operate near the national technological frontier. Based on a very conservative estimate, in most sectors the "adjusted" technological frontier is about five times the mean for all firms (World Bank, 2007). For small formal enterprises, average productivity is even lower, at about one-sixth the level at the technological frontier for each sector, and average productivity is roughly 15% of top local performers. Smaller informal enterprises are likely to be even less productive. The skewed distribution of enterprise productivity implies potentially huge productivity and output increases if knowledge existing in the country were to diffuse from top performers to the rest of the economy. Assuming the availability (or creation) of domestic competences to employ efficiently the technologies used in enterprises at the national frontier, Indian GDP could be 4.8 times higher if those technologies were successfully applied by their less productive rivals. Similarly, the productivity of innovative firms with more than 10 employees in Brazil, which account for 26% of total sales, is on average 6.5 times higher than that of firms of similar size classified as weakly innovative (World Bank, 2007b) (they account for 11% of sales but 38% of employment).

Measuring technological progress

Overall, the extent to which developing countries have effectively exploited technology over the past 15 years has improved substantially. The level of technology achieved in most developing countries has improved and the rate of change in both low- and middle-income countries has exceeded that in high-income countries. As a result, developing countries have closed the relative gap with respect to high-income countries. However, the gap remains large, partly because the technological frontier has also been moving (Figure 8.12). Moreover, the relatively strong performance of low-income countries reflects very large improvements in technological achievement by some, but much more modest advances in the majority. As a result, many are only maintaining pace or even losing ground with respect to high-income countries.

In general the level of technology achieved by a country is positively correlated with income levels. However, there is considerable variation within income groups (Figure 8.13). Among other things, this reflects the nature of the technology observed, the extent to which governments have given priority to and succeed in delivering services with a strong technology component, and the ease with which technologically sophisticated firms have been able to grow and expand their weight in the overall economy. These factors, which are summarised by the concept of technological absorptive capacity presented earlier, determine to a significant degree the level of technological achievement to which a country is converging. Differences in absorptive capacity help to explain why countries at similar income levels can have such different levels of technological achievement.

Figure 8.12. Technological achievement: converging, but the gap remains large

Rapid technological progress

Percentages change in technological achievement 2000s vs. 1990s

Relative convergence

Index: High-income countries = 100

Absolute changes in achievement

Technological achievement, index

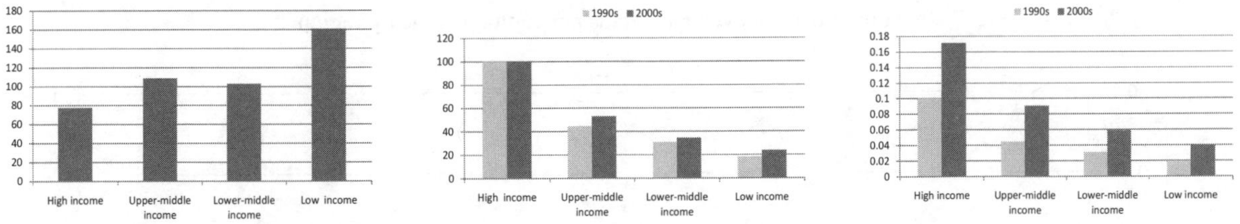

Source: The World Bank.

Figure 8.13. Overall index of technological achievement, 2000s

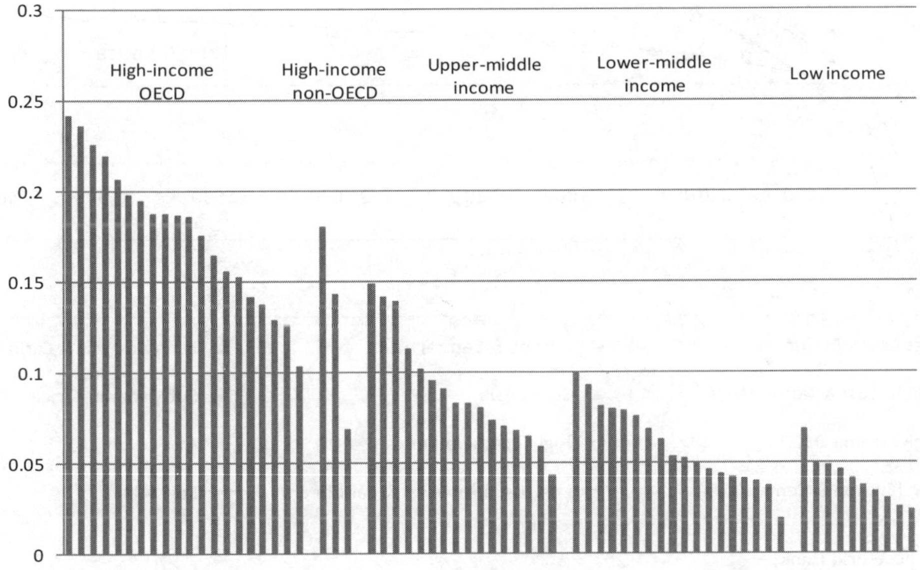

Source: The World Bank.

The higher the underlying level of technological absorptive capacity, the higher the level of technological achievement to which a country is converging over time. Thus, while technological achievement tends to rise with income, the relationship is non-linear and shows a tendency to level off. Moreover, it is not uniform across regions (Figure 8.14). Countries in Europe and Central Asia tend to have somewhat higher levels of achievement than would be expected on the basis of income alone, but the overall relationship between technological achievement and income tracks relatively well that of all countries. In contrast, technological achievement in Latin America tends to be lower than what would be expected given incomes, and the overall relation suggests that other factors appear to be restraining overall achievement even as incomes progress. This result is consistent with the view that problems of absorptive capacity (the uneven distribution of income and educational opportunities and a history of weak linkages between the R&D and business communities) may be constraining technological achievement.

Figure 8.14. Technological achievement tends to level off at different income levels in different regions

Technological achievement versus per capita income by region

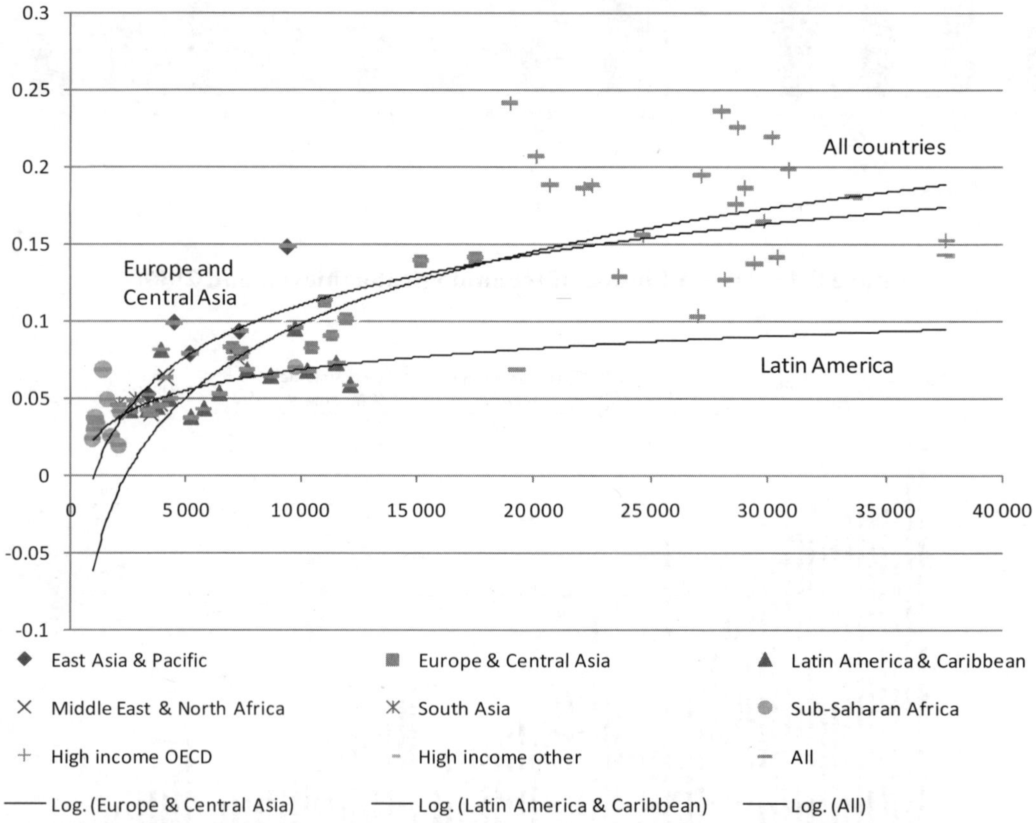

Source: The World Bank.

Conclusions

Technology diffusion in developing countries depends on access to foreign technology (through trade, foreign direct investment, international migration, and other networks) and on the ability to absorb technology (as determined by the quality of government policy and institutions, the stock of human capital, R&D efforts, and the financial system). One implication of the analysis and data presented here is that prospects for further technological progress in low- and middle-income countries are good. Over the past 15 years the main international channels through which technology is transferred have spread. Developing countries' imports of high-technology goods and of capital goods have risen relative to GDP, and their share of global high-technology export markets has increased. Inflows of foreign direct investment have increased six-fold relative to developing countries' output, and opportunities to purchase technology have risen along with FDI outflows.

Simultaneously, the absorptive capacity of developing countries has been increasing – albeit more slowly. Youth literacy rates are as much as 15 percentage points higher than for the adult population. As a result, the basic technical literacy of the population has been increasing, and it should continue to do so for many decades. The macroeconomic instability that plagued developing countries during the 1970s and 1980s has declined, and the business climate has improved, although not by as much or as uniformly as one might have hoped. Given a continuation of these trends, and assuming no major disruptions to global trade and financial systems, developing countries' access to, and ability to absorb, new technology should continue to rise over the medium term.

Of particular note is the speed with which communications technologies are evolving and diffusing in the developing world. Only 27 years after the introduction of cell phone technology, mobile phones are active in virtually every country and penetration rates are rising rapidly. Moreover, the range of economic activities that were once heavily dependent on infrastructure and that are now being conducted using mobile phone technology is impressive and expanding daily. Already, mobile phones are bringing banking, remittances and arm's-length financial transactions to regions of the world that until recently were unserved. Given the pace at which things are changing, most developing countries should continue to see a rise in their ability to communicate and process information over the next few decades. This should help speed the diffusion of other technologies as well.

For middle-income countries the relatively rapid technological progress of the past few years and improvements in both openness and technological adaptive capability suggest that their level of technological sophistication should continue to converge with that of higher-income countries. However, even the most advanced of these countries will not be able to benefit fully from the new technologies that are expected to become both technically and economically viable over the next several years because of inadequacies in their infrastructure (unreliable power or communication systems), insufficient technical literacy, or the absence of a critical mass of scientists and engineers necessary to exploit the technology (Box 8.4). For some countries the relative slowness with which technological absorptive capacity has been advancing could slow the pace of convergence as missing competences become an increasingly binding constraint on the absorption of additional technologies.

For low-income countries prospects are more complex. On average, among the low-income countries for which sufficient data are available to calculate recent increases in technological achievement, convergence is occurring, and more quickly than in middle-income countries. However, this finding reflects rapid progress in a few countries and more modest performance in many others. Overall they are only maintaining their ground relative to high-income countries.

Notwithstanding very strong technological progress in some cities and greater openness to technological flows, the gap between existing competences and those needed to converge with technological progress in high-income countries is immense, especially in rural areas. Moreover, the pace at which absorptive capacity is rising is disappointing. While some countries have recorded significant increases, developing countries on average are not catching up to high-income countries, an indication that the gap in their technology potential is not narrowing. As a result, unless substantial steps are taken to raise basic competences and invest in local networks that successfully disseminate technologies and technological competences, many of these countries are not expected to be able to master anything more than the simplest of forthcoming technologies (see Box 8.4).

Box 8.4. Technology in 2020

A recent report by the Rand Corporation (Silberglitt *et al.*, 2007) examines some 56 emerging technologies expected to be real products by 2020 and evaluates in detail the 16 judged to be most important on the basis of technical feasibility, marketability and societal impact. These applications include, among others, improvements in health services (targeted drug delivery, improved diagnostic and surgical methods), in access to information (rural wireless communications, quantum cryptography), and in the environmental sustainability of products and services (improved water purification, green manufacturing and hybrid vehicles). It then examines in detail the technical base required by a country to make effective use of each technology and the likelihood that each of 80 representative economies, including high-income and developing countries from every region in the world, will be able to exploit these technologies by 2020.

Requires increased technological sophistication

Technology application	Most of Africa, Middle East, Oceania	Latin America, South Africa, Turkey, Indonesia	China, India, Russia, Eastern Europe	Industrial countries
	Technologies likely to be mastered by 2020 (x)			
Cheap solar energy	x	x	x	x
Rural wireless communication	x	x	x	x
Genetically modified crops	x	x	x	x
Filters and Catalysts	x	x	x	x
Cheap autonomous housing	x	x	x	x
Rapid bioassays		x	x	x
Green manufacturing		x	x	x
Ubiquitous RFID tagging		x	x	x
Hybrid vehicles		x	x	x
Targeted drug delivery			x	x
Improved diagnostic and surgical methods			x	x
Quantum cryptography			x	x
Ubiquitous information access				x
Tissue engineering				x
Pervasive sensors				x
Wearable computers				x

Source: Silberglitt *et al.* (2007).

While many countries are expected to be able to take advantage of some of the simpler-to-use technologies, a wide range of countries are not expected to be able to do so, because they lack the required technological infrastructure, because their population is not sufficiently technically literate, or because a critical mass of scientists and engineers is not present to exploit the technology (see table). The report finds that most high-income countries will be able to adopt and exploit all of the technologies effectively. A second group of countries – China, India, Russia and the countries of Eastern Europe – has a considerable level of scientific and technological proficiency in specific applications, but barriers to technology adaptation are likely to limit their ability to take advantage of the most sophisticated network applications. A third group of middle-income countries – several Latin American countries, South Africa, Turkey and Indonesia – lack more prerequisites and are thus expected to exploit fewer of these technologies. A final group, comprising most of the world's poorest countries – most of the countries of Africa, the Middle East, and Oceania – are projected to make use of only the simplest of the new technologies. This analysis provides a useful snapshot of the prospects for technological progress based on current data. However, it does not incorporate the potential for dynamic improvements in technological progress (for example, through the rapid dissemination of existing new technologies), which could rapidly improve developing countries' ability to absorb new technologies.

One bright spot is the relatively rapid diffusion of some new technologies in low-income countries. Declining computing costs and the prospects for rapid declines in the cost of wireless Internet connections may improve the efficiency of ongoing economic activities in low-income countries and enable them to leapfrog into more advanced technologies (Primo Braga *et al.*, 2003). Indeed, the development of simple, low-cost computers and the spread of open-source technology have already enhanced the affordability of new technologies for low-income countries. Successful exploitation of these new technologies will nonetheless require stepped-up investments in human capital and reforms in policy and regulation to provide an appropriate incentives structure for ICT investments.

A rigorous roadmap for achieving rapid technological progress does not exist. However, the evidence presented in this report points to a number of conclusions, principles and policy directions that appear likely to promote technological progress and may be able to guide policy makers. Exactly how much weight to give to each of these and how they interact depends on specific country circumstances and should be the subject of future research.

- Openness to external technologies through foreign trade, FDI, the diaspora and other international networks is a critical and necessary condition for technological progress for both low- and middle-income countries, in which most progress occurs through the adoption, assimilation and adaptation of existing but new-to-the-market or new-to-the-firm technologies.
- The capacity of a firm or an individual to use a technology depends critically on the basic technological literacy of workers and consumers. This in turn depends on the capacity of the government to deliver quality education to the widest number possible.
- In a market economy the preeminent vehicles for the dissemination and diffusion of technology are the firm and the entrepreneur. Their success in doing so depends on their ability to undertake and expand new activities. This requires a stable macroeconomic environment, together with a regulatory environment that effectively enforces property rights and the rule of law, does not excessively restrict firms' ability to hire and fire, and does not impose excessive regulatory or financial burdens.
- The capacity of a firm or individual to take advantage of a technology can be constrained by affordability and by liquidity, thereby placing a premium on the efficiency with which the financial system intermediates between savers and borrowers both domestically and abroad.
- Given market failures, government has a role to play in assisting firms to learn about, adopt and adapt new technologies. In addition to focusing on research and development in new-to-the-market technologies, applied R&D agencies need to emphasise outreach, testing, marketing and dissemination activities. The huge rural-urban divide in technology and absorptive capacity in many developing countries underlines the importance of such activities to inclusive development.
- The government also can have an important impact on economic progress by integrating new technology into its own operations, including in the provision of education, health and publicly provided infrastructure; in procurement of goods

and services; in the provision of information and fostering public dialogue; and in the definition of standards for commercial products.

- The principal challenge facing many low-income countries is not the country's access to technology but its absorptive capacity, including physical, human and institutional capacity; limited financial resources; and the extent to which its social and political environment is supportive of entrepreneurship, investment and technological progress.

These conclusions highlight the critical role of government in establishing the general conditions that support rapid technological progress, in helping to overcome market failures that constrain innovation by firms, and in providing (and purchasing) high-quality goods and services. Countries that have achieved sustained and rapid technological progress have generally benefited from committed national leadership following coherent development policies, although the nature of these policies, and in particular the degree of public sector intervention in private markets, has varied enormously.

Some policy directions

This chapter does not offer a comprehensive explanation of why technological progress occurs, nor does it focus on the policies that governments should follow to increase the rate of technological progress. Nevertheless, the preceding analysis suggests that some combination of openness to foreign technology, strong domestic technological competences, a motivated public sector and a well-financed private sector are key ingredients to success. In addition, several general policy directions suggest themselves.

First, much of the technological progress recorded by developing countries over the past 15 years has been associated with the increase in openness that occurred during the same period. This increased the exposure of developing countries to foreign technologies, but their capacity to absorb it has improved much less. To the extent that technological absorptive capacity sets a limit on the level of technology an economy can achieve, as Latin American and high-income countries' tendency to level off suggests, the relatively weak improvement in absorptive capacity may result in a slowing rate of technological progress in some countries, unless significant steps are taken to raise the quality of domestic human capital and improve the regulatory environment and the efficiency with which government services are delivered. This risk may be most marked for countries like Mexico and Indonesia, which have taken advantage of globalisation in a relatively passive way, by exploiting their low-wage comparative advantage without seeking to improve domestic competences.

Second, because of the complementarity of technologies and infrastructure, countries in which older technologies have yet to penetrate particularly deeply may also face limits on the extent to which other technologies are able to diffuse. Therefore, the authorities should give particular emphasis to ensuring that publicly supplied technological services are as widely and economically available as possible, whether delivered directly by the state or via private firms.

Third, it is indispensable to ensure that technologies diffuse throughout the country, not simply to major centres or top-performing firms. This implies paying attention to reinforcing absorptive capacity at the sub-national level and to dissemination channels within countries, including domestic transport infrastructure, and the essential role to be played by outreach, testing, marketing and dissemination activities of applied R&D agencies.

Fourth, notwithstanding the relatively strong improvement in the level of technology achieved by some low-income countries, many others have improved only marginally or not at all. In particular, improvements in technological absorptive capacity have been limited. Efforts must continue to concentrate on increasing the quality of human capital, both by ensuring that more students stay longer in school and also by raising standards, which in too many cases are too low.

References

Archibugi, Daniele and Alberto Coco (2005), "Measuring Technological Capabilities at the Country Level: A Survey and a Menu for Choice", *Research Policy* 34, pp. 175-94.

Bankable Frontier Associates (2007), "Financial Service Access and Usage in Southern and East Africa: What Do finscopetm Surveys Tell Us?", *www.bankablefrontier.com.*

Comin, Diego and Bart Hobijn (2004), "Cross-country technology adoption: making the theories face the facts", *Journal of Monetary Economics*. Vol. 51, Issue 1, January, pp. 39-83.

Docquier, Frederic and Abdeslam Marfouk (2004), "Measuring the international mobility of skilled workers (1990-2000): release 1.0", Policy Research Working Paper Series 3381, The World Bank, Washington, DC.

Friedman, Thomas (2007),"If IT merged with ET", *International Herald Tribune*, 31October.

Goldberg, Itzhak, Lee Branstetter, John Gabriel Goddard and Smita Kuriakose (2008), "Globalization and Technology Absorption in Europe and Central Asia: The Role of Trade, FDI, and Cross-border Knowledge Flows", World Bank Working Paper No. 15.

Gwatkin, D.R., S. Rustein, K. Johnson *et al.* (2000a), "Socio-economic differences in Brazil: Health, Nutrition, and Population in Brazil", HNP/Poverty Thematic Group of the World Bank, Washington, DC, *http://worldbank.org/poverty/health/data/index.htm#lcr*.

Gwatkin, D.R., S. Rustein, K. Johnson *et al.* (2000b), "Socio-economic differences in Ethiopia: Health, Nutrition, and Population in Ethiopia", HNP/Poverty Thematic Group of the World Bank, Washington, DC, *http://worldbank.org/poverty/health/data/index.htm#lcr*.

Gwatkin, D.R., S. Rustein, K. Johnson *et al.* (2000c), "Socio-economic differences in Nigeria. Health, Nutrition, and Population in Nigeria", HNP/Poverty Thematic Group of the World Bank, Washington, DC, *http://worldbank.org/poverty/health/data/index.htm#lcr*.

Jefferson, Gary H., Thomas G. Rawski and Yifan Zhang (2007), "Productivity Growth and Convergence across China's Industrial Economy", Draft, Brandeis University, Waltham, MA.

Kannankutty, Nirmala and Joan Burrelli (2007), "Why Did They Come to the United States? A Profile of Immigrant Scientists and Engineers", InfoBrief NSF07-324, Science Resources Statistics, Directorate for Social, Behavioral, and Economic Sciences, United States National Science Foundation, Arlington, VA, June.

Lederman, Daniel and Laura Saenz (2005), "Innovation and Development around the World, 1960-2000", World Bank Policy Research Working Paper No. 3774, November.

Maloney, William (2006), "Missed Opportunities: Innovation and Resource-Based Growth in Latin America", Background paper, The World Bank, Washington, DC.

Martin, W. and D. Mitra (2001), "Productivity Growth and Convergence in Agriculture and Manufacturing" *Economic Development and Cultural Change* 49(2), pp. 403-423, January.

Ministry of Communication and Information Technology, Government of India (2006), *E-Readiness Assessment Report of States/Union Territories 2005*, New Delhi.

Mitra, Raja M. (2007), "India's Emergence as a Global R&D Center", Swedish Institute for Growth Policy Studies, Stockholm.

McKenzie, D.J. (2003), "Measure Inequality with Asset Indicators", BREAD Working Paper No. 042. Bureau for Research and Economic Analysis of Development, Center for International Development, Harvard University, Cambridge, MA.

Nicoletti, Giuseppe, Stefano Scarpetta and Olivier Boylaud (1999), "Summary Indicators of Product Market Regulation with an Extension to Employment Protection Legislation", OECD Economics Department Working Papers, No. 226.

Poncet, Sandra (2006), "The Long Term Growth Prospects of the World Economy: Horizon 2050", Working Paper 2006–16, Centre d'Etudes Prospectives et d'Informations Internationales.

Porter, M. and S. Stern (2003), "Ranking National Innovative Capacity: Findings from the National Innovative Capacity Index", *Global Competitiveness Report*, World Economic Forum, Geneva.

Primo Braga, Carlos A., John Daly and Bimal Sareen (2003), "The Future of Information and Communication Technologies for Development", Presented at the ICT Development Forum, Petersburg, Germany, 21 May.

Sagasti, Francisco (2003), *Knowledge and Innovation for Development: The Sisyphus Challenge for the 21st Century*. Edward Elgar, Cheltenham.

Silberglitt, Richard, Philip S. Anton and David R. Howell (2007), *The Global Technology Revolution 2020*, Rand Corporation.

Soubbotina, Tatyana P. (2006), "Generic Models of Technological Learning by Developing Countries", World Bank Science and Technology Program.

Sullivan, Nicholas P. (2007), *You Can Hear Me Now: How Microloans and cell Phones Are Connecting the World's Poor to the Global Economy,* Jossey-Bass, San Fransisco, CA.

United Nations Conference on Trade and Development (UNCTAD) (2003), "Transfer of Technology for Successful Integration into the Global Economy", UNCTAD/ITE/IPC/2003/6.

UNCTAD (2005), *World Investment Report 2005: Transnational Corporation and Internationalization of R&D,*: United Nations, New York and Geneva.

Watkins, Alfred and Anubha Verma (eds.) (2007), "Building Science, Technology and Innovation Capacity in Rwanda: Developing Practical Solutions to Practical Problems, The World Bank, Washington, DC.

World Bank (2005), "Sri Lanka: Offshoring Professional Services A Development Opportunity", The World Bank, Washington, DC.

World Bank (2006), *Transport for Development*, The World Bank, Washington, D.C.

World Bank (2007a), *Making Finance Work for Africa*, The World Bank, Washington, D.C.

World Bank (2007b), *Unleashing India's Innovation Potential*, The World Bank, Washington, D.C.

World Bank (2007c), *Global Trade Connections: Connecting to Compete*, The World Bank, Washington, D.C.

Vyas, Seema and Lilani Kumaranayake (2006), "Constructing Socio-economic Status Indices: How to Use Principal Components Analysis", *Health Policy and Planning 2006*, 21(6):], pp. 459-468, *doi:10.1093/heapol/czl029.*

Chapter 9

Foreign investment and the development of telecommunications in Latin America

Juan R. de Laiglesia[1]

Latin America leads developing countries by a significant margin in cumulative FDI flows to the telecommunications sector. Access to communications services has progressed very positively in the region in the past decade, accompanied by the expansion in foreign investment and innovation in both technology and service delivery. Both foreign investments – including from home-grown Latin multinationals – and mobile telephony, especially in the form of prepaid lines, have contributed to increasing the density of telecommunication services. However, service provision for relatively disadvantaged groups remains a major challenge that current investment and innovation trends have only begun to address. Two policy concerns dominate the communications sector: access to services remains very unequal – for example, only 25% of the poorer households have a phone at home – and innovations in network expansion are slow. This chapter makes a case for a stronger regulatory network, more competition and policies to support innovation.

1. Juan de Laiglesia is Economist, Poverty Reduction and Social Development Group, OECD Development Centre.

Introduction

Since the early 1990s, foreign direct investment (FDI) has increased rapidly all over the world. In Latin America, the 1990s were a period of significant FDI inflows, led by OECD-based multinational corporate investments in newly privatised or liberalised sectors in the region.

The telecommunications sector lies at the intersection of patterns of inward and outward FDI in Latin America. Two multinationals dominate FDI in the sector: Spain's Telefónica and Mexico's Carso group. The former came into the region during the wave of privatisations in the 1990s. The latter, owner of Telmex and América Móvil, moved, in little more than five years, from being a national monopoly to becoming a major regional player.

The performance of the telecommunications sector is also important because of its implications for aggregate economic performance in Latin America. Telecommunications services are a key means of information transmission. As such, not only does the rollout of telecommunications infrastructure profoundly change the way business is done, it has the potential to significantly improve the efficiency of markets as the cost of obtaining information is radically lowered, and thus also to contribute sizeably to economic growth and development. Moreover, these benefits for the economy as a whole display increasing returns, so that as service coverage approaches universality, the benefits increase more than proportionately (Röller and Waverman, 2001; Waverman *et al.*, 2005). The degree of people's and firms' access to telecommunications services is therefore crucial.

This chapter analyses the factors that have driven the evolution of telecommunications service coverage and measures the sector's progress both in terms of overall coverage and in terms of providing access to different segments of the population. It first describes investment patterns in telecommunications and discusses the role of FDI in the sector. It then summarises the performance of the region in terms of increasing density and of inequality in access to service. The next section assesses the drivers of that performance. A conclusion follows.

Investment in public services and multinationals: the case of telecommunications

Infrastructure plays a key role in economic development. By way of illustration, if the countries in Latin America with the largest gaps in infrastructure development – Bolivia, Guatemala, Honduras, Nicaragua and Peru – caught up with the regional leader, Costa Rica, their long-term growth performance is predicted to speed up by at least 5 percentage points and their income Gini coefficients (a common measure of income inequality) to drop by a significant 8 to 10 percentage points (Calderón and Servén, 2004).

Telecommunications are a particularly important part of a country's infrastructure. Not only do they open doors to new information and business opportunities and shape how firms do business. They can deeply affect economic and political life as a whole, because information and communications technology (ICT) influences the transmission of information for all kinds of purposes. ICT development has been shown to be associated with lower levels of corruption, for example, as well as with lower inequality (Bandyopadhyay, 2006).[2]

2. Bandyopadhyay (2006) finds inequality to be negatively related to ICT development for her whole sample but positively for the developing country sample, suggesting a non-linear relationship.

The telecommunications sector has been, since the early 1990s, witness to the increasing internationalisation of Latin American economies. FDI flows to Latin America increased dramatically throughout the 1990s, in conjunction with widespread privatisation in the region, from less than USD 10 billion in 1990 to a record USD 89 billion in 1999 (averaging USD 39 billion a year for the decade). Despite a notable fall in the early 2000s, the annual flow rose again in the middle of the decade to around USD 70 billion. Still, Latin America's share of global FDI flows, at 8% in 2006, remains well below the 14% peak reached during the 1970s and again in 1997 (ECLAC, 2007).

Not only that, the increasingly rapid pace of internationalisation of a relatively small number of Latin American enterprises has given rise to a major increase in outward FDI from Latin American countries, a pattern that has also affected the telecommunications industry through the rise of sister companies América Móvil and Telmex. Indeed, the telecommunications sector provides a particularly important example of the recent wave of developed-country multinational corporate investment in the region together with the rise of Latin American multinationals and their interaction.

Like other key areas of infrastructure, the telecommunications sector requires substantial capital investment. Among developing regions, Latin America has been at the forefront in allowing private capital into its telecommunications industry. Following the path opened by Chile in 1987, other countries, including Jamaica (1989), Argentina (1990) and Mexico (1990), privatised their incumbent operators. These privatisations brought in substantial amounts of private capital – the privatisation of Brazil's Telebras system alone raised USD 34 billion in 1998, for example. In the majority of cases privatisation was also an avenue for foreign capital to enter the industry. Of course, such has not always been the case, as in Mexico – where regulatory restrictions on FDI in the sector have prevented entry, and foreign participation remains limited to 49% outside of mobile telephony – and in a few countries, notably Ecuador and Paraguay, which have failed to secure the interest of private investors (see Rozas Balbontín, 2005, on successes and failures in privatisation).

With ICT greatly facilitating the internationalisation of firms and economies, companies that supply ICT services should gain significantly from internationalising their activity. The drive of telecommunications enterprises to become global players, accompanied in the last five years by a process of consolidation in the ICT industry, has given rise to a battle for supremacy in the Latin American telecommunications industry.

Privatisation, multinationals and FDI in telecommunications

Between 1990 and 2003, FDI in the telecommunications sector in Latin America amounted to USD 109.8 billion, well above that in any other developing region; 70% of this amount is the result of privatisation transactions (Guislain and Qiang, 2006). Despite a slowdown since the beginning of the 2000s, FDI in the sector picked up again in 2003 and 2004, with new FDI and investment by affiliates of foreign multinationals adding up to over USD 5 billion in 2004 and 2005 (Figure 9.1). But the objectives of FDI in the sector have changed. While consolidation has driven a series of corporate mergers and acquisitions, on the one hand, mobile telephony accounts for more than half of FDI in the sector since 2000, and the rise of mobile telephony has led to a growing share of greenfield investments, on the other.

Figure 9.1. Foreign investment[1] in telecommunications in Latin America

USD millions

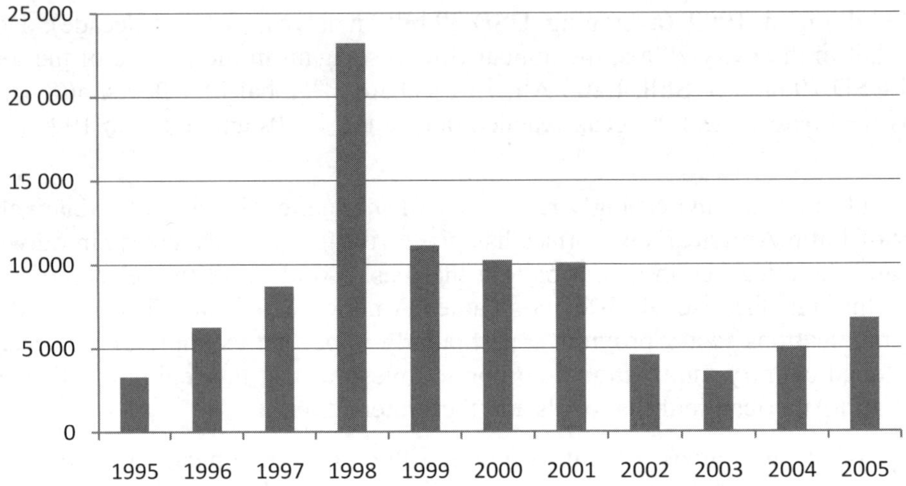

1. FDI and investment by affiliates of foreign multinationals.

Source: OECD (2007), *Latin American Economic Outlook*, OECD Development Centre, Paris; based on World Bank (2007), *PPI Database*.

Indeed, the telecommunications sector has witnessed increased internationalisation and consolidation worldwide since the early 1990s, and Latin America is no exception (Figure 9.2 and Table 9.1). A number of foreign investors, including Spain's Telefónica, entered Latin America following the privatisation of the landline incumbents during the 1990s. Despite difficult financing conditions after the burst of the Internet bubble in 2000, the large expenses incurred in purchasing third-generation licences and devaluations in Argentina and Brazil, the rise in demand from 1995 to 2005 for telecommunications services, especially mobile telecommunications, kept private enterprises in the game. At the same time, the region witnessed a gradual process of consolidation, led by Telefónica in 2000 and followed by the growth of América Móvil – a Telmex spin-off operating in the mobile segment – from 2002 to 2005. Reinforcing this trend, Telmex, the privatised Mexican landline incumbent, invested USD 4 750 million between 2003 and 2005 to compete with Telefónica in the landline and data-transmission markets.

The performance of the telecommunications sector in Latin America

Latin Americans do not have a positive perception of the privatisation of public utilities in their region. According to 2005 survey data (Latinobarómetro, 2005), only one-third of the region's population is satisfied with it. These opinion data bundle together the privatisation of water and electricity, where the entry of foreign private capital has often been more controversial, as well as telecommunications, where privatisation and liberalisation were widely preceded or accompanied by a rebalancing of user tariffs. Whereas previously, very costly long-distance calls often "paid for" extremely low-cost local calls and the sector's infrastructure, operators moved to eliminate this cross-subsidisation, thereby raising the cost of telephone services for those who make less use of international services, and hurting the poorer segments of the population.

Figure 9.2. Share of FDI in telecommunications in developing countries, by region, 1990-2003

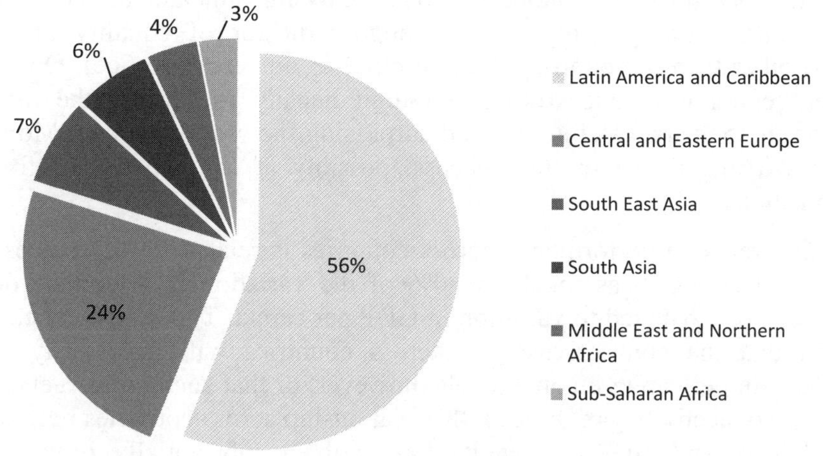

Source: World Bank (2006), *Information and Communications for Development 2006*.

Table 9.1. Main telecommunications operators in Latin America, March 2007

Thousands of serviced lines

	Telefónica	America Móvil	Telmex	Telecom Italia	Millicom
Land lines	23 810	2 913	20 374	5 173	--
Mobile customers[1]	85 637	122 434	--	38 656	10 438
-- *of which*					
Mexico	9 320	44 946	18 284	0	0
Central America	4 042	9 231	0	0	5 917
Brazil	41 064	24 608	2 068	26 300	0
Argentina	16 441	10 927	n/a	13 426	0
South America	96 085	71 170	n/a	56 109	4 521
Total	109 447	125 347	20 374[2]	43 829	10 438

1. Includes fixed mobile.

2. Total available data.

n/a not available; -- no operations.

Source: OECD (2007), *Latin American Economic Outlook 2008*, OECD Development Centre, Paris, based on enterprise reports for first quarter 2007, complemented by regulator websites.

The performance of Latin America's telecommunications sector has nevertheless been exemplary since privatisation started in the late 1980s in terms of increased aggregate availability and quality of services. The most commonly used aggregate indicator of progress in telecommunications is telephone density (or teledensity) in lines per 100 inhabitants. Given the substitutability of land and mobile lines for voice communications (OECD, 2007b), the sum of the two is taken as a measure of access to private telephone lines. Telephone density is a useful indicator of the connectivity gap across countries (though only a very rough indicator of equality of access within countries, other than in countries that are close to universal service). On average, since 1990, the region has gone from single-digit density to serving the majority of the population in 15 years, catching up and surpassing the world average. Mobile telephony has played a major role in this success, possibly at the expense of fixed telephone communications.

The difference in performance across countries is related to differences in aggregate economic performance: as much as 80% of the variation in telephone density across countries can be attributed to variation in GDP per capita. This statistical correlation does not mean that the connectivity gap across countries will necessarily be closed by convergence in per capita income levels, however, or that such convergence is necessary to close the connectivity gap. Indeed, the relationship across countries between telephone-service density and GDP per capita has evolved substantially over time, as more developed markets have reached saturation levels and middle-income countries have extended service more quickly than would have been predicted by income growth alone (Figure 9.3). While the relation between teledensity and per capita income was essentially linear until 1995, by 2005 that linearity had disappeared (Figure 9.4), a disappearance that is consistent, among other interpretations, with threshold effects in network extension. The relation between income levels and teledensity has limits, in other words, both in terms of its causal interpretation, and in terms of the degree to which it can be used to infer policy recommendations.

Figure 9.3. Telecommunications performance in Latin America 1990-2007

In lines/subscriptions per 100 inhabitants

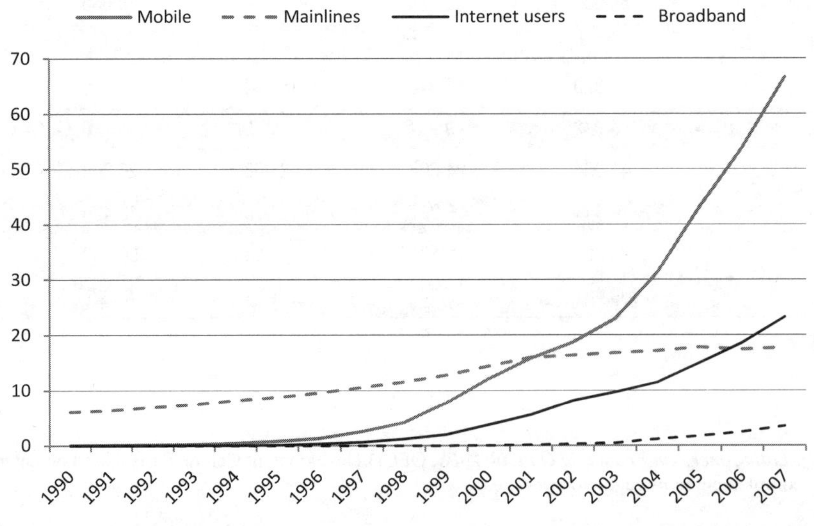

Source: Author's calculation based on ITU (2008), ICT Eye Regulatory Information Centre data.

Figure 9.4. Income per capita and telephone density: an evolving relationship

GDP per capita (PPP, in constant 2000 USD) and telephone density (lines per 100 inhabitants)

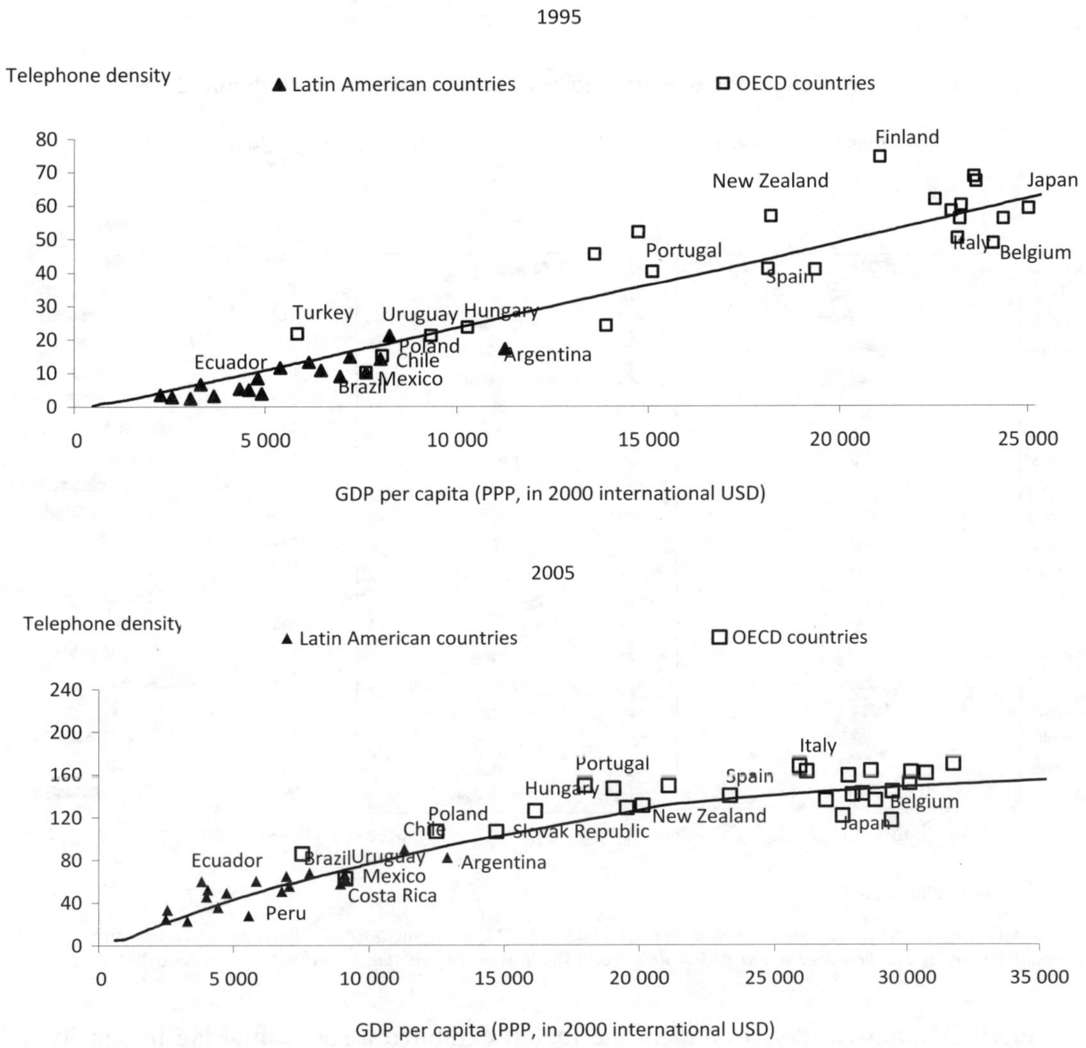

Note: The solid line is a non-parametric smoother based on data from all available countries at each date.

Source: OECD (2007), *Latin American Economic Outlook 2008*, OECD Development Centre, Paris, based on data from ITU, *World Telecommunications/ICT Indicators Database* 2006 and World Bank, *World Development Indicators Database* 2006.

Latin America's aggregate success in catching up with the world average in telephone density hides a diverse picture among countries in the region. Chile and Argentina are far ahead, with 90 and 82 telephone lines per 100 inhabitants, respectively. On the other hand, Haiti and Cuba have yet to experience take-off in telephone access and uptake, while the poorest Central American and Andean countries have only recently started their catch-up phase: in Ecuador, for instance, annual growth in teledensity accelerated from around 30% in the early 2000s to 53% in 2005.

Figure 9.5 attempts to account for both the effect of GDP per capita and the evolution of the changing relation between per capita income and telephone density. Country-level results are presented for 1995, 2000 and 2005, with each graph showing the deviation of each country's actual telephone density from the value predicted by the country's GDP per capita. Negative figures mean that a country is below the teledensity expected, given its level of income; positive figures mean that the country's teledensity is above the expected level.[3]

Figure 9.5. Telephone density and per capita GDP, 1995, 2000 and 2005

Deviations from non-parametric prediction of telephone penetration based on GDP

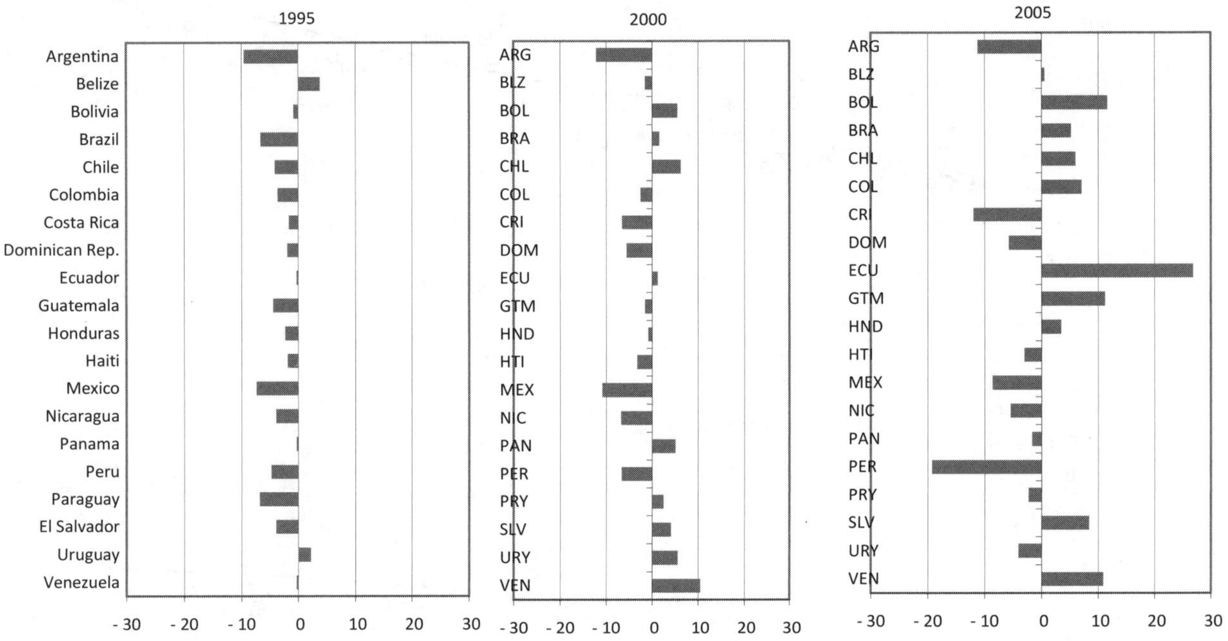

Note: Data for Haiti 2005 from 2004.

Source: OECD (2007), *Latin American Economic Outlook 2008*, OECD Development Centre, Paris, based on data from ITU, *World Telecommunications/ICT Indicators Database* 2006 and World Bank, *World Development Indicators Database* 2006.

In 1995, almost every country in the region exhibited a substantial lag in density of telephone services relative to the level its per capita income would predict. The only exceptions were Belize, one of the pioneers in the privatisation process, and Uruguay, which maintained the incumbent public operator; other early privatisers, including Argentina, Guyana, Mexico and Venezuela, had not yet achieved significantly better results. Five years later the picture was brighter. Chile, one of the first countries to liberalise the market, had more than caught up with the level of connectivity predicted by its level of income, and Brazil had also reduced the difference between actual and predicted connectivity. In both countries, foreign operators played a leading role.

Yet the latest data still present a mixed picture for the region. While several countries that privatised their incumbent telephone provider perform well, even when accounting

3. The actual model fitted is a non-parametric local linear regression of telephone density over GDP per capita in PPP terms. The results are broadly unchanged if a parametric approach is used by applying the best-fitting Box-Cox transformation to per capita GDP for each year.

for the rise in income levels (Brazil, Chile, El Salvador and Guatemala all exhibit more connectivity than expected), others, including Argentina, Mexico and Nicaragua, which have also privatised their incumbent operator, are behind their expected levels of connectivity. The remarkable case of Ecuador, which is clearly outperforming its peers both within and beyond the region, is due to a doubling of mobile subscribers in 2005 – probably the result of intense competition between the two major players, Telefónica's Movistar and América Móvil's Porta, spurred by the entry in 2003 and rapid expansion in coverage of a third mobile operator, Alegro.

Despite good aggregate performance, the degree to which the poor have benefited from the region's increased access to telecommunications varies greatly across countries. Figure 9.6 presents the proportion of individuals in the highest and lowest quintiles who have a telephone at home. While increases in average connectivity are indeed leading to increased availability of service for lower income quintiles, the variation is striking, even among countries where essentially all of the better-off have telephone lines. Moreover, the degree of inequality is very high, even when compared to provision of other utilities, as shown by the fact that in the poorer countries, virtually no one in the lowest quintiles has his/her own telephone line. The picture is not much brighter for Internet use (Figure 9.7). Despite reasonable rates of Internet use among the well-off, the poorest do not use the Internet in the poorer countries of the region.

Figure 9.6. Proportion of the population with a telephone at home, selected countries

By income quintile

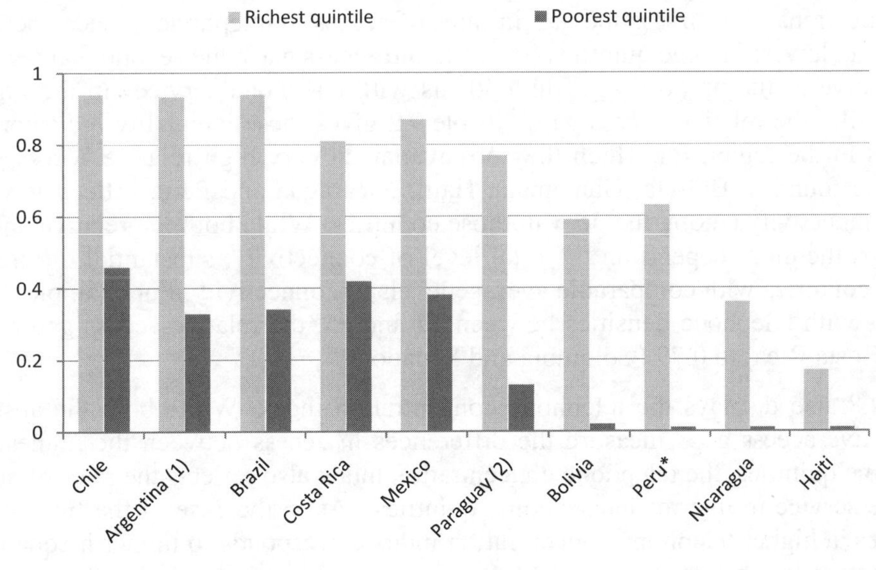

Note: Definitions as per national surveys.

1. Argentina data for urban areas only.

2. Fixed lines only.

Source: Author's calculation, based on SEDLAC data.

Figure 9.7. Proportion of Internet users, poorest and richest quintiles

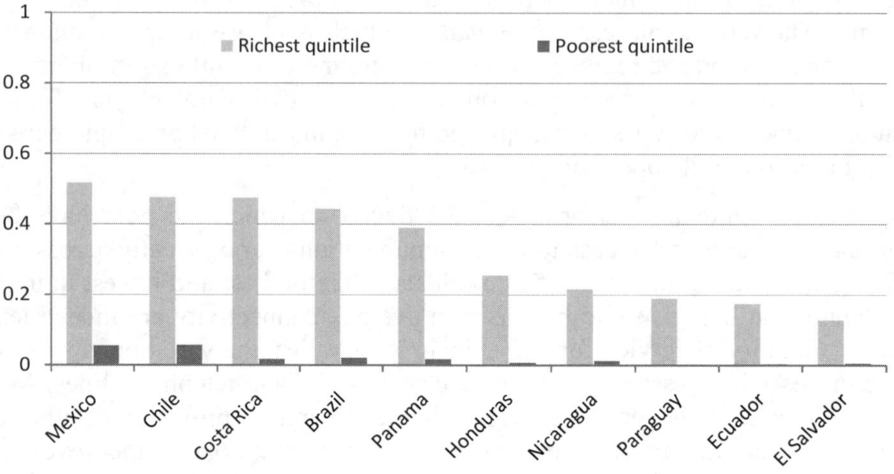

Note: Definitions as per national surveys.

Source: Author's calculation, based on OSILAC/ECLAC data.

Differences between top quintiles are a visually compelling but somewhat rough measure of inequality. Figure 9.8 displays three measures of inequality in access to telecommunications. The first is the gap in rates of access to telephone services between the highest and lowest income quintiles (the absolute access gap); the second is the size of this gap relative to the proportion of individuals with telephone services in the highest income quintile (the relative access gap).[4] Table 9.2 gives these inequality indicators for all countries in the region for which they are available. Very high relative access gaps, such as those found in Bolivia, Guatemala, Haiti, Nicaragua and Peru, reflect very low telephone connectivity among the poor in those countries. While both the relative and the absolute gaps therefore depend on the total level of connectivity, substantial differences exist across countries with comparable average levels of connectivity. For example, in the six countries with telephone densities between 57 and 65, the relative access gap ranges from 0.48 (Costa Rica) to 0.79 (Colombia and Ecuador).

Figure 9.8 also displays the telephone concentration index. While both the absolute and the relative access gaps measure the differences in access between the highest and lowest income quintiles, the telephone concentration index also reflects the rates of access to telephone service in intermediate income quintiles.[5] As in the case of the two access-gap measures, a higher telephone concentration index corresponds to higher inequality in access. Importantly, the data in Figure 9.8 reveal an inverse correlation between the

4. The relative gap is *(Q5-Q1)/Q5* where *Qi* is the proportion of individuals in income quintile *i* with access to a telephone. This measure is therefore between 0 (perfect equality) and 1 (if no one in the lowest quintile has a telephone).

5. The telephone concentration index measures the concentration of telephone ownership by the area over the Lorenz curve, plotting the cumulative share of owned telephones against income quintiles. It can be interpreted like a Gini coefficient: 0 is perfect equality (all individuals have the same probability of owning a telephone) and 1 perfect inequality (only individuals in the highest income quintile have a telephone).

concentration index and total density, which suggests that as telephone density increases, households in the poor (but not the poorest) segments of a country's population gain access.

Figure 9.8. Access gap in telephony for selected Latin American countries

Latest available data

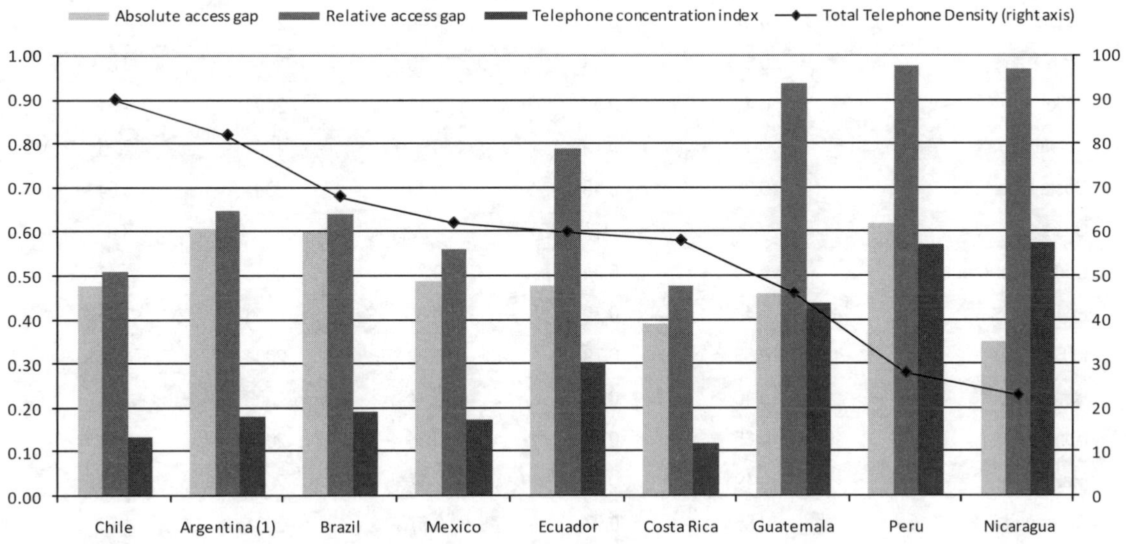

1. Urban areas only.

Source: OECD (2007), *Latin America Economic Outlook*, OECD Development Centre, Paris, based on SEDLAC (2007) data, except Argentina, based on IADB (2007), *Equity and Social Indicators* data.

These measures of telephone access inequality rely on household data and are based on telephone-line ownership, rather than access strictly speaking. While it can be argued that having access to one's own telephone line is a qualitative improvement over public access, the main avenue for the poor to gain access to telephone and other information technology services is often through public payphones or communal telecentres. Data on the density of public payphone networks show significant differences even within the group of countries with the most unequal access. While Peru and Nicaragua exhibit similar relative access gaps of close to 1 (Figure 9.8), Peru boasts 5.42 payphones for every 1 000 inhabitants to Nicaragua's mere 1.27, and Bolivia's 1.75.

Although access gaps are generally high, relative access gaps exhibit a decreasing trend in almost every country for which a long enough data series exists. Brazil provides an important illustration: until 1997, the country's relative access gap was higher than 0.9 – similar to those found in the most unequal countries in Latin America today – but by 2004 it had fallen significantly, to 0.6, thanks to an increase in the access rate of the poorest quintile from 5 telephones per 100 persons to more than 30.

Table 9.2. Inequality and access to telephone service in Latin America

	Telephone line subscription inequality				Public payphones per 100 inhabitants
	Absolute access gap	Relative access gap	Telephone concentration index	Year	2005 or latest
Argentina[1]	0.61	0.65	0.18	2001	5.91
Bolivia	0.57	0.97	0.48	2002	1.75
Brazil	0.60	0.64	0.19	2004	6.83
Chile	0.48	0.51	0.13	2003	4.51
Colombia	0.44	0.79	0.27	2004	2.44
Costa Rica	0.39	0.48	0.12	2004	5.04
Cuba	..	..	..	..	3.07
Dominican Republic	0.49	0.84	0.34	2005	1.47
Ecuador	0.48	0.79	0.30	2003	2.35
El Salvador	0.53	0.79	0.29	2004	4.04
Guatemala	0.46	0.94	0.44	2003	4.31
Honduras	0.65	0.89	0.36	2005	0.48
Mexico	0.49	0.56	0.17	2004	7.12
Nicaragua	0.35	0.97	0.57	2001	1.27
Panama	0.79	0.87	0.28	2003	2.54
Paraguay[2]	0.64	0.83	0.30	2004	2.54
Peru[2]	0.62	0.98	0.57	2003	5.42
Uruguay	0.57	0.60	0.16	2005	5.44
Venezuela	0.36	0.72	0.26	2003	4.57
Latin America average	0.49	0.75	0.3	--	3.74

Notes: .. not available ; -- not applicable. Definitions as in national surveys.

1. Urban areas only.

2. Fixed lines only.

Source: OECD (2007), *Latin American Economic Outlook 2008*, OECD Development Centre, Paris. Author's calculations, based on SEDLAC (2007), except Argentina, based on IADB (2007), *Equity and Social Indicators*.

Explaining the performance of the telecommunications sector in Latin America

This section reviews the drivers of the performance of telecommunications in Latin America. Two common explanations of good aggregate performance (privatisation and technology) are briefly presented; the specific roles of multinationals and regulatory frameworks are then examined with the help of the new performance indicators described in the preceding section.

Privatisation and performance in telecommunications

The good aggregate connectivity performance of the telecommunications sector in Latin America has often been attributed to the successful privatisation of incumbent operators (IADB, 2001). Figure 9.9 shows that indeed, on average, the rate of growth of connectivity per capita increased substantially, from 10% to 25% a year after privatisation. While these data may be interpreted as suggesting that privatisation played a key role in enhancing performance as measured by telephone density, such an interpretation is subject to important caveats. In addition to those noted earlier, it is important to recall that the 1990s saw the implementation of a series of reforms that accompanied privatisation: in some cases, as for example in Brazil, telecommunications markets were liberalised immediately after privatisation of the incumbent. Moreover, as the trends shown reveal, a second break has tended to occur in the upward trend of teledensity around the fifth year after privatisation, probably owing to the end of exclusivity periods and the subsequent opening of the market. Indeed, exclusivity periods have been shown to limit the benefits of privatisation substantially in terms of network extension (Wallsten, 2003).

Figure 9.9. Telephone connectivity and privatisation experiences

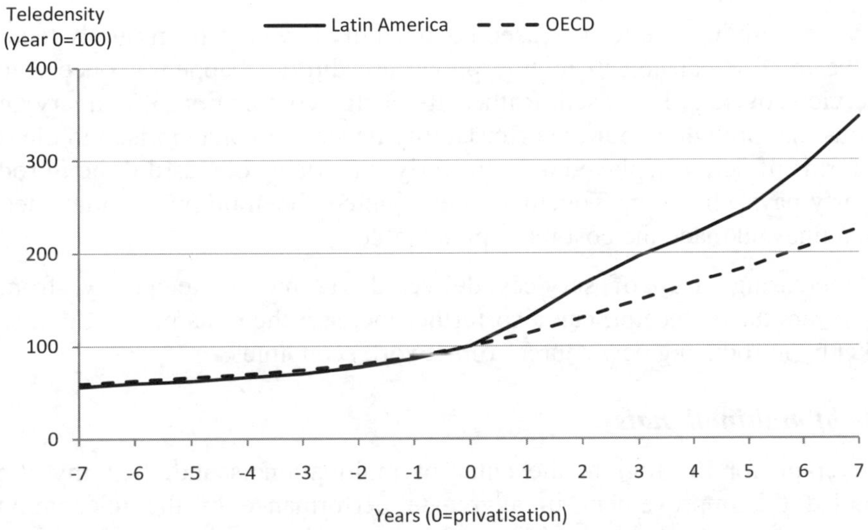

Note: Time 0 is the year of privatisation of the incumbent operator; the scale of telephone density is normalised to be 100 at the time of privatisation. The plot is the average of data for countries in which privatisation had occurred at least seven years previously and had available data for the post-privatisation period (Argentina, Belize, Bolivia, Brazil, Chile, El Salvador, Guatemala, Guyana, Mexico, Panama, Peru and Venezuela).

Source: OECD (2007), *Latin American Economic Outlook 2008*, OECD Development Centre, Paris, based on ITU (2008), ICT Eye Regulatory Information Centre data.

The mobile factor: technology and performance

Technology and innovation have played a major role in the successful spread of telephony in Latin America. As noted earlier (Figure 9.3), while landlines have become steadily more available, the great advances since the late 1990s correspond to rapid increases in mobile telephony.[6]

6. Comparable data on mobile phone coverage of the population for the countries considered that would allow a comparative perspective are not available.

The rising importance of mobile communications, and the segmentation of land and mobile telephony that telecommunications markets have witnessed as a consequence, make assessments of the performance of telecommunications policies more difficult. Indeed, mobile communications have typically been less stringently regulated and led to more competitive markets than landline communications. This difference reflects two factors: the lower initial investment necessary in mobile telephony, which reduces entry costs, and the absence of established public monopolistic providers.

The interaction between the diffusion of mobile technology and the consolidation process in the sector is one of the main determinants of the sector's competitive structure. The spread of mobile communications was accompanied by a number of new entrants into the sub-sector and a segmentation of the market. Moreover, the absence of an incumbent in most cases was a unique opportunity to create competitive markets from the start. Yet subsequent technological advances have tended, on the contrary, to blur the lines between landline, mobile and data transmission services, thereby reinforcing a process of consolidation and giving rise to fierce oligopolistic competition among a reduced number of actors.

Innovations giving rise to increased connectivity have not been purely technological. Indeed, the rapid expansion in mobile phone use did not happen at once when mobile phone service coverage increased. Rather, it was the conjunction of delivery and pricing innovations that brought about the affordability needed to democratise mobile telephony. Prepaid forms of delivery played a particularly important role, as did the introduction of calling-party-pays charging. Together, they limited the minimum outlay necessary to maintain a line and made the cost more predictable.

The increasing array of services delivered via mobile telephony, from Internet banking to remittance receipt, can also further increase the dynamism of that segment of the market by introducing new sources of network economies.

The role of multinationals

On average for the region, the entry of multinationals and their investment have accompanied the improvement in aggregate performance of the telecommunications sector, as measured by teledensity (Figure 9.10).

The entry and expansion of multinational firms in the telecommunications sector in Latin America have been driven by their market-seeking strategies. These are observable not only in their cross-border activity, but also within countries, as competitors strive to acquire customers beyond the affluent classes. Indeed, providers, and especially mobile providers, spend substantial resources to acquire new customers. To attract customers, mobile-telephony enterprises have even shown themselves willing to bear a substantial part of the cost of new telephones to make the acquisition of a telephone very attractive to new customers – essentially subsidising the price of terminals for customers. The average enterprise subsidy for the purchase of a mobile telephone is four times the average monthly revenue per customer in Colombia, Ecuador and Peru, for example, and six times the average revenue in Argentina and Brazil – meaning that the cost of the telephones is only recovered, on average, four and six months, respectively, after acquisition by the customer (Fundación Telefónica, 2007).

Figure 9.10. Foreign investment in telecommunications and the increase in connectivity

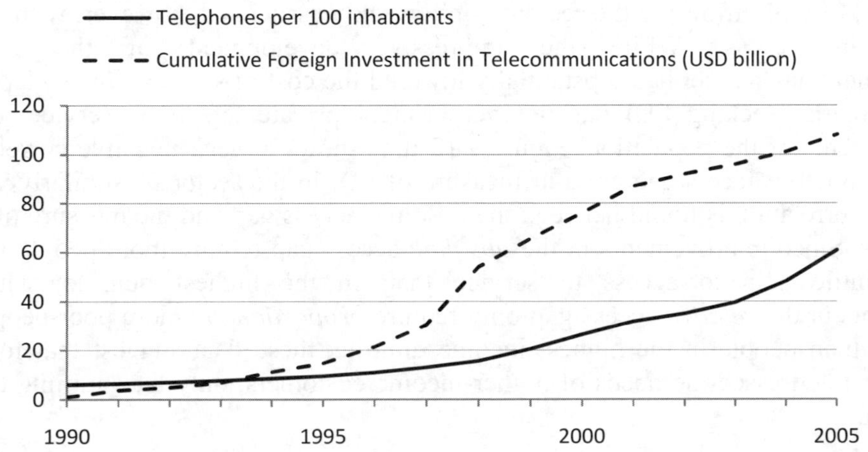

Source: Author's calculations, based on World Bank (2007), *PPI Database* and ITU (2008), ICT Eye Regulatory Information Centre data.

Figure 9.11. Foreign investment and changes in teledensity, 1988-2005

Cumulative foreign investment in telecommunications per capita (in USD)

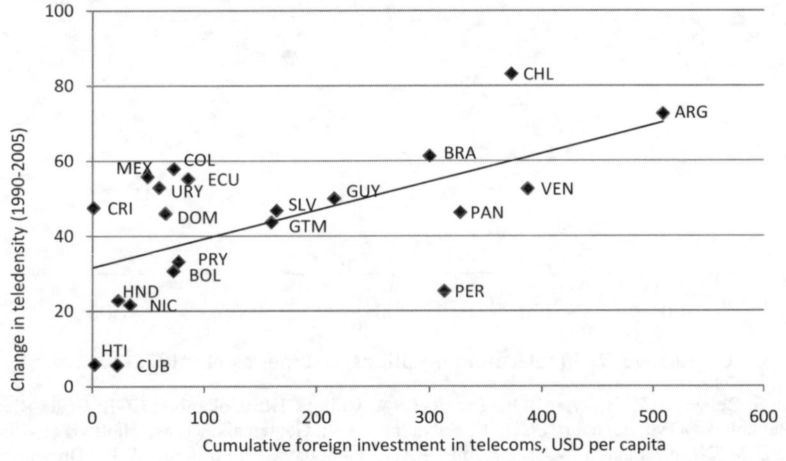

Argentina: ARG; Belize: BLZ; Bolivia: BOL; Brazil: BRA; Chile: CHL; Colombia: COL; Costa Rica: CRI; Dominican Republic: DOM; Ecuador: ECU; El Salvador: SLV; Guatemala: GTM; Haiti: HTI; Honduras: HND; Mexico: MEX; Nicaragua: NIC; Panama: PAN; Paraguay. PRY; Peru: PER; Uruguay: URY; Venezuela: VEN.

Source: OECD (2007a), *Latin American Economic Outlook 2008*, OECD Development Centre, Paris. Teledensity information from ITU (2006), World Telecommunications/ICT Indicators Database; investment information calculated from World Bank PPI Database.

Such market-seeking multinational corporate behaviour explains the positive correlation visible in Figure 9.10 between the level of FDI in a country's telecommunications sector and the change in the country's telephone density between

1990 and 2005. While the FDI data in Figure 9.11 are limited in terms of coverage and comprehensiveness, alternative data sources provide a similar aggregate picture.[7]

The level of FDI inflows to the sector is only weakly associated, however, with lower inequality in access. While the impressive development of the mobile telecommunications market has substantially lowered the cost of serving voice telephony customers, market-seeking FDI has not yet brought private telephone service to the poorest segments of the population. Figure 9.12 thus shows a weak negative correlation between the relative access gap and our measure of FDI in the sector. A similarly weak, but positive correlation is found between the absolute access gap and the measure of FDI in the sector. Since improvements in the *absolute* access gap require more people in the poorest quintile gain to access to service than in the highest quintile, whereas improvements in the *relative* access gap only require *proportionally* more poor people to gain access than people in the highest income quintile, these data suggest that foreign entry has first addressed the needs of higher-income customers, only later turning to the less well-off.

Figure 9.12. Foreign actors are not associated with lower inequality

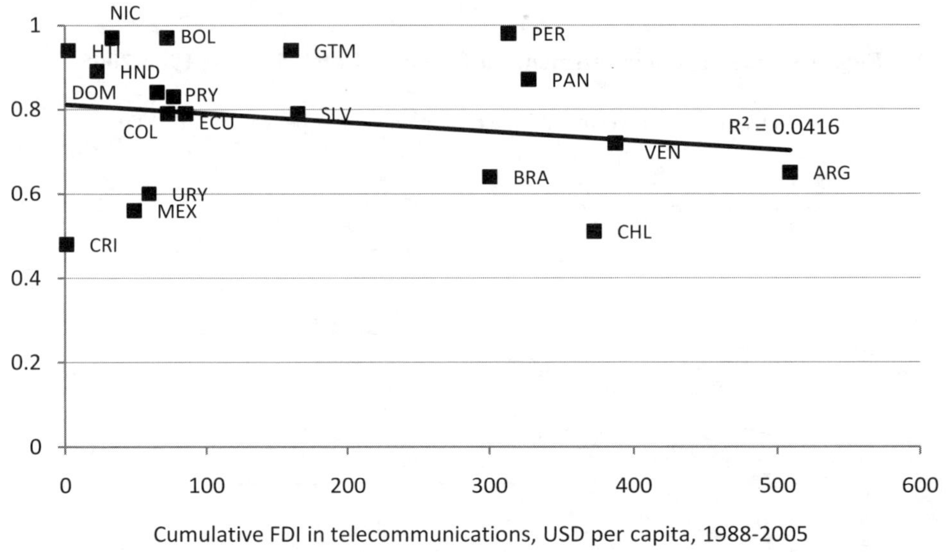

Argentina: ARG; Belize: BLZ; Bolivia: BOL; Brazil: BRA; Chile: CHL; Colombia: COL; Costa Rica: CRI; Dominican Republic: DOM; Ecuador: ECU; El Salvador: SLV; Guatemala: GTM; Haiti: HTI; Honduras: HND; Mexico: MEX; Nicaragua: NIC; Panama: PAN; Paraguay: PRY; Peru: PER; Uruguay: URY; Venezuela: VEN.

Source: OECD (2007), *Latin American Economic Outlook 2008*, OECD Development Centre, Paris, based on World Bank (2007), *PPI Database* and SEDLAC (2007) data.

7. Sector-disaggregated data are not available from official sources in a comparable fashion. The data used are drawn from the World Bank's *Private Participation in Infrastructure (PPI) Database.* They do not therefore strictly correspond to the OECD definition of foreign direct investment (OECD, 1999) and should be interpreted as a proxy. This database records total investment per project as well as the participation of foreign actors but seldom the participation (share) of foreign actors in each project and never its change over time. All investment data are weighted by the share in the joint venture when available. They therefore underestimate (sometimes substantially) actual investment, especially in the case of acquisitions. Moreover, the data do not account for losses or repatriation of profits. For further details on the advantages and shortcomings of the database for this purpose, see Guislain and Qiang (2006).

Still, the increased connectivity rate displayed by countries in which the telecommunications sector has received substantial FDI has benefited the population as a whole, if not the lowest income groups. Figure 9.13 plots FDI in the sector against the index of concentration presented in the preceding section. This measure of the concentration of telephone ownership across income groups puts less weight on the lowest income group. Increased access by the less poor and the middle class tends to lower the concentration index, which can still be interpreted as signalling less inequality in access. Again, the size of foreign investment inflows is weakly associated with less inequality in access. While one cannot determine causality from this exercise, a probable explanation of the relationship is the attractiveness, for foreign capital in the sector, of markets with sizeable middle classes.

Figure 9.13. Telephone concentration and FDI in telecommunications, 1990-2005

Index of telephone concentration (latest available) versus cumulated FDI in telecommunications (in USD)

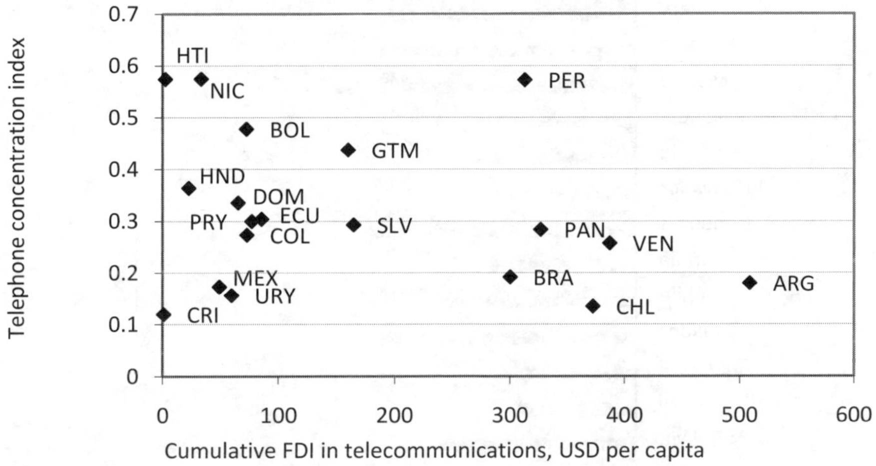

Argentina: ARG; Belize: BLZ; Bolivia: BOL; Brazil: BRA; Chile: CHL; Colombia: COL; Costa Rica: CRI; Dominican Republic: DOM; Ecuador: ECU; El Salvador: SLV; Guatemala: GTM; Haiti: HTI; Honduras: HND; Mexico: MEX; Nicaragua: NIC; Panama: PAN; Paraguay: PRY; Peru: PER; Uruguay: URY; Venezuela: VEN.

Source: OECD (2007), *Latin American Economic Outlook 2008*, OECD Development Centre, Paris, based on SEDLAC (2007) data and World Bank (2007), PPI Database.

Consolidation, competition and market structure

The consolidation process in the industry has raised concerns about an evolution towards an increasingly duopolistic market. This process notably accelerated in 2004, with Telefónica's acquisition of Bell South's Latin American mobile operations, on the one hand, and Telmex's purchase of AT&T's operations in Argentina, Chile, Colombia and Peru on the other. But the land and mobile segments of the market remain quite different, in terms of their competitive structures, and there is also wide dispersion across countries, as indicated by the Herfindahl-Hirschman Index (HHI) values presented in Figure 9.14. Mobile telephony markets tend to be less concentrated than landline markets (an observation that is reinforced by the fact that the low values of the HHI for Colombia and Bolivia hide substantial local market power in the hands of local telephone co-operatives in those countries). Significantly, Brazil and Argentina, which have received large inflows of FDI and are the main battlegrounds for supremacy, exhibit HHI values consistent with effective competition. But it also true that among the major markets, Mexico displays very high concentration indexes – well above those of Argentina, Brazil, Chile and Colombia – and that the HHI tends to underestimate market power especially when competition takes place at the sub-national level.[8]

Figure 9.14. Supply concentration in telephony, end 2005

Herfindahl-Hirschman Index, by segment

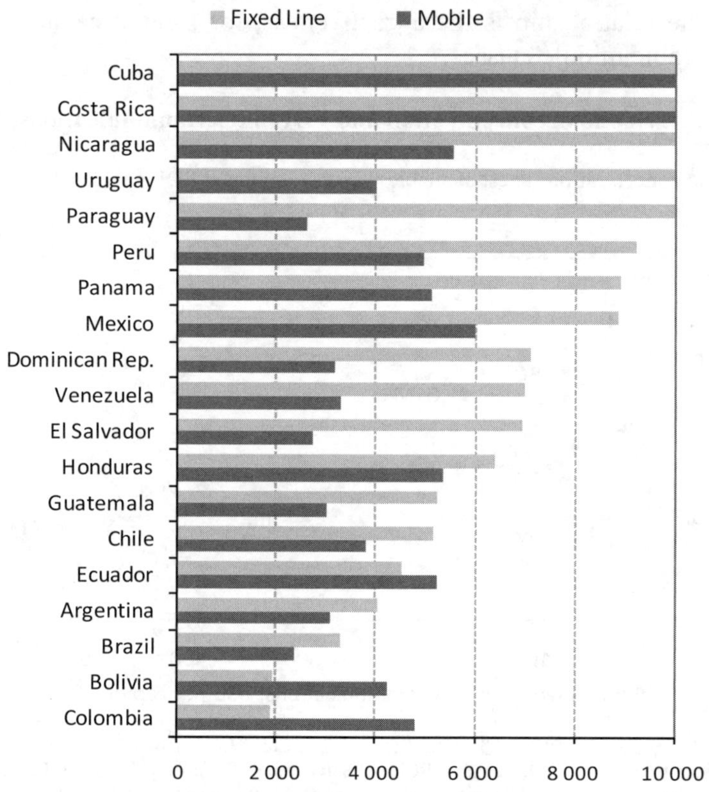

Note: The Herfindahl-Hirschman concentration index is constructed as the sum of market shares in each segment expressed in percentages: 0 corresponds to an atomistic market, 10 000 to a monopoly.

Source: OECD (2007), *Latin American Economic Outlook 2008*, OECD Development Centre, Paris, based on regulators' and operators' annual reports.

The relationships between supplier concentration and the sector's performance, both in terms of density of coverage and in terms of equality of access, are fairly weak. Figure 9.15 presents the relationships in each segment – mobile and landline – between supplier concentration on the one hand and performance outcomes in terms of density and total inequality (access indicators are not available by segment) on the other.

In the mobile segment, which is much more important than the landline segment for almost all countries, there is a weak negative correlation between telephone density and

8. Costa Rica stands out as the only country in the region where not only landline but also mobile telephony remain state monopolies.

supplier concentration (Panel A). The best performers in terms of total density (Argentina, Brazil, Chile) all exhibit fairly low supplier concentration. On the other end, the Costa Rican monopoly clearly stands out as having missed the great expansion of mobile telephony witnessed in the region since the mid-1990s, despite having achieved the highest density of landlines and the most equitable access in the region. Other relatively weak performers in mobile density include Honduras, Nicaragua and Peru, all of which have, in contrast to the previous group, very concentrated landline markets. These patterns suggest that dominant positions in the landline market may be hampering entry into the mobile market, while within the landline segment, supply concentration and teledensity are not strongly associated.

Figure 9.15. Market concentration and performance

Market concentration at end 2005, performance at end 2005 or latest available

Panel A: Market concentration and telephone density

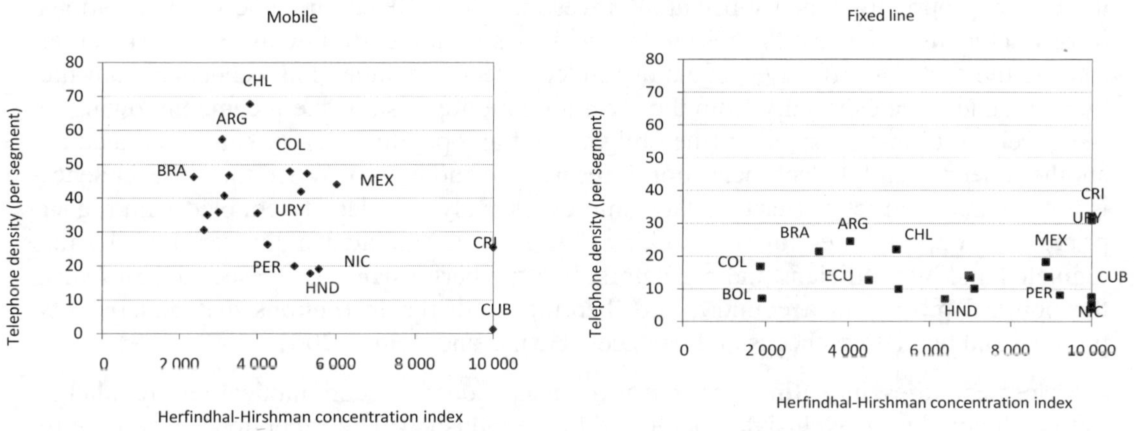

Panel B: Market concentration and telephone concentration

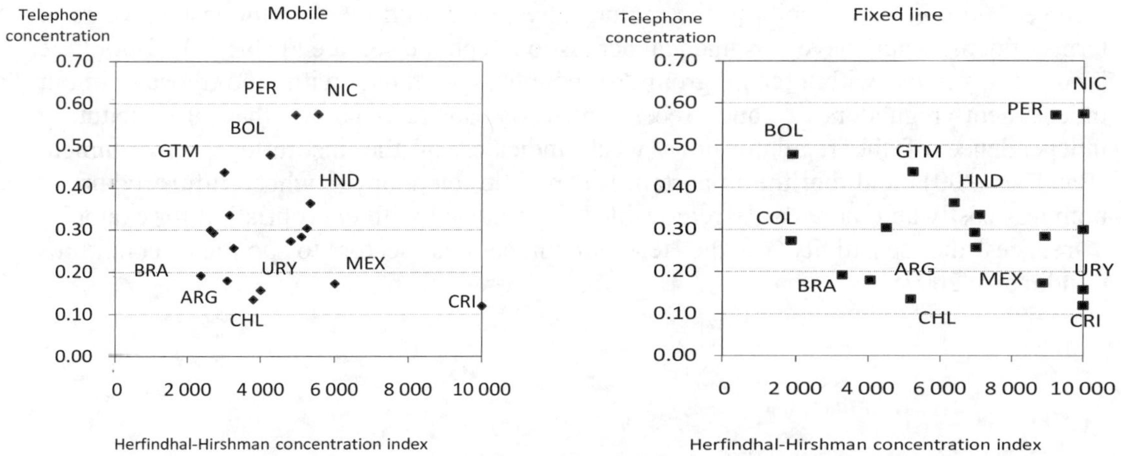

Argentina: ARG; Belize: BLZ; Bolivia: BOL; Brazil: BRA; Chile: CHL; Colombia: COL; Costa Rica: CRI; Dominican Republic: DOM; Ecuador: ECU; El Salvador: SLV; Guatemala: GTM; Haiti: HTI; Honduras: HND; Mexico: MEX; Nicaragua: NIC; Panama: PAN; Paraguay: PRY; Peru: PER; Uruguay: URY; Venezuela: VEN.

Source: OECD (2007), *Latin American Economic Outlook 2008*, OECD Development Centre, Paris.

Panel B presents the relation between inequality in telephone access and concentration of supply. Once the particularities of the Bolivian and Colombian markets are taken into account (supply in both these markets appears fairly dispersed, but local suppliers have significant local market shares), lower concentration in the supply of land telephony is associated with more equal access (in Argentina, Brazil and Chile). The sector's performance in terms of equality when the landline incumbent remains a monopoly ranges from the lowest inequality (in the state monopoly cases of Uruguay and Costa Rica) to the highest inequality (in the case of Nicaragua's monopoly, ENITEL, whose privatisation was finalised in 2004). This shows that monopolies, as such, neither guarantee nor preclude good results in terms of equality of access.[9]

Regulation is crucial

The quality of regulation plays a key role in the link among market concentration, market power and sector performance. Independent regulators are needed to address two major commitment problems. One derives from the state's difficulty to commit credibly to not expropriating the substantial investments involved in telecommunications infrastructure after the cost of those investments has been incurred by investors. The other reflects the fact that effective regulatory independence includes independence from the industry, and often especially from the local-loop monopolist, or the incumbent, owing to issues related to interconnection (the obligation of an operator to carry calls generated by another operator and the compensation it receives for doing so). When this independence is not ensured, market power translates into excessively low density coupled with higher prices. The opportunities for rent-seeking that exist in regulated industries, both by the regulated and by politicians and regulators, link the performance of these industries with the political process in a country, and therefore with the institutions that underpin its political and regulatory checks and balances (Henisz and Zelner, 2001).

Most countries in Latin America have independent telecommunications regulators (although not always exclusively dedicated to the industry). The exceptions, according to the criteria set by the International Telecommunications Union (ITU), are Chile, Haiti, Peru, Suriname and Uruguay. On average countries with independent regulators have received more FDI per capita in the sector, have progressed more in the last 15 years in terms of density, and have less unequal access to telephone service (Table 9.3). But, there is great diversity within each group of countries – those with and those without independent regulators – and recent analysis suggests both that the statutory independence of the regulator is a weak indicator of the institutional environment (Baudrier, 2001) and that the importance of mobile telephony – whose infrastructure is both less costly and more easily removable (if threatened with expropriation for example) – reduces the sensitivity of the telecommunications sector to political conditions (Andonova, 2007).

9. ENITEL was privatised in two stages. It was controlled by the Honduran consortium Megatel between 2001 and 2004, and since 2004, by América Móvil.

Table 9.3. Regulator independence and telecommunications performance indicators

Simple average of selected indicators

	Number of countries	Telephone density 2005	Change in telephone density 1990-2005	Telephone concentration index	Cumulative FDI in the sector per capita
With autonomous regulator	16	55.6	50.4	0.29	151
No autonomous regulator	5	49.6	43.0	0.35	58

Source: OECD (2007a), *Latin American Economic Outlook 2008*, OECD Development Centre, Paris, based on ITU data.

The heterogeneity of regulatory regimes and market structures, even among countries whose regulators are categorised as independent, explains the relatively weak differences in teledensity between countries with and without independent regulators. Still, cross-country analyses of the importance of regulatory independence find that countries that have independent regulators at the time of privatisation tend to have lower prices (Estache *et al.*, 2006). An important reason is that independent regulators help ensure that an increase in prices due to tariff rebalancing in conjunction with privatisation does not erode the potential gains for users from technological progress brought about by more dynamic providers. Regulatory performance in turn depends both on the governance of regulation itself, and on the political environment that the regulator faces. Chile, which does not have an independent regulator but has a political environment characterised by relatively high levels of transparency and accountability, illustrates the latter point. Studies also suggest that political accountability improves regulatory performance (Gasmi *et al.*, 2006), and that while even corruption can lead to performance improvements in the face of resistance to needed change and red tape (Estache *et al.*, 2006), reform policies can lead to better performance.

Beyond ensuring competition, Latin American regulators have shown great commitment to extending service. The models used vary greatly, ranging from a commitment to market liberalisation, to the creation of funds to finance universal-access projects, to state-mandated command-and-control mechanisms (Stern *et al.*, 2006). Countries with strong landline incumbents, including Bolivia, Panama and Mexico, have relied on universal-access obligations imposed on the incumbent with varying degrees of success and stringency. In Costa Rica and Uruguay, state or corporative objectives have achieved high levels of landline connectivity. Other countries, including El Salvador and Guatemala, have created very liberal regulatory regimes that have achieved increased teledensity without reducing significant regional disparities. More balanced approaches, notably Brazil's combination of a liberal licensing regime and universal-service obligations enshrined as regional development targets, have proved very successful.

The most noteworthy experiments have been in so-called universal access funds (UAFs). Most countries in the region have created such funds, although in some countries they have not begun their action (*e.g.* the Argentine fund was legislated in 2000 but was not yet in operation as of June 2007) and the accumulation of funds by the Brazilian Fundo Fiduciario do Serviço de Telecomunicações has raised questions about their future use. Chile's UAF is particularly innovative, and interesting, because of its competitive bidding mechanism, in which enterprises bid for universal access projects with the one requesting the lowest subsidy being awarded the project, and because the fund has been very successful: within five years of its establishment in 1995, the fund had succeeded in extending access to basic service to the majority of rural Chileans (Xavier, 2006). The majority of universal-access actions have been to provide public payphones and community telecentres that offer a wider array of telecommunications services, although Peru's UAF (FITEL) has also financed pilot projects that extend individual access to the local network.

Looking ahead, it is clear from the experiences in the region that, while an access policy with clear and stable rules is necessary and can be very successful, well-regulated open and contestable markets can do much to provide access on commercial terms to a large part of the population. Given the degree of concentration of supply in much of Latin America's landline segment, and just as eyes turn to Brazil's and Chile's successes in extending coverage, Brazil is in the process of shifting to new interconnection regulations, with rates based on a fully allocated cost model (OECD, 2007b).

Conclusions

The evolution in Latin America of FDI inflows and outflows and of the strategies of multinational corporations, home-grown as well as from outside the region, provides a lens through which to assess the progress and impact of the region's opening to the global economy. Since 2000, the irruption of emerging-market multinationals, including Latin American multinationals – the so-called *multilatinas* – has significantly altered this panorama.

The challenges ahead: inclusion and mobile service

Latin America's telecommunications sector has received substantial FDI flows in conjunction with three sectoral phenomena – privatisation, mobile telephony and industry consolidation – of varying importance across countries. The market-seeking nature of these flows, the arrival of mobile technology, and an emerging political commitment to foster universal service have created a major opportunity to provide better service to more people. Despite recent progress, however, the gap in access to telephone services between the rich and the poor remains substantial in most countries.

Providing voice service can go a long way towards enhancing communications to strengthen social ties and increase physical, economic and social mobility. It can also improve the efficiency of markets by allowing timely communication between potential buyers and sellers (Jensen, 2007).

However, voice communication is only a first step towards bridging the communication and digital divide. With mobile-phone-based Internet still far from maturity in terms of coverage and expansion, there is dire need for the mobile-phone-based services that can reduce the immense disparity that still exists in such areas as e-banking and e-government (paying taxes or voting through the Internet, for instance). Moreover, while Internet access is outpacing the growth of landline expansion thanks to communal forms of access, broadband access remains limited by the restricted expansion of the land network.

Indeed, notwithstanding great progress in most countries, access to land telephone lines remains difficult for many people. As both a medium of communication and a source of content, the Internet holds great promise for enhancing transparency and governance. Yet to play this role for all, it needs to be accessible across all segments of the population. The rise of mobile banking in southern Africa, for example, shows how process innovations in business organisation can be as important as technological innovations – mobile banking actually started with customers using prepaid telephone-card numbers as a vehicle for money transfers, and was then picked up by operators and the banking industry. Similarly, in Latin America, initiatives that allow individuals to use their mobile telephones to retrieve remittances from migrant family members have created a new way for providers of banking services to establish and maintain relationships with low-income or other previously inaccessible populations.

Lessons beyond telecommunications

The prevailing institutional environment plays a crucial role not only in determining the incentives for both incoming and outgoing investment, but in determining the effective contribution of multinational corporations to development. The example of telecommunications shows that their contribution has been and will undoubtedly continue to be significant in this sector of vital importance for aggregate economic performance and development – albeit initially mainly to the benefit of the better-off segments of local populations, thereby also increasing access inequalities – in countries across the region.

The example of telecommunications also draws attention to the crucial importance of regulatory regimes, certainly in key public services. While some countries, most notably Costa Rica, have restricted the entry of private actors in telecommunications services and performed relatively well in terms of equity of access to existing services, those same countries have not performed well in terms of service extension, letting crucial opportunities pass them by. The experiences of other countries in the region show that a regulatory regime that allows foreign actors into the sector has great potential to accelerate service extension. The combined effects of significant new FDI, the competitive market-seeking behaviour of the main investors, the spread of digital mobile technology and market liberalisation have been to greatly increase connectivity in the region – faster than in other regions. Effective access-promotion policies with clear and stable rules are nevertheless required to ensure access to the poorer segments of the population, access which is vitally important for economic and political development. Well-regulated, open and competitive markets that encourage innovation from within as well as from without will induce and facilitate corporate strategies that maximise the contribution of multinationals to development throughout the region.

References

Andonova, V. (2007), "La importancia de las instituciones políticas y el impacto en las telecomunicaciones", mimeo, Universidad de los Andes, Bogotá.

Bandyopadhyay, S. (2006), "Knowledge-Driven Economic Development", Discussion Paper No. 267, University of Oxford, Oxford.

Baudrier, A. (2001), "Independent Regulation and Telecommunications Performance in Developing Countries", paper presented at the Annual Conference of the International Society for New Institutional Economics (ISNIE), *www.regulationbodyofknowledge.org/documents/08/003.pdf.*

Calderón, C. and L. Servén (2004), "The Effects of Infrastructure Development on Growth and Income Distribution", World Bank Policy Research Working Paper No. 3400, The World Bank, Washington, DC.

ECLAC (United Nations Economic Commission for Latin America and the Caribbean) (2007), *Foreign Investment in Latin America and the Caribbean 2006*, ECLAC, Santiago de Chile.

Estache, A., A. Goicoechea and M. Manacorda (2006), "Telecommunications Performance, Reforms and Governance", World Bank Policy Research Working Paper No. 3822, The World Bank, Washington, DC.

Fundación Telefónica (2007), "El impacto socio-económico de la telefonía móvil en Latinoamérica", presentation at the seminar, Desarrollo Económico, Desarrollo Social y Comunicaciones Móviles en América Latina, Buenos Aires, 20-21 April, *www.fundacion.telefonica.com/noticias/pdf/Estudio%20Impacto%20socioeconomico%20telefonia%20movil%20210407.pdf*, accessed 12 July 2007.

Gasmi, F., P. Noumba and L. Recuero Virto (2006), "Political Accountability and Regulatory Performance in Infrastructure Industries: An Empirical Analysis", World Bank Policy Research Working Paper No. 4101, The World Bank, Washington, DC.

Guislain, P. and C.Z.-W. Qiang (2006), "Foreign Direct Investment in Telecommunications in Developing Countries", *Information and Communications for Development: Global Trends and Policies*, The World Bank, Washington, DC.

Henisz, W. and B.A. Zelner (2001), "The Institutional Environment for Telecommunications Investment", *Journal of Economics and Management Strategy*, Vol. 10, No. 1, The MIT Press, Cambridge, MA.

IADB (Inter-American Development Bank) (2001), "Competitiveness: The Business of Growth", *Economic and Social Progress in Latin America 2001*, IADB, Washington, DC.

IADB (2007), *Equity and Social Indicators (EQxIS)*, *www.iadb.org/sds/xindicators/*, IADB, Washington, DC.

ITU (International Telecommunications Union) (2006), *World Telecommunications/ICT Indicators Database*, 2006, ITU, Geneva.

ITU (2008), ICT Eye Regulatory Information Centre, *www.itu.int/ITU-D/icteye/Regulators/Regulators.aspx*, accessed May 2008, ITU, Geneva.

Jensen, R. (2007), "The Digital Provide: Information (Technology), Market Performance and Welfare in the South Indian Fisheries Sector", *Quarterly Journal of Economics*, The MIT Press, Cambridge MA, forthcoming.

Latinobarómetro (2005), *Summary Report*, www.*latinobarometro.org*.

OECD (1999), *OECD Benchmark Definition of Foreign Direct Investment*, Third Edition, *www.oecd.org/dataoecd/10/16/2090148.pdf*, OECD, Paris.

OECD (2004), "Mexico: Progress in Implementing Regulatory Reform", *OECD Reviews of Regulatory Reform*, OECD, Paris.

OECD (2005), *OECD Economic Surveys: Mexico*, OECD, Paris.

OECD (2007a), *Latin American Economic Outlook 2008*, OECD Development Centre, Paris.

OECD (2007b), *OECD Communications Outlook 2007*, OECD, Paris.

Röller, L.-H. and L. Waverman (2001), "Telecommunications Infrastructure and Economic Development: A Simultaneous Approach", *The American Economic Review*, Vol. 91, No. 4. American Economic Association, Pittsburgh, PA.

Rozas Balbontín, P. (2005), "Privatización, reestructuración industrial y prácticas regulatorias en el sector telecomunicaciones", *Serie Recursos Naturales e Infraestructura*, Vol. 93, ECLAC, Santiago de Chile.

SEDLAC (Socio-Economic Database for Latin America and the Caribbean) (2007), CEDLAS and The World Bank, La Plata and Washington, DC, *www.depeco.econo.unlp.edu.ar/cedlas/sedlac/default.html*.

Stern, P.A., J. Monedero and D. Townsend (2006), "New Models for Universal Access in Latin America", report for the Regulatel/World Bank/ECLAC Project on Universal Access for Telecommunications in Latin America, Montreal, Boston and Madrid.

UNCTAD (United Nations Conference on Trade and Development) (2006), *World Investment Report, 2006: FDI from Developing and Transition Economies: Implications for Development*, UNCTAD, New York and Geneva.

Wallsten, S.J. (2003), "Privatizing Monopolies in Developing Countries: The Real Effects of Exclusivity Periods in Telecommunications", *Related Publication* 03-17, AEI-Brookings Joint Center for Regulatory Studies, Washington, DC.

Waverman, L., M. Meschi and M. Fuss (2005), "The Impact of Telecoms on Economic Growth in Developing Countries", *Africa: The Impact of Mobile Phones*, The Vodafone Policy Paper Series No. 3, Vodafone, Berkshire.

World Bank (2006), *World Development Indicators Database 2006*, The World Bank, Washington, DC.

World Bank (2007), *Private Participation in Infrastructure Database*, *http://ppi.worldbank.org*.

Xavier, P. (2006), "Universal Access to Telecommunications in a Competitive Environment", *Liberalisation and Universal Access to Basic Services: Telecommunications, Water and Sanitation, Financial Services, and Electricity*, OECD Trade Policy Studies, OECD and The World Bank, Paris and Washington, DC.

Chapter 10

Broadband as a platform for economic, social and cultural development: lessons from Asia

Tim Kelly[1]

Far from "playing catch-up", Asian economies have been setting the pace in the development of broadband networks, both on fixed and mobile networks. Korea was an early leader in fixed broadband, and Japan has been leading in the early stages of mobile broadband deployment. Singapore is one of the world leaders in urban fibre deployment while Hong Kong, China, is a pioneer in the provision of Internet Protocol Television (IPTV). Among the developing countries of the region, China now has the largest installed base of broadband users. India has recognised the critical importance of broadband for its bourgeoning software outsourcing industry.

1. Tim Kelly is Lead ICT Policy Specialist, *info*Dev/World Bank.

Introduction

Among Asian economies, there has been keen competition and a close interest in global comparisons of broadband performance, such as penetration rates, speeds and prices. There has also been considerable government support for broadband promotion. The implicit assumption is that broadband can drive growth in the rest of the economy and the development of an information society. But can it and, if so, how? How does broadband create jobs, how can it spur innovation and how does it reduce costs for businesses? In a similar vein, does broadband simply imply faster video downloads and longer hours spent gaming or can it be a platform for broader-based social and cultural development? Furthermore, can broadband help with wider problems of society, such as climate change, rising fuel prices or food insecurity? This chapter looks at the lessons that can be learned from the Asian broadband superstars in terms of the changes that broadband can bring, both positive and negative, and how they might be passed on to the developing world.

Why broadband?

Students of long-term trends in economic, social and cultural development have long been fascinated by the emergence of new "drivers" that will provide for a sustained upturn in the global economy. The study of so-called "Kondratieff long waves" has identified previous drivers, or general purpose technology (GPT) enablers, such as steam power, oil, the motor vehicle, plastics, etc., which have sustained long-term waves of innovation and growth. More recently, the development of information and communication technologies (ICTs), which date from the development of microprocessors (semiconductor chips) in the late 1960s, has given rise to a new period of sustained development that has been termed the "Information Society".

If we assume that the average lifecycle of such long waves is 40-60 years, and that, with technological progress, this is speeding up over time, the world is now overdue for another significant technological driver of economic progress. There are many candidates – ranging from mobile phones and the Internet in the ICT sector to nanotechnology or bioengineering – and it is unlikely that a single technology would be capable of supporting a sustained rise in global GDP. But a good case can be made to argue that the next large-scale drive will come from the ICT sector and that it will encompass the three major strands of that sector – telecommunications, computing and broadcasting. The leading candidate of the moment is broadband.

What is broadband?

Conventional definitions of broadband focus on what it is *not* rather than what it *is*. For instance, broadband is not narrowband, in that there is general consensus on the low-end cut-off speed for broadband as offering a transmission capacity equal to or above 256 kbit/s. But there is no upper limit placed on what broadband can become, and the evidence points to the fact that broadband speeds, and its performance/price ratio, are tending to double every 12-15 months. This trajectory makes it tempting to make comparisons with semiconductor chips for which, following Moore's Law, performance/price has been doubling on average every 18-24 months since the late 1960s.

Similarly, broadband can be defined as not being a "scarce" service, by contrast with the telephone which is used on average for less than 30 minutes a day, and which is priced per minute, per mile and per megabyte. Rather, broadband offers an "always-on" service, which is distance-independent and which, for the most part, is available without limitations on capacity use. Again, by stating what broadband is not, there remains much to discover about what it can become, particularly as the main users of broadband in the future will not necessarily be humans but machines and objects that are interconnected and controlled in a ubiquitous network.

Finally, a third way of looking at broadband would be to say that it is neither fixed nor mobile but rather that it can be both. In other words, in a next-generation network (NGN) environment, the term broadband will refer to the quality of the access environment, and the possibilities this allows, rather than whether a particular user is connected to a fixed-line or wireless environment at any particular time. Similarly, broadband is not limited to a telecommunications environment, but can be used to support broadcast audio and video and to provide access to so-called "cloud computing".

In summary, although existing definitions focus on what broadband is not, there are no real limitations on what it can become. It is therefore better to avoid rigid definitions. There are lessons to be learned from an analogy with the application of microprocessors. In the 1970s, they were used in machine tools and in hand-built personal computers, such as the Altair or the Apple 1. In the 1980s, PCs were mass-produced for the first time and microprocessors started appearing in devices such as washing machines and elevators. By the 1990s, miniaturisation and rising speeds saw them appear in mobile phones and automobiles. In the 2000s the most prolific form of microprocessors is in RFID chips and smart cards. In other words, the simultaneous processes of increasing power, shrinking size and falling price has allowed microprocessors to be embedded in all sorts of devices and to be used in many more applications than were ever envisaged when they were first developed.

Can we expect broadband to follow a similar trajectory? At present, the relatively high price of broadband means that it is generally restricted to applications for which consumers are willing to pay a premium price, such as entertainment or business. But in the future, as broadband capability becomes embedded in a wider range of devices, and as machines and objects rather than humans become the main users, broadband will give birth to a wide range of devices and applications as yet unimagined. Again with reference to the development of microprocessors, improvements in performance/price can continue at a regular level over a long period, but it is only once some kind of threshold is passed that this incremental progress becomes revolutionary. Atkinson and Castro (2008), in their newly released report on *Digital Quality of Life* (Atkinson et Castro, 2008), argue that the inflection point that triggered the ICT revolution did not come until the mid-1990s, some 30 years after Moore's Law was first coined. The price of chips and of memory storage had fallen so far by the mid-1990s that it became viable to interlink websites on the Internet or to manufacture digital mobile phones that could be carried in the palm of a hand. If we date the birth of broadband to the late 1990s, and if it follows a similar path to the microprocessor, by the late 2020s we can expect speeds measured in terms of Gigabits per second for prices lower than USD 20 per month. At that point, a revised definition of broadband will be long overdue.

Where is broadband?

Broadband has been one of the fastest-growing ICT services ever seen in terms of its diffusion worldwide. Korea was the first country to deploy fixed-line broadband on a commercial scale in the late 1990s. By 2000, some 36 economies had launched broadband service, of which two had a penetration of fixed-line broadband subscribers per 100 inhabitants greater than 5: Korea (8.2) and Hong Kong, China (6.67).

Two years later, in 2002, the number of economies having launched service had reached 71 and three had above 10 subscribers per 100 inhabitants: Canada (11.2); Hong Kong, China (15.4); and Korea (21.9). Korea was the clear leader and was able to showcase its technology in that year when co-hosting the FIFA Football World Cup (ITU, 2003). Elsewhere in the world, 2002 was something of a nadir for the Internet industry, following the bursting of the dot.com bubble at the start of the decade. Indeed, it was the Asia region, which had been largely unaffected by the dot.com bubble and the inflated prices paid for 3G licences, that was leading the global broadband economy. In that year, in addition to the market leaders, broadband penetration in a further three Asian economies exceeded 5 per 100 inhabitants: Chinese Taipei (9.4), Japan (7.4) and Singapore (6.5).

By 2004, as the industry as a whole began to emerge from the downturn, a much greater percentage of the broadband map was coloured in, with some 131 economies having commenced service. In that year, some 22 economies had passed the significant barrier of 10 subscribers per 100 inhabitants, which corresponds to roughly one-third of all households. Of these, six were in the Asia-Pacific region, including the two with the highest level of penetration: Hong Kong, China (22.4) and Korea (24.8). By contrast, only three economies in the American hemisphere had passed this milestone: Barbados (10.2), the United States (12.7) and Canada (16.9). The others were European. By 2004, there were also stirrings among the developing countries of the Asia-Pacific region as China reached the 25 million subscriber mark, second only to the United States with 37 million in that year.

By 2006, as broadband subscribers globally surpassed 300 million, broadband had been launched in some 166 economies, including most of the Asia-Pacific region. By this stage, the early leadership of the Asian tigers had begun to erode as Europe had caught up, with Denmark the first major economy to cross the 30 subscribers per 100 inhabitants mark, just behind the minnows of Bermuda (37.1) and Monaco (33.4).

By 2008, more than 180 economies worldwide had launched broadband and global subscriber numbers had surpassed 500 million, if fixed and mobile broadband subscribers are added together. The leading economies, in terms of penetration, are now mainly the smaller economies of Europe – such as Denmark, Iceland, the Netherlands, Norway and Switzerland – though Asian economies are still leading in the newer field of mobile broadband. However, as these economies approach the point at which most households that want broadband already have it, other factors become more important in differentiating between performance, such as speed of service, pricing and level of market choice. In addition, other factors – such as level of urbanisation, size of local loops or degree of inter-modal competition between fixed-line and wireless services – come into play in explaining small differences in terms of penetration rate among the leading economies. These are explored below.

Why Asia?

Asian pioneers

Although the lead established by the leading Asian economies over the rest of the world on broadband penetration has narrowed over time, and indeed some of the smaller European economies have now overtaken them, it is nonetheless instructive to look at

Asia for lessons on promoting broadband and on the impact it is having on development for a number of reasons:

- the early start that Asian economies had on their broadband adventure, both for fixed-line and mobile services;
- the fact that Asian economies frequently have broadband service with higher performance, especially where it is based on fibre, and which is generally cheaper than elsewhere;
- the diversity of Asia, which offers many different models for broadband, including state-driven and private-sector-driven, fixed-line or wireless-based, and including both developing and developed economies.

The leading broadband economies, by penetration rate, are shown in Figure 10.1 and the leading economies in terms of number of subscribers, at the start of 2008, are shown in Figure 10.2.

Figure 10.1. Top 20 fixed-line broadband economies, in subscribers per 100 inhabitants, by technology, year-end 2007

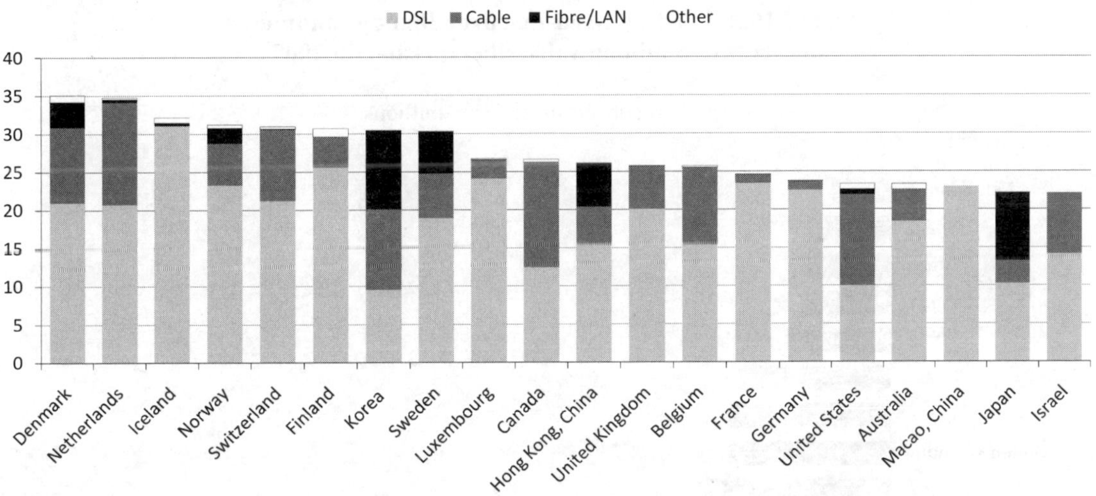

Source: ITU/OECD/World Bank.

The Asia-Pacific economies that feature among the world leaders have a number of distinguishing factors which help explain their success:

- They are more likely to have inter-modal competition. This is well illustrated by Korea, whose total of almost 15 million fixed-line broadband subscribers (at year end 2007) were almost equally split between DSL, cable modem and LAN/fibre optics platforms. Furthermore, if the mobile broadband subscribers were added, the variety of platforms is even greater with 3G mobile (both Wideband CDMA and CDMA 2000 1x EVDO), Wi-Fi subscribers and WiBro (a local Korean implementation of WiMAX) added to the mix. Inter-modal platform competition seems to be just as important for promoting market growth in broadband as competition between companies using the same technology.

- The Asia-Pacific economies among the leaders are more likely to have a high percentage of fibre optic users – fibre to the home (FTTH) or fibre-to-the-kerb – among total broadband users. Japan and Hong Kong, China, both illustrate this point. Japan is second only to Korea in its penetration of FTTH (around 9% per 100 inhabitants) with Hong Kong, China, not far behind with 6%. In both countries, fibre and apartment LANs have served as a way of introducing greater competition to the market, through Yahoo! BB and Hong Kong Broadband Network, respectively. Fibre is also used as the basis for IPTV services in both countries.
- Third, and partly as a result of these two factors, Asian broadband users frequently pay lower prices for broadband than their counterparts in the rest of the world. As Table 10.1 shows, Japan and the Korea, both of which offer speeds of up to 51 Mbit/s download for broadband, charge users less than USD 1 per Mbit/s per month. Three other Asian economies – Viet Nam, Singapore and Chinese Taipei – also feature among the ten lowest-priced broadband economies. The case of Viet Nam is particularly noteworthy because, although typical speeds on offer are just 1.5 Mbit/s, it is able to offer some of the lowest entry-level prices, at below USD 5 per month.

Figure 10.2. Leading fixed-line broadband economies, in excess of 5 million subscribers, year-end 2007

Total broadband subscribers, millions

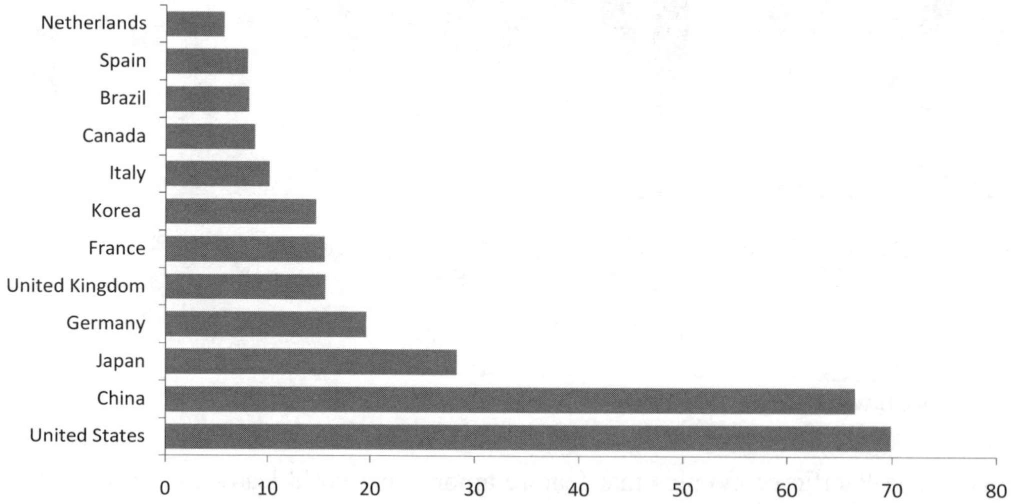

Source: ITU/OECD/World Bank.

Developing Asia

The leading Asian economies have become role models for developing Asia. This is exemplified by the fact that entry-level prices for broadband among developing Asian economies are typically among the lowest in the world. As Table 10.2 shows, in addition to Viet Nam, six other Asian economies have entry-level broadband prices below USD 10 a month, less than a tenth of the global average in mid-2007 of USD 108 (this is a simple, unweighted average with each country with broadband counting as one case). Overall,

Asian prices are below the world average in every price category (low-speed, high-speed, per Mbit/s and as a percentage of monthly GNI per capita) even though average broadband speeds are higher in Asia than in the rest of the world.

Table 10.1. Ten economies with lowest broadband prices worldwide, mid-2007

USD per 1 Mbit/s per month

	Lower speed		Higher speed		Lowest sampled cost		
Rank	Monthly charge USD 2007	Speed(kbit/s) Down 2007	Monthly charge USD 2007	Speed (kbit/s) down 2007	USD per 1 Mbit/s 2007	As a % of monthly income (GNI)	Internet service provider
1 Japan	28.57	8 192	35.70	51 200	0.70	0.00	Yahoo BB
2 Korea	29.94	10 240	48.11	51 200	0.94	0.01	KT
3 Netherlands	12.25	4 000	27.30	20 000	1.36	0.00	Orange
4 Chinese Taipei	6.08	256	22.20	12 288	1.81	0.02	Chunghwa
5 Sweden	14.65	1 000	44.23	24 000	1.84	0.01	Tele2
6 France	20.39	2 000	40.91	20 000	2.05	0.01	Neuf telecom
7 Singapore	13.49	512	80.81	30 720	2.63	0.01	StarHub
8 Viet Nam	4.34	1 536	4.34	1 536	2.82	0.62	FPT Communications
9 Finland	34.07	1 024	68.28	24 000	2.84	0.01	Elisa-Lajaakaista
10 Switzerland	40.82	3 500	57.48	20 000	2.87	1.05	Bluewin
World average	107.95	759	278.18	4 392	299.21	97.43	

Source: ITU *World Telecommunication Indicators Database*.

Table 10.2. Asian economies with entry-level broadband prices below USD 10 per month, mid-2007

	Lower speed		Higher speed		Lowest sampled cost		
Rank	Monthly charge USD 2007	Speed (kbit/s) down 2007	Monthly charge USD 2007	Speed (kbit/s) down 2007	USD per 1 Mbit/s 2007	As a % of monthly income (GNI)	Internet service provider
1 Viet Nam	4.34	1 536	4.34	1 536	2.82	0.62	FPT Communications
2 Malaysia	5.80	384	5.80	384	15.11	0.39	TMNET
3 Chinese Taipei	6.08	256	22.20	12 288	1.81	0.02	Chunghwa
4 China	7.88	512	7.88	512	15.38	1.43	E-NET
5 Sri Lanka	9.01	512	9.01	512	17.60	2.09	SLTnet
6 Maldives	9.27	256	15.38	512	30.03	1.44	Dhiraagu
7 India	9.65	256	37.48	2 048	18.30	3.54	Tata Indicom
Asia	70.06	818	222.36	4 587	278.89	60.85	
World	107.95	759	278.18	4 392	299.21	97.43	

Source: ITU *World Telecommunication Indicators Database*.

The low-priced Asian economies include China and India, the world's most populous economies. No prior models exist for economies that have gone from developing to developed status with the benefits of low-priced access to information, and it is entirely possible that a new model of economic development may emerge as a result. China has experienced a remarkable compound annual growth rate in broadband penetration of 81% over 2002-07, with 66 million subscribers at the end of 2007. Yet it still has the slowest growth rate among the developing Asian economies highlighted in Figure 10.3. Malaysia, growing at 131% a year, overtook China in terms of penetration rate in 2007, and even India, lingering towards the bottom of the figure, is doubling every year, on average. But Viet Nam is the standout case, with a CAGR over 2002-07 of more than 300%, which shows what can be achieved when broadband prices are among the lowest in the world.

Figure 10.3. Broadband penetration rates, per 100 inhabitants, in selected developing Asian economies, 2002-07

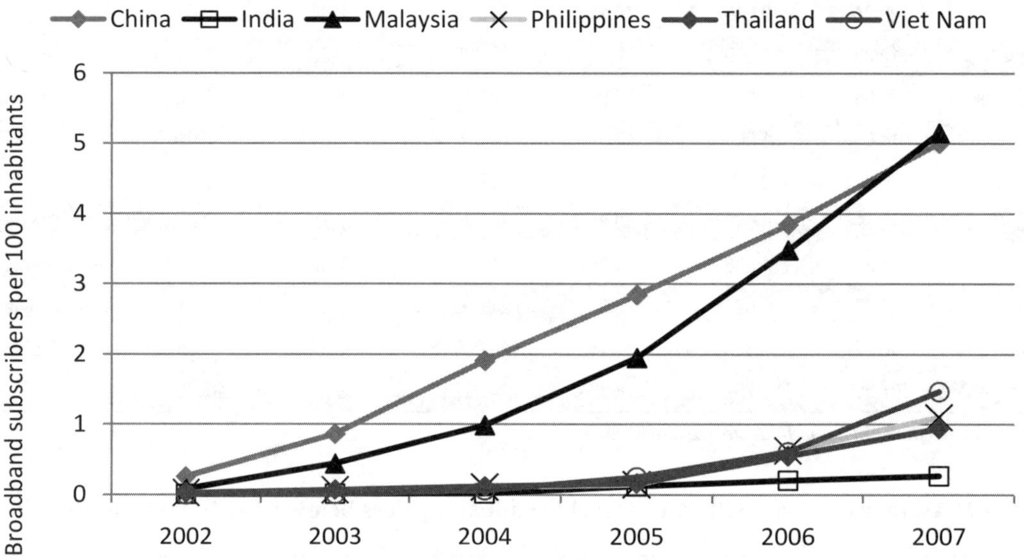

Source: ITU *World Telecommunication Indicators Database.*

To return to China, it is instructive to compare its growth with that of the other economic powerhouse, the United States. It is possible to compare the year in which both economies reached a penetration level of 5 per 100 inhabitants for different ICT services (Table 10.3). For fixed line telephone, the United States reached that level in 1912, but it took China a further 85 years to reach it. For PCs, the catching-up process took 22 years and for Internet users it was 8 years. But for mobile phones and broadband, the catching up process took just seven and five years, respectively. Furthermore, although China still has fewer PCs or Internet users than the United States, by the end of 2008 it nevertheless had more broadband users. Although broadband speeds in China do not yet match those in the United States, the entry-level price for low-speed services is just one-third of that charged in the United States, suggesting the potential for narrowing the gap further.

Table 10.3. The narrowing gap in ICTs between the United States and China

Year in which each country reached 5% penetration (per 100 inhabitants) for selected ICTs

	Fixed-line	PC	Internet users	Mobile	Broadband
United States	1912	1983	1994	1993	2002
China	1997	2005	2002	2000	2007
Gap (years)	85	22	8	7	5
Year when China overtook USA in subscribers	2002	n/a	n/a	2000	2008

Source: World Bank, adapted from *ITU World Telecommunication Indicators Database* and AT&T.

Broadband and economic growth

The example of China catching up with the United States, and of other developing Asian nations prospering in their broadband development, raises the question of how broadband will affect future trajectories of economic growth.

Growing evidence of the benefits of broadband

There is a growing body of literature which points to the potential of broadband for promoting development. The theoretical basis for this takes into account:

- The shift from intermittent to "always on" access to information that broadband permits.
- The availability of higher speeds and lower unit costs. Closely associated with this is the reduction in waiting time for access to information and waiting for a website to load.
- The enhancement of user experience through multimedia, both at home (*e.g.* the rise of video on demand services, such as YouTube) and in business (*e.g.* the facilitation of e-commerce).
- The access that broadband provides to global markets, for instance for outsourcing, and for encouraging global, real-time, transparent competition.
- The ability of broadband to "cannibalise" other services, for instance by substituting downloaded video-on-demand for live TV shows, or substituting voice over broadband for regular telephone service. Broadband's ability to offer multiple play should generate cost savings and efficiencies.

A number of studies have attempted to document how broadband can promote growth at the microeconomic level. For instance:

- By contrast with narrowband users, broadband users tend to visit a wider range of sites and to make greater use of content-rich or socially interactive websites (although the faster download speeds also mean that users tend to spend less time at a particular site) (Rappoport *et al.*, 2002).
- An early attempt to study the business impact of the Internet found that, by 2002, it had already achieved cumulative business savings of USD 155 billion in the United States, which would presumably be much greater with broadband (Varian *et al.*, 2002).

- Broadband is credited with boosting e-commerce sales, for instance for sales of movies both for online delivery and for online ordering (Smith and Telang, 2006).
- Mobile broadband can improve business processes even further by generating productivity savings, up to USD 6.9 billion in the case of the US healthcare sector (Entner, 2008).
- Broadband can also help communities, as was shown in the study of its impact on economic activities in ten US communities between 1998 and 2002 where it had added 1 to 1.4 percentage points to job growth (Gillett *et al.*, 2006).

The benefits of broadband are not limited to the commercial arena. It can help with social and cultural development across a wide range of fields. For instance:

- A study conducted by Climate Risk for the Australian incumbent, Telstra, shows how broadband creates opportunities for low-carbon development (Climate Risk pty, 2007). The report identifies seven opportunities for carbon reduction or abatement, all of which are based on the use of broadband networks. These include remote management of power appliances and presence-based power, telecommuting, real-time freight management and on-live high definition video-conferencing.
- In a similar vein, the Report of the Global eSustainability Initiative (GeSI) entitled *Smart 2020: Enabling the Low Carbon Economy in the Information Age* (GeSI, 2008) looks at the enabling effect of ICTs, especially broadband, in terms of generating opportunities for carbon abatement that are five times greater than the direct carbon emissions of the ICT industry itself.
- Broadband can contribute to improvements in the quality of life, as illustrated by a BT study of flexible working patterns among small businesses, which showed that 82% of small businesses that had adopted broadband reported a better work-life balance (BT Business, 2007).

Performance/price ratios

In analysing the developmental impact of broadband, it is important to use the most recent data. The reason for this can be illustrated by the rapidly improving performance/price ratio for broadband worldwide. Figures tracked by Biggs and Kelly (forthcoming) show worldwide trends over time in broadband price reduction and speed increase. Between 2004 and 2007, for some 170 economies, median broadband prices, per Mbit/s per month, fell by a compound rate of 25% a year, while average broadband speeds rose by 26% a year (Figure 10.4). Taking these two trends together, it can be seen that the performance/price ratio of broadband is doubling in less than 18 months.

Furthermore, in some parts of Asia, the rate of improvement is even greater. In Korea and Malaysia, for instance, the performance to price ratio doubled in just 12 months over 2003-07. At such exponential rates of improvement, it is clear that the market environment will continue to evolve rapidly, with products and services that are not commercially viable to offer at the moment, because of speed or price constraints, becoming viable within just a few years. For instance, the idea of downloading full-length movies was a pipedream just a few years ago but has now become commonplace in many countries, even in high-definition format. Similarly, because broadband is always on and additional bandwidth use is close to zero marginal cost, voice over broadband is increasingly substituting for telephone service.

Figure 10.4. Trends in broadband speeds and pricing: worldwide 2004-07, selected Asian economies

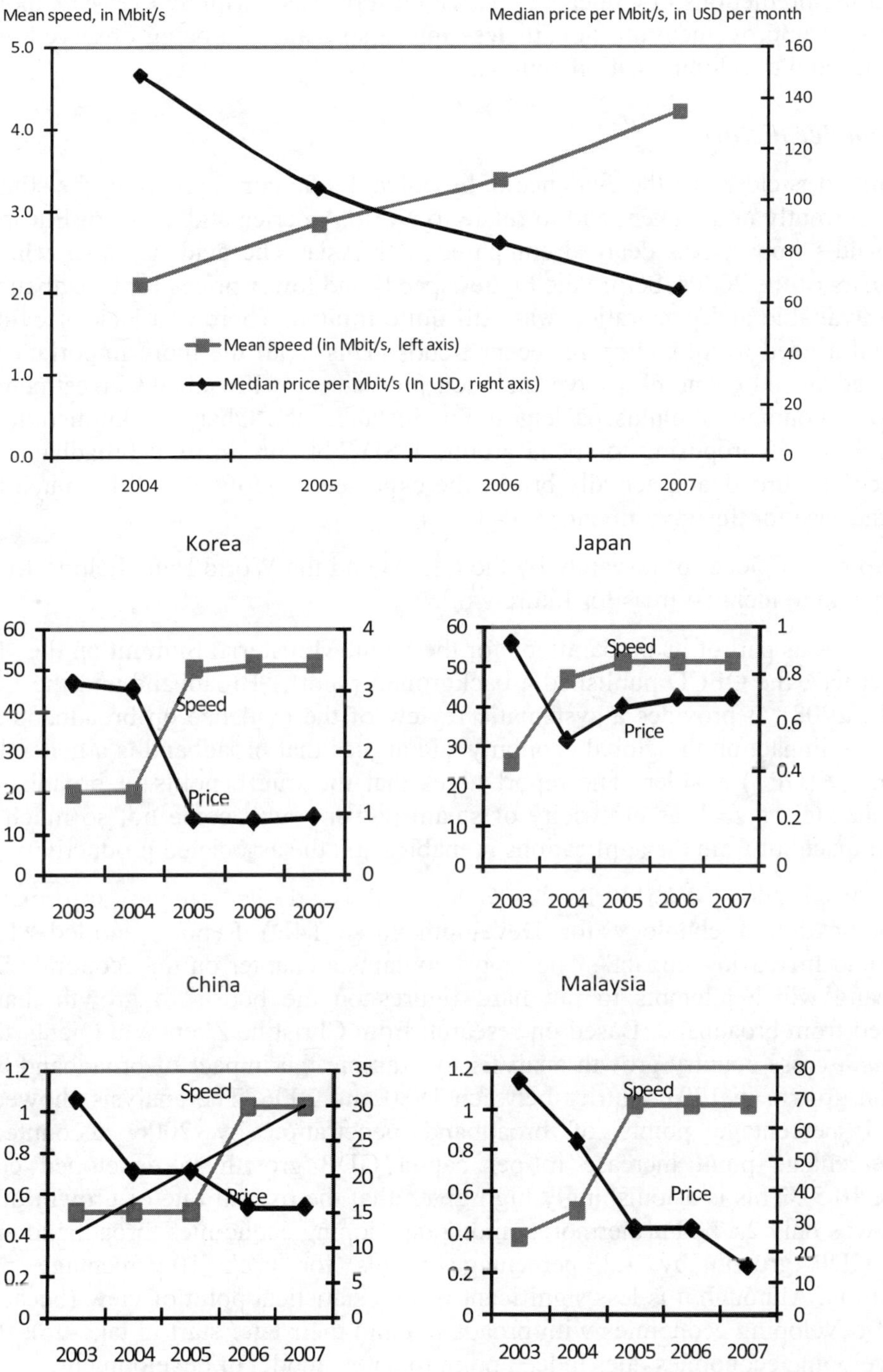

Source: Biggs and Kelly (forthcoming).

Of course, it is not possible to forecast whether these trends will persist over a long period, as was the case for Moore's Law for semiconductor price performance, which has been sustained since the late 1960s. And the rate of growth in bandwidth availability falls short of the predictions in Gilder's Law, which forecasts a tripling every 12 months. But the rates of improvement are nonetheless impressive and are being observed equally in developed and developing nation contexts.

The broader picture

While it is clear that the evidence in broadband's favour is growing, the studies of its impact currently undertaken tend to relate to North America and Western Europe, where broadband's impact was delayed compared with Asia. The evidence also relates to the early years of the 2000s, before the higher speeds and lower prices for broadband became widely available and penetration was still quite limited. There is a lack of evidence for Asia and a need to look at more recent trends. This is all the more important now that several advanced economies have included public-sector broadband investment in their proposed economic stimulus packages. For instance, the Obama Administration in the United States is proposing to spend around USD 7 billion on rural broadband. Without evidence that broadband actually brings the expected benefits, it will be much harder to make the case for this investment.

Two recent pieces of research, by the OECD and the World Bank help to fill some of this gap and to identify areas for future work.

In 2008, as part of its preparations for the Seoul Ministerial Summit on the "Future of the Internet", the OECD published a background report, "Broadband and the Economy" (OECD, 2008). It provides a systematic review of the evidence on broadband's current and future impact on the global economy and argues that broadband is a general purpose technology (GPT) enabler. The report notes that the true benefits of broadband – like forerunner GPTs such as electricity or steam power – will come not so much from its direct impact but from the applications it enables and the associated productivity gains.

In May 2009, World Bank/*info*Dev published its second Information and Communication Technology for Development (ICT4D) Report, entitled "Extending Reach and Increasing Impact". The report contains a chapter on the economic impact of broadband which attempts to put hard figures on the boosts to growth that can be expected from broadband. Based on research from Christine Zhen-Wei Qiang, the report presents a cross-country growth analysis to examine the impact of broadband and other ICTs on growth in 120 countries between 1980 and 2006. The analysis showed that an extra 10 percentage points of broadband penetration by 2006 accounted for a 1.21 percentage point increase in per capita GDP growth in developed economies (Figure 10.5). This is astonishingly high given that the overall rate of growth during this period was only 2.1%. Furthermore, among developing economies, broadband appears to boost GDP growth by 1.38 percentage points for each 10 percentage points of penetration. Although it is less significant from a statistical point of view (because of the lack of developing economies with broadband and their later start in take-up), this result for developing economies does indeed point to a new model of development.

What is also remarkable is the fact that the impact of broadband is so much higher than that observed for other ICTs, such as fixed-line, mobile or narrowband Internet. Broadband's impact in developed countries, for instance, is found to be twice the impact of mobile phones, even though the latter have penetrated more widely. Much more has been written about the economic impact of these other services; broadband is still relatively under-researched.

Figure 10.5. Growth effects of ICTs: percentage point increase in GDP per capita for every 10 percentage point increase in ICT penetration, 1980-2006

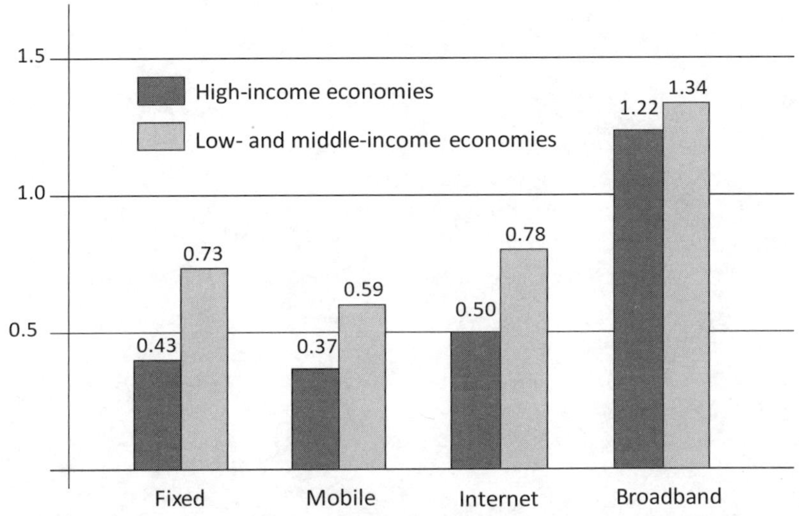

Note: All results are statistically significant at the 1% level except for that of broadband in developing countries, which is at the 10% level.

Source: Qiang, Christine Zhen-Wei and Carlo Rossotto (2009), "Economic Benefits of Broadband", Chapter 3 in *Information and Communications for Development 2009: Extending Reach and Increasing Impact*, The World Bank, Washington, DC..

Some evidence of the beneficial impact of broadband on the economy can be seen in the rising share of the contribution of ICT to GDP in broadband-intensive economies. In the case of Korea, the percentage contribution of telecommunication services (including broadband) to GDP more that doubled, from 2.05% to 4.99%, between 1995 and 2005, the decade of broadband's expansion in the Korean economy.

Of course, there is a difficulty in interpreting causality, in that wealth and ICT use rise concurrently. Therefore, it is possible that it is rising wealth which is leading to rising broadband use (*i.e.* because Internet use becomes more affordable as incomes rise and because affluence brings more leisure time). Statistically, this reverse causality cannot be rejected. But, given the growing literature at the microeconomic level about the benefits of broadband, it must be assumed that the causality works both ways and that the relation between broadband and economic growth is mutually beneficial.

Conclusions

Clearly, more work is needed to investigate the nature of the link between broadband and economic growth, at both microeconomic and macroeconomic levels, and especially to understand how this works in developing economies where the use of broadband is still in nascent stages. To date, too little research has been done on Asia, which has historically led the way. It is also too early to trumpet the argument that the improvements in the performance/price ratio that have been observed for broadband, especially in Asia, can be sustained in the long term with the kind of longevity demonstrated by Moore's Law for semiconductors. Nevertheless, the evidence points to the fact that broadband, although still in its infancy, is having a profound and positive effect on economic, social and cultural development. What is more, the vectors of future change, with rising speeds and falling prices, suggest that the best is yet to come.

References

Atkinson, Robert and Daniel Castro (2008), "Digital Quality of Life: Understanding the Personal and Social Benefits of the Information Technology Revolution", IITF, Washington DC, *www.itif.org/index.php?id=179.*

Biggs, Phillippa and Tim Kelly (forthcoming, 2009), "Trends in Broadband Pricing", *Info.*

BT Business (April 2007) "State of the Small Business Nation", *http://businessclub.bt.com/stateofthenation.pdf.*

Climate Risk pty (2007), "Towards a high-bandwidth, low-carbon future: Telecommunication-based opportunities to reduce greenhouse gas emissions", *www.itu.int/dms_pub/itu-t/oth/06/0F/T060F0060080023PDFE.pdf.*

Entner, Roger (2008), "The Increasingly Important Impact of Wireless Broadband Technology and Services on the U.S. Economy", an Ovum Study for CTIA-The Wireless Association, *http://files.ctia.org/pdf/Final_OvumEconomicImpact_Report_5_21_08.pdf.*

GeSI (2008) "Smart 2020: Enabling the low carbon economy in the information age", *www.smart2020.org/.*

Gillett, Sharon E. *et al.* (2006), "Measuring the Impact of Broadband Deployment", Prepared for the US Department of Commerce, Economic Development Administration, Washington, DC, *www.eda.gov/ImageCache/EDAPublic/documents/pdfdocs2006/mitcmubbimpactreport_2epdf/v1/mitcmubbimpactreport.pdf.*

International Telecommunications Union (ITU) (2003), "Broadband Korea: Internet Case Study", March, *www.itu.int/ITU-D/ict/cs/korea/material/CS_KOR.pdf.*

OECD (2008), "Broadband and the Economy: Ministerial Background Report", OECD Ministerial Meeting on the Future of the Internet Economy, Seoul, 17-18 June, *www.oecd.org/dataoecd/62/7/40781696.pdf.*

Qiang, Christine Zhen-Wei and Carlo Rossotto (2008), "Economic Benefits of Broadband", Chapter 3 in *Information and Communications for Development 2009: Extending Reach and Increasing Impact*, The World Bank, Washington, DC.

Rappoport, Paul N., Donald. J. Kridel and Lester D. Taylor (2002), "The Demand for Broadband: Access, Content and the Value of Time", Washington, DC, *http://aei-brookings.org/admin/authorpdfs/redirect-safely.php?fname=../pdffiles/phpvw.pdf.*

Smith, Michael D. and Rahul Telang (2006), "Piracy or Promotion? The Impact of Broadband Internet Penetration on DVD Sales", *http://archive.nyu.edu/bitstream/2451/14958/2/USEDBOOK16.pdf.*

Varian, Hal *et al.* (2002), "The Net Impact Study: The projected economic benefits of the Internet in USA, UK, France and Germany", *www.netimpactstudy.com/NetImpact_Study_Report.pdf.*

Chapter 11

User-driven innovation and communications development

Sam Paltridge[1]

Improving access to communication networks in developing countries is a driver of their overall economic and social development. Over the past decade greater reliance on the market in many low-income countries has made owning and using a telephone increasingly affordable and accessible. This has helped create new sources of income and employment as well as encouraged innovation aimed at meeting local requirements. This chapter looks at some of the innovation surrounding wireless communications in developing countries and at how network effects and network externalities influence innovation in developing countries.

1. Sam Paltridge is Communication Analyst, Directorate for Science, Technology and Industry, OECD. The views are those of the author and may or may not reflect those of the OECD or its member countries.

Users and innovation

As access to networks has expanded, the use to which communication devices and services are being put shows extraordinary creativity. In Kenya, for example:

- elephants are being tagged with SIM cards and GPS, fitted to collars, so that when they stray from reservations into populated areas rangers receive alerts.
- inexpensive personal accident cover for 24-hour periods can be activated by sending an SMS with the premium deducted from a customer's mobile phone credit.
- more than 3.8 million users had begun using mobile banking services within 18 months of service initiation, a level it took the traditional banking sector more than a century to accomplish in that country.

Some of these services are further explored in this chapter, but it is first worth asking what role users played in the development of each. The example of geotagging animals shows that if users have the tools available they will develop applications that suit local requirements. The example of the insurance service shows that entrepreneurs will develop commercial services even where few existed. In both the benefit of expanded network coverage and access is clear, though no one could foresee that these applications would be developed following the liberalisation of telecommunication markets. Both are also examples of innovation stemming from users rather than network operators.

Mobile banking, which is growing rapidly in developing countries and is a service innovation offered by operators, also has its roots in user-driven innovation. The first to recognise the utility of mobile phones for transferring funds were users whose ingenuity found a way to send and receive money that was cheaper than other ways. A user purchases prepaid airtime and calls an airtime vendor in the location of the recipient. Following negotiation of the commission, the local vendor adds the airtime value purchased by the sender to its account and then pays cash to the recipient.

The success of this informal system soon attracted the attention of operators who recognised that users were obtaining value (*i.e.* a positive externality) which was not reflected in their prices. They realised they could internalise this externality and add value to their customer proposition by creating a commercial service. To do so they developed a mobile banking system with trusted agents, transparent pricing, service promotion, and so forth. The net effect was to create banking and payment services for the unbanked. To explore this phenomenon further it is necessary to step back and see how regulatory reform made it possible for the market to deliver innovation applicable to local needs and to better understand the role of network effects and network externalities.

Regulatory reform and network growth

Until the end of 1997 monopolies provided telecommunication services in the majority of OECD countries, and most developing countries. The following decade witnessed rapid liberalisation. Performance in both eras can be compared by considering the following global developments:

- The first billion telephone subscriptions, reached by the end of 1997, took more than a century to accomplish. It took just four years to add the second billion, three years for the third billion and two years for the fourth.

- In 1997 teledensity for low-income countries was just 1.5 fixed and mobile telephone subscriptions per 100 inhabitants. By 2007 this had increased to 23.9 per 100 inhabitants (Figure 11.1).
- Telecommunication revenue in low-income countries grew from USD 10 billion in 1996 to USD 60 billion in 2006. Over the period total foreign direct investment (FDI) in telecommunications, in developing and transition economies, was USD 100 billion.
- Between 1996 and 2006 annual investment in telecommunication infrastructure more than doubled in low-income countries, from USD 4.4 billion to USD 9.6 billion in an environment of falling equipment costs and a shift to less expensive wireless-based systems.

After a century of low growth in access in developing economies, progress in recent years has been little short of breathtaking, something of which those countries can be justifiably proud. It can, of course, be readily acknowledged that there is not a one-to-one relationship between the number of subscriptions and individual access. In fact, this is the case in both eras (monopoly and liberal) and in developed and developing countries alike. Some historical and contemporary factors to take into consideration, when thinking about overall access, include the number of second lines on fixed networks, the level of ownership of dual SIM cards, resale and shared access. They may, however, offset each other in terms of assessing whether policy objectives are being met. The ownership of dual SIM cards means that individual access is overstated when simply looking at subscriptions. On the other hand, particularly in developing economies, there is significant resale of prepaid subscriptions by micro-entrepreneurs as well as shared use by families, friends and neighbours. This means that access may be understated when only considering subscriptions. Be that as it may, it is clear that competitive markets have far outperformed monopolies in terms of expansion. For low-income countries almost half of the access growth to the end of 2007 had occurred in the previous two years (Figure 11.1).

Figure 11.1. Low-income countries (teledensity)

Fixed and mobile subscriptions per 100 inhabitants

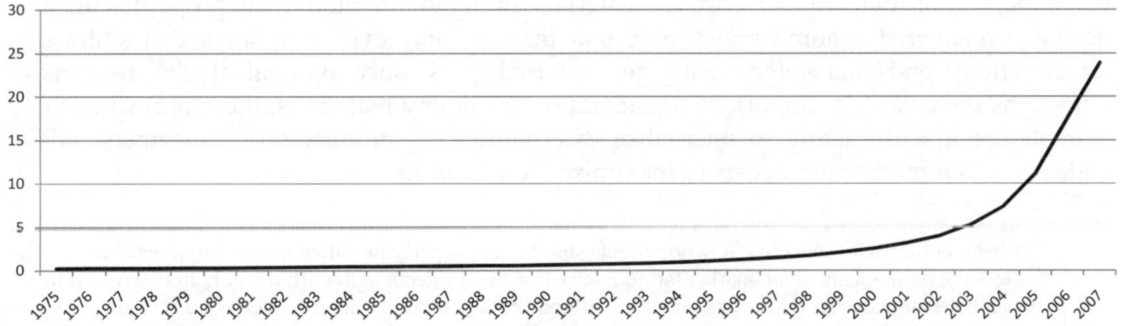

Source: OECD based on International Telecommunications Union.

Policy and regulatory reform surrounding ICTs have been chiefly responsible for the gains over recent years. Liberalisation of communication markets was the critical step, along with separation of policy and operational responsibilities, the creation of independent regulators, and privatisation. All of these initiatives aimed at providing the right incentives for operators to expand service, introduce new technologies and attract investment from capital markets. Just as important have been reforms to the principles underpinning international traffic exchange. These include recommended best practices, such as cost orientation and non-discrimination, as well as efforts towards greater transparency.

Network effects and externalities

At the individual level the benefits of expanded access to communication networks is easily grasped. By joining a network a user can communicate with all other users on that network and, beyond that, on interconnected networks. In addition, the network itself becomes more valuable to participants as the number of users, or opportunities for communication, increases (*i.e.* the network effect). The network effect is sometimes termed a "network externality" though the two terms, strictly speaking, are not interchangeable.[2]

Externalities are side effects of economic activities in which those affected, positively or negatively, have not participated directly in the decisions which led to those outcomes (Baumol and Oates, 1998). They are sometimes categorised as market outcomes that affect parties other than the direct producers and consumers of a good or service. Textbook examples include a factory that pollutes the surrounding environment (a negative externality for those affected) or householders, who by painting their residence or tending their garden, improve the value of surrounding properties (a positive externality for neighbours). Side effects such as these, when not taken into account in the pricing of economic transactions, are a form of market imperfection. As the costs and benefits, for third parties, were not part of the calculations of those deciding to undertake these activities, too much or too little of this activity may be produced.[3]

For economists the solution to the problem of externalities is to seek a way to bring them into the financial considerations of decision makers. The aim is for the costs or benefits of actions to be accounted for in economic transactions (*i.e.* internalisation). A polluter, for example, may cease to pollute if it faces financial penalties or if consumers shun their products. A householder may be more inclined to paint his/her house if neighbours contribute to the cost of materials or maintain their own properties in a desirable manner. Economists also note that markets may evolve in ways that address imperfections and that internalising an externality is only optimal if the benefits outweighs the costs. An important consideration for policy makers is, therefore, whether markets are already acting to internalise externalities, as producers or consumers will endeavour to minimise their costs or maximise their benefits.

---

2. "First a definitional concern: Network effects should not properly be called network externalities unless the participants in the market fail to internalize these effects. After all, it would not be useful to have the term 'externality' mean something different in this literature than it does in the rest of economics. Unfortunately, the term externality has been used somewhat carelessly in this literature." (Liebowitz and Margolis, n.d.)

3. Farrell and Klemperer (2001) note that when a network effect is not internalised (*i.e.* it remains an externality) it can be argued that there would be under-adoption of a good (service). Under these conditions it has been argued that the network will be too small and adoption will tend to be too slow so that optimality requires subsidising the marginal adopter to the extent of his external contribution to others.

The potential for network expansion can be viewed as a positive externality. Those who make this case argue that when a user joins a network he/she takes into account his/her benefit but does not consider the benefit for others. The corollary is that if this additional benefit is not captured in the pricing of the service, network expansion will be less than optimal (*i.e.* there will be too little production of a positive externality). Their suggested solutions may range from premiums paid by existing users or subsidies from other sources (*e.g.* public revenue) to align the interests of the prospective subscriber with the benefit to others already on the network.

There are several pitfalls in thinking about network externalities instead of simply using the concept of network effects. All can agree that the expansion of networks (the network effect) is desirable, and policies can be crafted and assessed with that in mind. By way of contrast, a policy presupposing an externality starts from the premise that an imperfection needs to be addressed. It does not necessarily take into account whether the market is already acting to meet policy goals. In other words, even if an imperfection exists and can be measured (many externalities are widely considered not to be measurable), it is far from clear than any action will expand the network faster than pursuing market-based growth. This is why economists urge caution in using the concept of externalities because "to some it may imply the need for a public policy intervention when none is called for" (Crandall and Waverman, 2000).

Producers and consumers often do take externalities into account, even if not always at the aggregate level, and market solutions frequently evolve. By addressing network externalities, rather than network effects, actions may be contrary to policy objectives or impose greater costs than benefits. It leaves open the possibility, for example, that an externality will be paid for twice – once voluntarily and directly by the actors directly affected and a second time by a top-down subsidy scheme to which they may be forced to contribute. At the same time, analysis of network externalities has not kept pace with technological and service developments. Markets take externalities into account and interact with them. A further weakness is that the underlying academic literature is almost entirely predicated on domestic rather than international communication.

Internalisation of externalities

The decision of each user to join a network is, for the most part, independent of any other user's decision, although, if both do so, they share a common benefit. On the other hand, the existence of a "network externality" hinges on perceptions of value by individual users in relation to others. This includes the value users perceive from being part of a network and their willingness to compensate others for joining a network. The United Kingdom's Competition Commission has defined a "network externality" as the benefit users derive from calling and being called by new users, which people often do not take into account when deciding whether to subscribe to a service (United Kingdom Competition Commission, 2003, pp. 225-226). They point out that social welfare is maximised if the price of joining a network takes into account externalities (*i.e.* internalises them) in a way that optimises the network effect.

One of the challenges in dealing with externalities is the difficulty in identifying those affected. In contrast to the network effect (where it can be argued that all users benefit equally from an expansion in calling opportunities), the strength of network externalities will be relative to each user. In defining the network externality, Crandall and Waverman (2000) observed:

> "The network externality in telephone networks may be characterised fairly simply. What I am willing to pay to join a network is a function of how many others I can call (and can call me). The standard economic analysis of this externality concludes that the nth person to join a network should pay a price below her costs to induce her to subscribe. She does not, in evaluating the costs and benefits of a telephone, consider the benefits she provides to others. As a result the externalities [relate to] the impossibility of identifying the beneficiaries, who otherwise might be induced to compensate her directly."

If producers or consumers make decisions, which incorporate the costs or benefits of side effects on others, these are not properly speaking externalities. As Liebowitz and Margolis (n.d.) point out:

> "Although the individual consumers of a product are not likely to internalize the effect of their joining a network on other members of a network, the owner of a network may very well internalize such effects. When the owner of a network (or technology) is able to internalize such network effects, they are no longer externalities."

In other words, "The difference between a network effect and a network externality lies in whether the impact of an additional user on other users is somehow internalized." (Liebowitz and Margolis, n.d.) In seeking to address one phenomenon (*i.e.* the network effect) operators may alter related factors (*i.e.* network externalities). Network operators, for example, in recognition of the network effect, co-operate on the creation and use of standards to permit communication across different networks. They do so because the benefits (interconnectivity) outweigh the costs (making standards). Their actions reflect demand from their customers, who would be quick to complain if their communication options were limited and who ultimately bear the cost of the standardisation. An important point, which needs to be recognised in all discussions of network externalities, is that network operators have incentives to internalise the phenomenon as they strive to enhance the network effect.

Internalisation of network externalities by network owners

The introduction of Short Message Service (SMS) provides a good example of operators internalising externalities and taking the network effect into account. Initially, SMS was not planned as a commercial service. The original purpose was for operators to share information with customers as well as staff (Wray, 2002). The unplanned side effect was to enable users to text message each other but for operators not to be able to bill users for this service. This meant that the economic value of a positive externality, for consumers, was not being captured by producers. As an unplanned side effect, rather than a commercial service, SMS also had limitations. These included users on one network not being able to text users on another network (BBC Online, 2002). Struck by the growing popularity of the phenomenon, operators internalised the externality by introducing pricing and billing systems for SMS and added value by enabling transmission across networks to enhance the network effect.

The solution to the positive externality generated by SMS would not have surprised Ronald Coase. Professor Coase was the economist who pointed out, in the 1960s, that property rights provided incentives for their owners to tackle inefficient outcomes resulting from externalities (Coase, 1960; Crandall and Waverman, 2000, p. 24). The problem for universal application of this principle, as recognised by Coase, occurred in situations where high transaction costs acted as a barrier to efficient resource allocation (Coase, 1960, p. 7). Among these costs is the identification of those affected by any externality and creating a solution (*e.g.* negotiations, contracts, etc.). Network operators, it can be argued, played that role in the case of SMS. This raises the question of whether they perform the same role with other externalities and whether markets are already acting to internalise network externalities.

The classic problem for network externalities is the challenge of identifying beneficiaries that might otherwise directly contribute to the cost of adding new telephony subscribers. In the monopoly era, network externalities tended to be dealt with obliquely. Universal service polices were largely justified on the basis of social externalities associated with having a telephone service (*e.g.* ensuring widespread availability of the ability to call emergency services) or, more broadly, the economic and social benefits of the network effect. In fostering cross-subsidies from one broad group to another to meet these goals, network externalities were addressed to an extent but arguably in a haphazard fashion. In other words very broad groups (business/consumer, urban/rural, long distance/local) met the costs or reaped the benefits. If such policies had been applied only in respect to network externalities there would have been more limited intervention. As Crandall and Waverman (2000, p. 25) point out, it would only have been necessary to intervene for those individuals whose private benefits did not exceed the cost of serving them, but who generated sufficient external benefits to make up the difference.

Enter the innovation of prepaid cards

There is, however, a different way of looking at the problem of network externalities in a market-based environment. Consider a situation in which a mechanism is introduced with the effect that the private benefits to the user almost always exceed the price they are willing to pay. Consider further that, while someone joining the network does not internalise the benefit to others, this benefit is well recognised by network operators. The solution, therefore, is a system that enables a network operator to charge users whatever they are willing to pay. In fact, this is what a market-based approach has delivered through prepaid cards.

The introduction of prepaid cards has also gone a long way toward internalising, and in some countries eliminating, network externalities. Few of those writing about network externalities in the previous century envisaged that teledensities would exceed 100%, let alone 200%.[4] Fewer still would have expected developing countries to be surpassing 100%.[5] This does not, of course, mean that everyone in these economies has a telephone. It does show, however, that competitive markets, in contrast to monopolies, are rapidly expanding access and, in the course of doing so, are internalising network externalities.

During the monopoly era operators typically charged a connection fee, a fixed monthly line rental and usage charges. There was little flexibility in these arrangements. In this context, the network externalities for a traditional telephone service, as described by Crandall and Waverman (2000), represented a larger consideration or barrier than they do today for users considering joining a network. Markets are better able to determine

4. At the close of 2007 Hong Kong, China, recorded teledensity of 200% as reported by the ITU.

5. Gabon may have been the first Sub-Saharan African country to surpass 100% mobile penetration. Gabon's regulatory authority reports quarterly updates at *www.artel.ga*. Based on 2007 population data, the country exceeded 100% during 2008. These data include users with dual SIM cards and operators may not always remove inactive subscribers from their databases.

price sensitivities, and network operators, like any business, will endeavour to charge consumers the maximum amount they can for any good or service (Harford, 2008, p. 35). Consumers have, of course, different propensities to pay. Prepaid cards not only eliminated credit barriers to joining a network but enabled operators to introduce a means for consumers to signal how price-sensitive they were to paying the fixed fees of traditional telephone service. Moreover, the new strategy had the advantage of internalising network externalities (*i.e.* people who would not otherwise have joined a network became subscribers).

For price-sensitive consumers, the innovation of prepaid cards as a result of competition turned the traditional pricing model on its head. Not only were connection fees commonly waived but joining a network could be considered in terms of what may be called the minimum cost of subscription (MCS). The MCS is the lowest possible charge a user can pay and still remain connected to a network. In some developing countries this can now be less than USD 2 a month for wireless service (OECD, 2007, p. 24). This is why in developed countries it is increasingly common to see nominal mobile penetration rates greater than 100%, indicating that at least some consumers, attracted by the MCS, have more than one mobile service (Sutherland, 2008). The advantage for the low-usage or low-income user, whose private benefit may not have justified subscription under a traditional pricing model, is that they can now rationalise having service. Their private benefit, in other words, will always exceed the cost of joining the network (*i.e.* what they are willing to pay).

At the same time, many mobile users continue to be billed on a post-pay basis. They generally pay fixed monthly fees, have higher network usage and a reduced unit price (a lower per minute rate or unlimited service). In this way post-pay users, who are less price-sensitive, make a greater individual contribution to meeting a network's joint and common costs. The issue here, however, is not one of cross-subsidisation. Both prepaid and post-paid users pay their way but the market has found a better way for them to signal how price-sensitive they are. Users may give little or no thought to the network externalities being addressed but the effect is the same. Both post-paid and prepaid users benefit from the internalisation of network externalities and the resulting growth of the network effect.

The prepaid model is also being applied to fixed network services in some countries. In India, Bharat Sanchar Nigam Ltd (BSNL) has introduced a scheme whereby subscribers can have a fixed line telephone connection on a prepaid basis (Thomas, 2006).[6] The service differs from telephone cards in that no PIN code is needed, making it simpler to use. The scheme aims at making fixed lines more attractive relative to mobile services. The MCS for this service is less than USD 0.50 per month.[7] Further discounts are available in rural areas for exchanges serving fewer than 1 000 users, effectively setting the price of the MCS at zero.

---

6. Brazil Telecom offers a prepaid service that integrates cellular, fixed line and public telephony. Users buy credit which can be used on mobile phones and make calls from fixed lines, including public telephones. Brazil Telecom, "Brasil Telecom GSM announces launch of operations", 2004, *www.secinfo.com/d17EG1.197.htm*.

7. Users have two options. Under Option 1 they pay an upfront fee of USD 5.44 for 24 months. Under Option 2 they pay USD 0.43 per month for 24 months. Option 2 includes USD 0.43 of included calls. Calls under both options are charged at USD 0.02 per unit. The length of units varies according to several distance bands with local calls including 180 seconds.

India's highly competitive access market has also delivered the innovation of lifetime prepaid cards for wireless service. Aimed at low-income people the prepaid cards have no fixed validity period. In other words the cards enable users to continue to receive calls even in months when they cannot afford to top up their cards with credits for making calls. Competition, therefore, not only acts to expand access for low-income users but also to retain them as customers when they might otherwise leave the network.

Network investment and externalities

It is not just in pricing that operators may take network externalities, or the network effect, into consideration. Operators invest in networks, for the most part, in anticipation of future growth. The economics of building and maintaining networks in relation to that growth is also a significant factor. The cost of putting in place capacity for future requirements may be marginal compared to adding that capacity, or replacing a system, at a later stage. Nevertheless, the investment must be paid for by shareholders, in anticipation of future growth and returns, or increasingly, as the network grows, by subscribers. Thus, network operators already act to internalise network externalities by addressing network effects. By getting today's subscribers to contribute to a network's joint and common costs, they lay the foundation for new subscribers. Operators have arguably identified the beneficiaries of future growth, at least in part, as people who are already on a network. As such, existing subscribers are compensating new subscribers even if they may not directly benefit from any particular subscriber joining a network or take the benefit for others into account in their own decision to subscribe.

Internalisation of network externalities by consumers

Consumers also internalise what may otherwise remain externalities when they perceive a direct benefit for those they call and those who call them. In some countries fixed lines may be retained by people with mobile phones because of lower termination charges and consequently, lower call charges for other users who call them. On mobile networks individuals who share a relation may select the same network to benefit from reduced "on-net" calls. Some may purchase SIM cards for more than one network for the same purpose. Consumers also buy telephones or telephone service for each other. Family members routinely buy service for each other in recognition of the benefits of calling and being called. While this is well recognised at the national level, a growing international element acts to internalise network externalities. Needless to say, the strongest positive network externalities are geographically local, with family and friends, or with local business intermediaries, accounting for many tariff plans which take this into account.

Migrants and offshore workers provide an international dimension to the issue of network externalities, particularly for developing countries. Diasporas are making increasing use of services which enable them to purchase telephone service for others in countries in which they do not themselves reside (Table 11.1). By inputting a user's telephone number, such services enable customers to send credit to that mobile phone.

Anywhere in the world, for example, people can use "mamamikes" to purchase telephones and airtime for friends and relatives in Kenya and Uganda (*www.mamamikes.com*). A spin-off from the same company, Kikwe, offers a top-up service in Kenya, Ghana and Sierra Leone and offers users the chance to send a complimentary SMS to the recipient when they remit the airtime (*www.kikwe.com/*). Another service provider, aryty, enables people in Canada and the United States to purchase airtime for users in the Philippines through the Internet or their mobile phones (*www.aryty.com/*). The aryty service, started in 2007, is currently being expanded to India and the Gulf Region (*The Economic Times*, 2008). Global Topup already enables users around the world to purchase prepaid airtime for people in India. In 2007, Telecom Malaysia also began a service enabling migrant workers in that country to top up the airtime of people in Indonesia and Bangladesh.

Table 11.1. Selected services offering direct international payments for telephone service

Name of Service	Description	URL
mamamikes	A service for anyone around the world to top up airtime for users in Kenya or Uganda. The recipient can chocse to collect their airtime from a Nairobi or Kampala office or it can be automatically loaded onto their phone. Airtime PIN numbers can also be sent via SMS to the mobile phone.	*www.mamamikes.com* *www.kikwe.com*
RechargeBrazil	Introduced in July 2008 and aimed at people the United States paying for airtime for users in Brazil. Real-time top-up via a website using PayPal in co-operation with mobile providers.	*www.recargabrasil.com*
Mexico En Linea	Introduced in 1998 and aimed at people in Canada and the United States paying for fixed lines for users in Mexico.	*www.telmexusa.com/us/ret_mexlinea.html*
aryty	Introduced in 2007 the service is aimed at people in Canada and the United States paying for airtime for users in the Philippines. Real-time top-ups can be made via a website or SMS from a previously credited card.	*www.aryty.com*
Paybox, Transfer To and Telecom Malaysia	Paybox and Transfer To provide tools for international remittances and mobile banking. Working with Telecom Malaysia they created a service which enables Indonesian or Bangladeshi workers, living in Malaysia, to send airtime to their relatives abroad by SMS. A similar service MoneyBox Africa has been announced for Nigeria.	*www.paybox.net* *www.transfer-to.com* *www.fixedandmobile.com*
Primissimo	A service for the Cameroonian diaspora to buy airtime for people in Cameroon.	*www.primissimo.net*
Global Topup	Enables users around the world to purchase prepaid airtime for users in India. Plans to expand to other markets.	*www.globaltopup.com*

Source: OECD.

In Latin America, Unpaid Systems launched a service (RechargeBrazil) in July 2008 which enables people in the United States to top up the phone cards of their friends and family in Brazil (*www.recargabrasil.com/recarga.do?method=load*). Working with PayPal, RechargeBrazil allows users to make direct payments, for others, on any of Brazil's mobile service providers. While the majority of schemes are recent and focus on wireless services, Telmex has offered "Mexico En Linea", aimed at fixed services, since 1998 (Luxner, 2002). People residing in Canada and the United States can pay for the installation of a Telmex line in Mexico and pay a portion of or the full amount of the recipient's monthly telephone bill (*www.telmexusa.com/us/ret_mexlinea.html*). There is no monthly subscription fee for the service. Between 1998 and 2002, TelmexUSA received about 1 million applications for connections under the scheme, of which around 70% were approved. Mexico's largest wireless operator offers "Telcel Amigo", a service which allows people in the United States to top up airtime for its subscribers in Mexico (*www.usatelcel.com/Recarga%20Aqui.html*).

In September 2008 Digicel, in partnership with Western Union, launched a service enabling US residents to buy airtime and send it to mobile phone accounts belonging to its users in Jamaica, Haiti, El Salvador and Guyana (CNN, 2008). In the future, the companies plan to add the service for Anguilla, Antigua, Dominica, Grenada, St. Kitts,

St. Lucia, St. Vincent, Barbados, and Trinidad and Tobago. Western Union has also partnered with Orascom Telecom which operates six mobile networks in Algeria, Pakistan, Egypt, Tunisia, Bangladesh and Zimbabwe to offer remittance services.

An example of an alternative or complementary service, which internalises international network externalities, is provided by TracFone Mobile. Tracfone, a network reseller owned by America Movil, offers low-cost post-pay and prepaid wireless service to users in the United States. TracFone's "International Neighbors" service allows users to obtain, at no cost, three Mexican or Canadian phone numbers (*www.tracfoneild.com/?AID=ILDHPB*). A user's family and friends in those countries can then call a TracFone user in the United States by dialling a local number in their own country, without paying for an international long distance call. The TracFone user in the United States pays for the incoming call at their standard airtime rate with no additional charge. The price they pay depends on their plan but may range, by way of example, from USD 0.17 to USD 0.30. The Canadian or Mexican user only pays the cost of a local call.

The advantage of the foregoing schemes is to enable beneficiaries to directly compensate other users. In turn, this provides demand-side encouragement for network expansion and use. As noted earlier, one of the challenges of externalities is to identify the parties affected and to apply an efficient means of internalisation. Services such as the foregoing enable the market to address both these challenges. They are more efficient than possible alternatives as there is far less risk that funds will be mislaid, misdirected or appropriated by an intermediary. In addition, the problem of corruption is vastly reduced or eliminated. For much the same reasons, mobile networks are now emerging as a platform for financial remittances of all types (Box 11.1). Policy makers and operators should encourage greater diffusion of such schemes because they directly and efficiently act to internalise positive externalities. Policy makers can also look to innovation in other areas to see if the market is addressing internalisation of network externalities.

Innovation and network externality internalisation

The literature dealing with externalities in communication markets has, for the most part, a 20th century perspective. It views telephones and networks solely as a means for making and receiving telephone calls. It largely does not take into account different ways of paying for telephone service (*e.g.* prepaid cards), although these have figured in discussions of network externalities in the United Kingdom. For the most part it locates the pricing of telephones in a single country. As a consequence, for an industry characterised by rapid technological and service innovation, the textbook examples do not always fit how telephones are used in practice. This is particularly true for network externalities in developing countries and the calculation of the private benefits of owning a telephone.

Box 11.1. International financial remittances

Some have proposed that international traffic between developed and developing countries should be subject to a network externality premium. The stated aim is to fund further network expansion. The effect of a network externality premium, imposed on users, would be to raise the price of financial remittances over mobile phones. It would penalise users who already contribute revenue to the terminating carrier and add costs to making more numerous small remittances. By way of contrast, a policy that aimed to reduce the price of telecommunication services through competition would enhance the opportunity for people without bank accounts, in both developed and developing countries, to increase the use of communication networks for financial remittances. The benefits for broader economic and social development stand to be far greater than the financial flows from the foreign purchase of airtime. Yet, both will contribute to network development through direct and indirect stimulation of demand.

The World Bank estimates that international financial remittances towards developing countries totalled USD 251 billion in 2007.[1] The phenomenon is challenging to measure for a variety of reasons including the harmonisation of definitions and data collection as well as financial remittances that take place outside the banking system.[2] It represents, however, a larger financial flow than foreign direct investment and a much higher total than official development aid (Gupta *et al.*, 2007).[3] As a ratio of GDP it typically ranges from 4% to 8% in African countries. In relation to export earnings it can range from 15% to 40% (Gupta *et al.*, 2007).[4] That being said, the flows of financial remittances to Africa are lower than towards other developing regions.[5] While undoubtedly reflecting a number of factors, the lack of an efficient means to remit funds and its high cost have acted as a barrier for Africa. ICTs can be used to address this barrier if the right policy and regulatory frameworks are in place.

Mobile phone networks are an emerging platform for international financial remittances. There are a number of advantages, including much less expensive transactions than traditional banking or remittance services, and the fact of being an enabling technology for people without bank accounts. Additionally, because transaction costs are lower, particularly for small sums, it enables users to send smaller amounts of money more regularly and ensure payment is made directly to the beneficiary. It also has the effect of drawing money that would otherwise have gone via informal remittances into systems that are observable and measurable. The service also generates revenue for operators. Each remittance is responsible for at least two SMS, between a remittance company and the sender and the receiver, to ensure the transaction.

The IMF has defined the remittance process using mobile phones as below:

Information flow: A sender visits a remittance company and fills out an information sheet. The remittance company in the sender's country submits to its partner bank in the recipient's country all relevant information provided by the sender. At the same time, the remittance company assigns an account number to the beneficiary's mobile phone and credits the number of the beneficiary. The service sends a text message to both the sender and the recipient, notifying them that the money has been transferred. The beneficiary can now get cash from the partnering institutions or through an ATM at a participating bank.

Funds flow: The remittance company in the sender's country maintains a pre-funded account at a local bank in the receiving country. When a sender remits, the funds are transferred to the beneficiary's money card. Meanwhile, the beneficiary receives a notification through a text message that the funds are credited to his money card. The beneficiary then claims his cash at accredited encashment centres by showing the text message in his mobile phone (Gupta *et al.*, 2007).

1. World Bank Migration and Development Brief 5, Development Prospects Group, Migration and Remittances Team, 10 July 2008, *http://siteresources.worldbank.org/INTPROSPECTS/Resources/334934-1110315015165/MD_Brief5.pdf.*

2. The IMF's Statistics Department, together with the Luxembourg Group, has completed a draft of the new "International Transactions in Remittances: Guide for Compilers and Users" (RCG), September 2008. See in particular pp. 19-20 at *www.imf.org/external/np/sta/bop/remitt.htm.*

3. For some caveats on comparisons, OECD, "Remittances as Development Finance", *www.oecd.org/dataoecd/62/17/34306846.pdf.*

4. The totals are higher relative to GDP and export earnings for Lesotho and Cape Verde.

5. Gupta *et al.* (2007), note that Africa received just 4% of total remittances, by far the smallest share to developing countries in 2005.

The theory of network externalities presupposes that users do not take into account the benefits they provide others when they join a network and therefore are not compensated. In fact, in developing countries, telephones are commonly shared by many individuals, and compensation for use is paid to the telephone's owner. These may be micro-entrepreneurs who resell airtime from street stalls or bicycles, or friends and neighbours who share the cost of a telephone. In an interview, Michael Joseph, chief executive officer of Kenya's Safaricom, says: "If you want to go and buy airtime now, for instance, you won't go to a big shop. You'll likely go to some shop on the side of the road. That person probably didn't get the airtime from one of our dealers. He bought it from another person, who bought it from another person, who bought it from another person. It's this creative distribution channel selling our products and services that's created a tremendous amount of work apart from all of the other things we do, like building our base stations, maintaining them, and things like that." (Gabel, 2008). In India, Pakistan, the Philippines and Thailand, for example, more than 50% of all mobile subscribers in rural areas share mobile phones with friends, relatives and acquaintances outside their households.[8]

Resale of telephone service also occurs in OECD countries. Prepaid cards selling below face value are routinely bought and sold on eBay. This may be because micro-entrepreneurs can buy cards, in bulk, at wholesale rates. In addition various promotional offers from wireless operators offer lower introductory prices for prepaid service. From the operator's perspective such promotions lower the barriers to joining a network particularly for low-income users. Once a user has purchased a card they may become a customer and purchase additional airtime. Other users may prefer to use such offers as a form of disposable service. They use the minutes for as long as the card is valid and then bid on or buy a new card. The benefit to low-income users may be relatively inexpensive rates compared to normal prepaid service (*e.g.* USD 0.03 or USD 0.04 instead of USD 0.10). Purchasing prepaid airtime through auctions offers a further avenue for some price-sensitive users to lower the minimum cost of telephone access. The downside of "disposable service", in terms of network externalities, is that each new card comes with a new telephone number. Even here, however, markets may act to externalise externalities as mobile phones interface with Internet-based call-forwarding services. In the United States "Grand Central" (*www.grandcentral.com*) currently offers a free service which provides users with a single number towards which they can direct up to six other numbers. Grand Central may, in future, charge for this service, which could limit its attractiveness to low-income users. However, during 2008, a prepaid user could purchase a single card with a longer duration (*e.g.* one year) and have a constant number which can be redirected to their Grand Central number. At the same time they could purchase multiple other "disposable" cards in online auctions with lower rates.

Nor is it a simple matter to determine the proportion of the overall benefit a user, as part of a network, derives from telephony. Making and receiving telephone calls is, of course, only one application on today's multi-functional devices. Examples of other applications include communicating text (SMS, email) receiving broadcast services (radio and television), playing games or personal entertainment (music, video), accessing information services (news, weather, commodity prices), taking and transmitting photographs or videos, social networking (*e.g.* Twitter) as well as interaction with other

8. LirneAsia has documented significant sharing of mobile phones outside the household by poor people in some developing countries: *http://lirneasia.net/wp-content/uploads/2008/07/shared-use-of-mobiles-in-sri-lanka-3.png*.

devices (*e.g.* from slingboxes to personal computers). Mobile phone users also interact with the Internet in a myriad of new ways. Examples include uploading photographs to websites such as Flik'r or video to Youtube or text to blogs. Mobile phones can be used to indicate presence (*e.g.* Loopt, Wrrl, Zintin) or determine location, using the cellular networks themselves, WiFi or GPS, and provide associated maps or information (*e.g. www.navizon.com, www.whrrl.com, www.loopt.com, jaiku.com, zintin.com*).

The uses to which mobile phones are being put show extraordinary creativity. One illustration of the variety of uses of such devices is illustrated by Apple's iPhone. In early October 2008, there were more than 2 600 applications for the iPhone available for downloading on Apple's website. Examples included applications which informed motorists of local fuel prices in their location; price checks for products through ISBN and UPC codes against online vendor alternatives; social networking; entertainment; as well as all manner of news and information services.

Innovation surrounding networks in developing countries is also readily apparent from the work of a growing number of "ICT anthropologists". Companies such as Nokia and Intel are devoting more resources to looking at how people are using ICTs in order to improve their design and applicability (Palmer, 2008; Corbett, 2008). Because people share mobile phones in developing countries, for example, Nokia introduced the option of multiple address books on devices to enhance privacy. The fact that users are highly price-sensitive and that others share and resell service prompted the introduction of prepaid and cost-tracking applications.[9] Increasingly, however, phones are put to uses that were not anticipated by manufacturers and service providers. The ICT anthropologists document the customisation (hacks) of phones for various purposes and local conditions and feed the information back to design teams.[10] Some uses may have little to do with telephony. Examples in developing countries include:

- In mid-2008, a Kenyan teenager received a great deal of publicity after he was reported to have developed a car immobiliser operated by a mobile phone.[11] The system can be used as an anti-theft or tracking device. If the ignition is started when the car is supposed to be locked, the device communicates via SMS with the owner's mobile phone requesting a PIN code to enable the vehicle to start. Similar services have emerged in other countries such as India (Parthasarathy, 2007).
- Mobile phones are being used as broadcast technologies for text services and document exchange. NGOs and other groups, for example, have used the Frontline SMS software to broadcast text messages to users for purposes ranging

9. "Nokia unveils two handsets that offer a range of useful features and colours aimed at consumers in emerging markets", 22 January 2008, *www.nokia.com/A4136001?newsid=1183854*; Johanna Liukkonen, Manager, Government and Industry Affairs, Emerging Markets, Nokia Corporation, "Innovating for the local needs of low-income consumers at the BoP", *www.hse.fi/NR/rdonlyres/FA89FD84-74F8-4BB5-90B1-9DC37B16A1E0/0/Johanna_Liukkonen_BoP_Helsinki.pdf.*

10. See for example the blog of Jan Chipchase, one of Nokia's anthropologists, *www.janchipchase.com/*. A selection of Chipchase's photos are also available at: *http://technology.newscientist.com/data/images/ns/av/mg19826602000V1/mg19826602000V1.html.* Intel has also recognised that local hacks of ICT equipment can sometimes make inexpensive substitute products available. Craig Barrett's interview with Dr Johnny Chung Lee shows how he uses a "hacked" Nintendo Wii to create an electronic whiteboard system for USD 50, *http://download.intel.com/pressroom/kits/events/idffall_2008/CraigBarrett_keynote_transcript.pdf.*

11. Kenyan television coverage of this story is available at: *www.youtube.com/watch?v=j_dR-nh5tvg.*

from monitoring elections, corruption and human rights to providing information on healthcare (*www.frontlinesms.com/*; Zuckerman, 2008). By using the Episurvey, for example, health authorities can collate information on the spread of disease by sending questionnaires across mobile networks, and data can be gathered from people on their phones (BBC, 2008). Perhaps the best example is the mPedigree service which enables users, after purchasing a drug and scratching a panel, to text the number to see if the product is genuine and prior to its date of expiry (http://mpedigree.org/home/). Authorities have also used text services to broadcast information about natural disasters.[12]

- Mobile phones are used as a substitute for the services and tools provided by financial intermediaries or institutions. This can range from acting as a store of value, a payment mechanism or a tool for financial remittances. Safaricom's M-PESA service, for example, enables users in Kenya to transfer money with the use of a mobile phone (*www.safaricom.co.ke/index.php?id=228*). In the first 18 months of service it transferred funds valued at over USD 500 million (Rosenberg, 2008). In that time the service registered 3.8 million customers, a figure the Kenyan banking system took a century to reach (Kinyanjui, 2008). These unanticipated services, pioneered in developing countries, are now being made available by mobile providers to "unbanked" people both in developing and developed countries.[13] At the same time application developers are taking successful models from developed countries and adapting them to local requirements such as services similar to PayPal for mobile phones.[14]

- Mobile phones are being used for the delivery of government services. In India, for example, pensions are delivered through a combination of RFID-enabled cards and mobile phones with Near Field Communication (NFC) (D'Monte, 2008; Menon, 2008). Under this scheme pensioners have a chip-enabled card which includes their personal information together with a fingerprint. On presentation of the card the information is scanned by the mobile phone and the agent is authorised to dispense the pension entitlement. The scheme is reported to have significantly reduced transaction costs and eliminated opportunities for corruption and fraud, as well as provided users a more efficient service. Accordingly, the Indian government aims to expand the use of mobile phones to provide services to people without bank accounts.

- Mobile phone networks have become a platform for commerce. CellBazaar in Bangladesh is a service which has been described as a Craigslist for mobile telephones. People advertise a wide variety of things from agricultural produce and real estate to jobs and services. The GSM Association has noted, "While common telephony establishes one-to-one communication, CellBazaar links many-to-many using the same basic mobile infrastructures."[15] In Africa, mobile phones can be used to access TradeNet, an agricultural market information and

12. In Sir Lanka, LirneAsia's research, for example, explores the use of mobile communications for disaster warnings in two types of models: community-based and public warning models, *http://lirneasia.net/projects/2008-2010/mobile20bop/vertical-aspects/mobiles-for-disaster-warning/*.

13. Trumpet Mobile, in co-operation with Western Union, offers remittance services, in the United States, aimed at users without bank accounts, *www.trumpetmobile.com/customerportal/index.aspx*.

14. *http://whiteafrican.com/2008/10/04/local-software-for-local-needs/*.

15. *www.globalmobileawards.com/winners.shtml#winner_cat5*.

trading service for farmers and their suppliers (*www.tradenet.biz/?lang=en*). The service, pioneered in Ghana but now available across Africa, mostly uses SMS to enable trade and information dissemination. Co-operatives and partners have the option of subsidising SMS messages for their constituents or customers. Senegal's Manobi offers a similar platform for farmers and fishermen (*www.manobi.net/worldwide/*).

The foregoing raises the question of how innovation surrounding the use of networks can inform debates about externalities. One obvious aspect is that the revenue opportunities for operators are no longer limited to telephony. Operators can share revenue with third-party service providers (as occurs with NTT's i-mode) or act as intermediaries themselves to sell products and services (terminal devices, ringtones, music, information, applications, financial remittances/mobile banking and so forth). They can also sell access to their subscriber base to advertisers. Some mobile service providers, for example, provide SMS and MMS advertising to their subscribers in return for lower telephony prices. For example, Blyk is a free European mobile operator for young people, funded by advertising. Advertisers pay Blyk to help them reach their customers and Blyk members get free texts and minutes (*www.blyk.co.uk/about*) (Moses, 2008). In the Philippines, U-mobile is a service owned by two network operators (*www.umobile.com.ph/*). It is aimed at 15-to-35-year-olds, by invitation only, who are willing to accept advertising in return for free calls and texts.

In India, mGinger offers advertising on SMS received by users. In return for signing up for the service and providing personal information users receive a payment for each text they receive and payment for texts read by the people they have referred to the service (*http://mginger.com/*). They are free to indicate the number of messages they are willing to receive and the time of day. Cheques are sent to these users when the amounts they accumulate exceed USD 6.11. Dubbed "permission-based advertising", mGinger has spawned a number of competitors offering advertising-supported SMS (*e.g. www.mgarlic.com/*, *www.admad.mobi/*, *www.m-earn.com/* and *www.160by2.com/*). Opportunities for revenue growth such as these may be taken into account when operators consider or address network externalities. The value of the customer to the operator is, in other words, no longer limited to their use of telephony.

In South Africa, Vodacom offers an SMS-based service that generates about 20 million messages a day (Morris, 2008). "Please Call Me" is Vodacom's free call-back service aimed at prepaid users. Users can send an advertiser-funded SMS for free to request a call back from another user. Vodacom also offers a service called "AdMe"; users give their permission to receive advertising with promotional offers from which the operator derives revenues (*www.adme.co.za/*).The genesis of the service is related to externalities. In developing countries highly price-sensitive users will signal others to call them back, for example through information on missed calls. This represents a positive externality for the user and generates a cost for the operator. Vodacom's solution internalises this externality by adding value to the caller (*i.e.* through improved information) and creating an opportunity for revenue (*e.g.* generating a new call, call termination or advertising).

Advertising on mobile phones still represents a relatively small share of the overall advertising market. However, it is a rapidly growing market segment with a potential reach exceeding 4 billion users in 2009. One of the largest firms serving advertisements to mobile devices is AdMob (*www.admob.com/s/home/*). The company reported that requests for advertisements increased from 1.5 billion in September 2007 to 5 billion in

September 2008 (Table 11.2). AdMob data cover only the mobile sites, applications and publishers that are part of its global network. It is, therefore, not representative of the entire market. Nonetheless, as one of the largest players in this market, the data reveal some interesting characteristics. One is that OECD countries' share of advertising requests for the top 20 countries was 51% of the total, down from 63% in 2007. Moreover a number of developing countries had many more advertisements served over mobile phones than OECD countries. These data suggest that advertising on mobile networks is developing rapidly outside the OECD area.

Table 11.2. AdMob mobile advertising

	September 2008		September 2007		Sept. 2007-Sept. 2008
Country	Requests	% of requests	Requests	% of requests	Request increase
United States	1 992 732 034	39.3	734 272 875	46.1	2.7x
Indonesia	1 135 977 186	22.4	84 390 452	5.3	13.5x
India	391 278 541	7.7	166 297 747	10.4	2.4x
United Kingdom	197 404 623	3.9	89 726 007	5.6	2.2x
Philippines	144 663 278	2.9	11 110 124	0.7	13.0x
South Africa	107 451 198	2.1	116 342 069	7.3	0.9x
Nigeria	89 848 935	1.8	2 884 122	0.2	31.2x
Romania	62 286 761	1.2	30 044 024	1.9	2.1x
China	54 403 922	1.1	7 471 365	1.1	7.3x
Malaysia	47 201 472	0.9	5 859 766	0.4	8.1x
Brunei	39 272 572	0.8	15 657 179	1.0	2.5x
Saudi Arabia	35 622 323	0.7	3 723 609	0.2	9.6x
Israel	33 906 466	0.7	21 748 162	1.4	1.6x
Australia	32 267 923	0.6	11 390 551	0.7	2.8x
Canada	30 846 195	0.6	26 110 308	1.6	1.2x
Kenya	30 547 361	0.6	10 343 288	0.6	3.0x
Italy	29 488 307	0.6	7 605 276	0.5	3.9x
Pakistan	28 849 805	0.6	8 127 478	0.5	3.5x
Bangladesh	26 267 624	0.5	24 051 999	1.5	1.1x
France	22 797 962	0.5	13 454 907	0.8	1.7x
Other countries	538 586 636	10.6	203 041 854	12.7	2.7x
Total	5 071 701 124	100	1 593 653 162	100	3.2x

Source: AdMob.

In competitive markets operators will seek to internalise externalities for particular groups of people. At the same time, the multi-dimensional nature of today's telephones is making adoption by these communities more rewarding. Africa's TradeNet provides one example. Another example comes from India where Bharti Airtel has a noteworthy agreement with the Indian Farmers Fertiliser Cooperative Limited (IFFCO). Bharti Airtel is a network operator in India with over 64 million subscribers. IFFCO is a co-operative of just under 40 000 societies which provides fertilisers and agricultural services to India's farming communities. The agreement, announced in May 2008, makes available to IFFCO members low-priced handsets bundled with Airtel mobile connection and discounts on calls to other IFFCO members (calls between IFFCO members cost around USD 0.01 a minute). The farmers also get access to five free voice messages on market prices, farming techniques, weather forecasts, animal husbandry, rural health initiatives, fertiliser availability, and so forth, on a daily basis. In addition, the farmer is able to call a dedicated helpline, staffed by people with expertise in various fields, for answers to specific queries. The stated aim is to promote community building in rural areas. There are, of course, other advantages for both parties. Bharti Airtel gets to set up towers at sites provided by IFFCO societies and has an advantage over competitors in marketing services to rural consumers. At the same time, IFFCO is able to strike a better deal for its members than might otherwise be available along with information and services targeted towards their needs that include, but also go beyond, telephony.

Today, calculating the benefits of owning a telephone, which was complex for telephony, takes on an air of impossibility. Some proportion of the private benefit and external benefit from having a telephone will obviously be telephony. But users interact with other people or machines through communication networks in increasingly multifaceted ways. This includes using mobile phones to interact with communication networks in ways not originally intended by the suppliers of services. This changing perception of the value of owning or using a mobile phone will be reflected in many things that relate to network expansion from word of mouth recommendations by consumers to the strategies of operators. In short network externalities are no longer a simple function of being able to call or be called but the result of the range of applications which users can use.

Why is this significant? Simply put, in a competitive market operators have an incentive to add value which internalises externalities whether in device functionality or in services. For their part consumers add value to services in ways that range from contributing user-generated content to social networking. The content may range from the Los Angeles Fire Department's use of Twitter to provide information on emergencies (*http://twitter.com/LAFD*) to a Bangladeshi farmer using CellBazaar to find the price of commodities. The user who was once unaware of and unrewarded for the external benefit they generated for others in joining a network may today be compensated financially (*e.g.* a lower price through acceptance of advertising or resale of their airtime) or in kind (*e.g.* information or content provided by other users which provides a financial or social benefit).

References

Baumol, William J. and Wallace E. Oates (1998), "Relevance and Theory of Externalities" in *The Theory of Environmental Policy*, Second Edition, Cambridge University Press, Cambridge.

BBC Online (2002), "Hppy bthdy txt!" 3 December, *http://news.bbc.co.uk/2/hi/uk_news/2538083.stm*.

BBC Online (2008), "Mobiles combat Kenyan polio outbreak", 18 September, *http://news.bbc.co.uk/2/hi/technology/7619473.stm*.

CNN (2008), "Digicel Teams with Western Union to Expand Diaspora Program in the U.S.", 9 September, *http://money.cnn.com/news/newsfeeds/articles/marketwire/0431767.htm*.

Coase, Ronald H. (1960), "The Problem of Social Cost", *Journal of Law and Economics*, Vol. 3, pp. 1–44, *www.sfu.ca/~allen/CoaseJLE1960.pdf*.

Corbett, Sarah (2008),"Can the Cellphone Help End Global Poverty?", *The New York Times*, 13 April, *www.nytimes.com/2008/04/13/magazine/13anthropology-t.html?_r=1&ref=world&oref=slogin*.

Crandall, Robert W. and Leonard Waverman (2000), *Who Pays for Universal Service?: When Telephone Subsidies Become Transparent*, Brookings Institution Press.

D'Monte, Leslie (2008), "Villagers across 12 States to Benefit from Mobile Banking", 20 February, *www.business-standard.com/india/storypage.php?autono=314350*.

Farrell, Joseph and Paul Klemperer (2001), "Coordination and Lock-In: Competition with Switching Costs and Network Effects", *www.nuff.ox.ac.uk/users/klemperer/lockinwebversion.pdf*.

Gabel, Katy (2008), "Kenya: Low-Cost Phones Revolutionize Africa", 18 September, *http://allafrica.com/stories/200809180964.html*

Gupta, Sanjeev, Catherine Pattillo and Smita Wagh (2007), "Making Remittances Work for Africa", *Finance and Development*, IMF, Vol. 44, No. 2, June, *www.imf.org/external/pubs/ft/fandd/2007/06/gupta.htm*.

Harford, Tim (2008), *The UnderCover Economist*, Abacus, London.

Kinyanjui, Kui (2008), "Safaricom Revolutionises Money Transfer Services", *Business Daily*, 4 September, *www.bdafrica.com/index.php?option=com_content&task=view&id=9805&Itemid=5812*.

Liebowitz S.J. and Stephen E. Margolis (n.d.), "Network Externalities (Effects)", *www.utdallas.edu/~liebowit/palgrave/network.html*.

Luxner, Larry (2002), "El gran negocio de la diáspora", *TelePress Latinoamérica*, April, *www.luxner.com/cgi-bin/view_article.cgi?articleID=633.*

Menon, Amarnath K. (2008), "Swiping Reforms", 17 January, *http://indiatoday.digitaltoday.in/index.php?issueid=&id=3657&option=com_content &task=view&sectionid=21.*

Morris, Iain (2008), "Please Ad Me", IT and Telecoms Analyst, Economist Intelligence Unit, May, *www.itudaily.com/home.asp?articleid=515200801.*

Moses, Asher (2008), "Cheap mobile deal to serve up ads", 26 August, *www.smh.com.au/news/articles/cheap-mobile-deal-to-serve-up-ads/2008/08/26/1219516447151.html.*

OECD (2007), "Global Opportunities for Internet Access Development", DSTI/ICCP/DCD(2007)3-FINAL.

Palmer, Jason (2008), "Interview: The Cellphone Anthropologist", 11 June, *www.newscientist.com/channel/opinion/mg19826602.000-interview-the-cellphone-anthropologist.html.*

Parthasarathy, Anand (2007), "Igniting IT with Embedded Systems", *Hindi Business Line*, 22 October, *www.thehindubusinessline.com/ew/2007/10/22/stories/2007102250030100.htm.*

Rosenberg, Jim (2008), "Why has M-PESA Become so Popular in Kenya?", 17 June, *http://technology.cgap.org/2008/06/17/why-has-m-pesa-become-so-popular-in-kenya/.*

Sutherland, Ewan (2008), "Counting Mobile Phones, SIM Cards & Customers", 19 April, *www.itu.int/ITU-D/ict/statistics/material/sutherland-mobile-numbers.pdf.*

The Economic Times (2008), "Aryty brings overseas prepaid phone credit service to India", 6 July, *http://economictimes.indiatimes.com/News/News_By_Industry/Telecom/Aryty_brings _overseas_prepaid_phone_credit_service_to_India/articleshow/3202745.cms.*

Thomas, Thomas K. (2006), "BSNL Offers Pre-paid Scheme on Fixed Line", 6 September, *www.thehindubusinessline.com/2006/09/07/stories/2006090704740100.htm.*

United Kingdom Competition Commission (2003), "Vodafone, 02, Orange and T-Mobile: Reports on References under Section 13 of the Telecommunications Act 1984 on the Charges Made by Vodafone, 02, Orange and T-Mobile for Terminating Calls from Fixed and Mobile Networks."

Wray, Richard (2002), "First with the Message, Interview: Cor Stutterheim, Executive Chairman, CMG", *The Guardian*, 16 March, *www.guardian.co.uk/business/2002/mar/16/5.*

Zuckerman, Ethan (2008), "Ken Banks on Frontline SMS", 8 February, *www.ethanzuckerman.com/blog/2008/02/08/ken-banks-on-frontline-sms/.*

OECD PUBLISHING, 2, rue André-Pascal, 75775 PARIS CEDEX 16
PRINTED IN FRANCE
(03 2009 07 1 P) ISBN 978-92-64-07396-8 – No. 56999 2009